Chromosomes: Eukaryotic, Prokaryotic, and Viral

Volume I

Repetitive Human DNA: The Shape of Things to Come. Sequence-Specific DNA-Binding Proteins Involved in Gene Transcription. Physical and Topological Properties of Closed Circular DNA. Structure of the 300A Chromatin Filament. Three-Dimensional Computer Reconstructions of Chromosomes in Human Mitotic Cells. The Kinetochore and its Role in Cell Division. X Inactivation in Mammals, An Update. The Y Chromosome of *Drosophila*.

Volume II

Meiosis. Chromosome Structure and Function During Oogenesis and Early Embryogenesis. Chromatin Organization in Sperm. Structure and Function of Polytene Chromosomes. *Saccharomyces cerevisiae:* Structure and Behavior of Natural and Artificial Chromosomes. DNA Replication in Higher Plants. Chromatin Structure of Plant Genes. Ploidy Manipulations in the Potato. The Chloroplast Genome and Regulation of its Expression.

Volume III

Bacterial Chromatin (A Critical Review of Structure-Function Relationships). The Chromosomal DNA Replication Origin, *oriC*, in Bacteria. Replication and Segregation Control of *Escherichia coli* Chromosomes. Termination of Replication in *Bacillus subtilis, Escherichia coli* and R6K. Polyoma and SV40 Chromosomes. The Genome of Cauliflower Mosaic Virus: Organization and General Characteristics. Reinitiation of DNA Replication in Bacteriophage Lambda. *In Vivo* Fate of Bacteriophage T4 DNA. Double-Stranded DNA Packaged in Bacteriophages: Conformation, Energetics and Packaging Pathway. Bacteriophage P1 DNA Packaging. Bacteriophage P22 DNA Packaging.

Chromosomes: Eukaryotic, Prokaryotic, and Viral

Volume III

Editor

Kenneth W. Adolph, Ph.D.

Associate Professor
Department of Biochemistry
University of Minnesota Medical School
Minneapolis, Minnesota

CRC Press, Inc.
Boca Raton, Florida

Library of Congress Cataloging-in-Publication Data

Chromosomes: eukaryotic, prokaryotic, and viral.

  Includes bibliographies and index.
  1. Chromosomes. I. Adolph, Kenneth W., 1944-
QH600.C498   1989   574.87'322   88-35340
ISBN 0-8493-4397-6 (v. 1)

Direct all inquiries to CRC Press, Inc., 2000 Corporate Blvd., N.W., Boca Raton, Florida, 33431.

© 1990 by CRC Press, Inc.

International Standard Book Number 0-8493-4397-6 (Volume I)
International Standard Book Number 0-8493-4398-4 (Volume II)
International Standard Book Number 0-8493-4399-2 (Volume III)

Library of Congress Card Number 88-35340
Printed in the United States

PREFACE

All animal cells, plant cells, and viruses capable of reproduction contain genetic material. For each living system, the genetic material exists as a complex of the coding molecules of DNA or RNA with proteins. The protein molecules serve to protect the DNA from degrading enzymes and shearing, and also control expression of the genetic information. Eukaryotic chromosomes are organized by the association of DNA with histones: these small, highly charged proteins coat the DNA to form the bead-like structures of the nucleosomes. Chromatin fibers are produced by the helical coiling of the beads-on-a-string filament of nucleosomes, and folding of the chromatin fiber results in the recognizable morphology of mitotic chromosomes. Nonhistone proteins further contribute to the structure and function of eukaryotic chromosomes, particularly in the regulation of gene expression.

The characteristic structures of mitotic chromosomes are observed during the process of normal cell division. But other processes and activities can influence DNA packaging in eukaryotic cells. The reduction of the diploid to haploid complement of chromosomes in meiosis is accompanied by special DNA packaging in the formation of eggs and sperm. In addition, chromosomes are active in replication and transcription, and this requires changes in the organization of chromosomes.

Prokaryotes, included bacteria and cyanobacteria (blue-green algae), have much smaller genomes than eukaryotes and their chromosomes are correspondingly less complex. The intestinal bacterium *E. coli* has 1000 × less DNA than human cells, lacks the histone proteins, and doesn't undergo mitosis or meiosis. Yet the DNA is contained in a defined nucleoid and is organized as supercoiled loops. This arrangement permits the efficient expression of genes and the segregation of replicated chromosomes to daughter cells. The lower complexity of prokaryotic chromosomes is an advantage for studies of the molecular biology and biochemistry of transcription and replication.

Viral chromosomes have even smaller genomes as a consequence of their parasitic reproduction cycles. For example, animal viruses such as polyoma and SV40 have DNA molecules that are 1000 × shorter than those of *E. coli*. Basic differences in the replication of animal, plant, and bacterial viruses have produced a variety of means of packaging viral genomes. Some animal viruses, included polyoma and SV40, have true minichromosomes with histones and nucleosomes. The heads of bacterial viruses such as T7 and P22 contain tighly folded or wound DNA strands that are free of bound protein. And most plant viruses (cauliflower mosaic virus is a notable exception) have RNA genomes enmeshed in protective coat protein subunits.

Major factors which determine the structures of chromosomes therefore include: the biochemical nature of DNA as a linear molecule composed of A, T, C and G subunits, the size of the genome and extent of DNA coiling and interaction with proteins, the nature of the histone proteins as globular and highly-charged, the activity of the genome in replication and transcription, and the requirement for special DNA packaging in mitosis and meiosis. These and additional topics will be examined in the following chapters, which will hopefully convey the variety and fascination of eukaryotic, prokaryotic, and viral chromosomes.

The chapters are divided into five sections; the first three of these are concerned with eukaryotic chromosomes, while the remaining two sections are devoted to prokaryotic and viral chromosomes. Each section begins with an introduction to give a brief overview of the subject matter and to relate it to the topics in other sections. Reviewing the results of research on different systems is important because, although chromosomes possess vastly different degrees of complexity, all chromosomes share similar features. Genetic information is encoded in the same DNA (or RNA) molecules and the sequence of genes must be protected and compacted by interacting with proteins. However, even though all chromosomes have the underlying unity of being protein-nucleic acid complexes, the special features observed for chromosomes of different sources makes studying chromosomes particularly challenging.

Chromosomes: Eukaryotic, Prokaryotic, and Viral should be a valuable resource for readers with a variety of interests and backgrounds. It is hoped that the information presented will be useful and that a sense is imparted of the excitement of research on chromosomes.

THE EDITOR

Kenneth W. Adolph, Ph.D., is presently a faculty member in the Department of Biochemistry of the University of Minnesota Medical School in Minneapolis. His research concerns two fundamental aspects of chromosomes: the structure of chromosomes determined by analysis of electron micrographs and the roles of nonhistone proteins in chromosome organization. He also maintains an interest in virus assembly. Kenneth W. Adolph has been a faculty member at the University of Minnesota since 1978 and is currently an Associate Professor. Postdoctoral training at Princeton University in the Department of Biochemical Sciences preceded this appointment; metaphase chromosome substructure and nuclear substructure were investigated in the laboratory of U.K. Laemmli. Earlier postdoctoral and graduate research was concerned with the assembly of viruses, protein-nucleic acid complexes much simpler than eukaryotic chromosomes and equally interesting. The structure of an icosahedral plant virus was studied during a postdoctoral year working with D. L. D. Caspar at the Rosenstiel Center, Brandeis University. Prior to this, the editor had a post-doctoral position at the Medical Research Council Laboratory of Molecular Biology in Cambridge, England. Research in the laboratory of Aaron Klug involved the *in vitro* reassembly of another simple, icosahedral plant virus. Kenneth W. Adolph received his Ph.D. from the Department of Biophysics, University of Chicago. His thesis concerned the isolation and characterization of cyanophages, and his advisor was R. Haselkorn. B.S. and M.S. degrees were received from the Department of Physics at the University of Wisconsin, Milwaukee.

CONTRIBUTORS

Kenneth W. Adolph, Ph.D.
Associate Professor
Department of Biochemistry
University of Minnesota Medical School
Minneapolis, Minnesota

Sherwood R. Casjens, Ph.D.
Professor
Department of Cellular, Viral, and
  Molecular Biology
University of Utah Medical Center
Salt Lake City, Utah

Charles E. Helmstetter, Ph.D.
Professor
Department of Biological Sciences
Florida Institute of Technology
Melbourne, Florida

Thomas M. Hill, Ph.D.
Assistant Professor
Department of Bioscience and
  Biotechnology
Drexel University
Philadelphia, Pennsylvania

Leon Hirth
Professor
Laboratory of Plant Viruses
Institut de Biologie Moleculaire des
  Plantes
Strasbourg, France

Ross B. Inman, Ph.D.
Professor
Institute for Molecular Virology
and
Department of Biochemistry
University of Wisconsin
Madison, Wisconsin

Eduard Kellenberger, Ph.D.
Professor
Department of Microbiology
Biozentrum
University of Basel
Basel, Switzerland

Andrzej W. Kozinski, Ph.D.
Professor
Department of Human Genetics
University of Pennsylvania
Philadelphia, Pennsylvania

Marie-Hélène Kryszke, Ph.D.
Department of Molecular Biology
Institut Pasteur
Paris, France

Peter L. Kuempel, Ph.D.
Professor
Department of Molecular, Cellular, and
  Developmental Biology
University of Colorado
Boulder, Colorado

Genevieve Lebeurier
Professor
Laboratory of Plant Viruses
Institut de Biologie Moleculaire des
  Plantes
Strasbourg, France

Alan C. Leonard, Ph.D.
Associate Professor
Department of Biological Sciences
Florida Institute of Technology
Melbourne, Florida

Jean-Michel Mesnard, Ph.D.
Laboratory of Plant Viruses
Institut de Biologie Moleculaire des
  Plantes
Strasbourg, France

Anthony J. Pelletier, Ph.D.
Department of Molecular, Cellular and
  Developmental Biology
University of Colorado
Boulder, Colorado

Philip Serwer, Ph.D.
Professor
Department of Biochemistry
The University of Texas Health Science
  Center at San Antonio
San Antonio, Texas

Douglas W. Smith, Ph.D.
Professor
Department of Biology
and
Center for Molecular Genetics
University of California, San Diego
La Jolla, California

Nat Sternberg, Ph.D.
Research Leader
Department of Central Research and
  Development
E. I. Dupont de Nemours
Wilmington, Delaware

Moshe Yaniv, Ph.D.
Professor
Department of Molecular Biology
Institut Pasteur
Paris, France

Judith W. Zyskind, Ph.D.
Professor
Department of Biology
San Diego State University
San Diego, California

TABLE OF CONTENTS

Volume 3

SECTION I. PROKARYOTIC CHROMOSOMES

SECTION II. VIRAL CHROMOSOMES

Section I. Prokaryotic Chromosomes

INTRODUCTION

Packaging of prokaryotic DNA is very different from DNA packaging in eukaryotes discussed in the previous volumes. Prokaryotic chromosomes are simpler, but have special aspects. The DNA does not interact with histones to form "30-nm" fibers and there are no structures analogous to mitotic or meiotic chromosomes. Major chromosomal rearrangements are not, therefore, a characteristic of cell division in prokaryotes such as the bacterium *Escherichia coli*. The DNA is contained in a nucleoid region, but the DNA is not constrained by a surrounding nuclear envelope. The absence of complicating features has made bacteria invaluable for research into the fundamental mechanisms of replication and transcription. The field of molecular biology was founded on studies of bacteria and their viruses.

How, then, is DNA organized in prokaryotes? The *E. coli* chromosome is a circular molecule of double-helical DNA, 1.3 mm in length. The folded *E. coli* chromosome (about 30 μm in diameter) is further condensed into the 1- to 2-μm nucleoid structure by supercoiling of the DNA. RNA contributes about 30% of the mass of the nucleoid, but protein, mainly RNA polymerase, is only about 1%. Transcriptional activity of the folded chromosome is not limited to a particular phase of the cell division cycle.

The chapters in Section I review basic aspects of the organization and activity of bacterial chromosomes. The ultrastructure of bacterial chromatin is difficult to investigate by electron microscopy because of the amorphous distribution of DNA in the nucleoid. But the results provide good evidence for the presence of DNA loops or domains and the absence of bulk nucleosomes. Because of the lack of nucleosomes, it is essential to obtain further information about the role of small anions and cations of the cell sap in condensing chromatin.

Bacterial chromosomes are particularly useful for studies of DNA replication. The genomes are not as large as eukaryotes nor as specialized as viruses. This has allowed details of the three stages of replication (initiation, elongation, and termination) to be determined. Replication of the circular DNA molecule is initiated at a sequence of nucleotides termed *oriC*. DNA synthesis proceeds bidirectionally from *oriC* as the strands of the parental DNA duplex separate. The precise sequence of nucleotides of *oriC* has been determined for *E. coli* and other bacterial strains. *In vitro* replication that initiates at *oriC* was found to require proteins that include DNA polymerase, primase, and gyrase.

Chromosome segregation to daughter cells following DNA replication is another topic covered. At cell division, *E. coli* chromosomal DNA is not randomly partitioned. That is, the chance of a DNA strand segregating toward the same pole at successive divisions is greater than 50%. A mechanism to explain this fact is based on the concept that multiple attachment sites for the *oriC* replication origin are present in the cell envelope. Thus, the cell envelope is not merely a container for the bacterial chromosome, but is closely involved with chromosome organization.

After initiation and elongation have duplicated the original chromosome, termination occurs to release the chromosomes for segregation to daughter cells. Termination has a unique character for *E. coli* and other bacteria because the chromosome is circular and replication is bidirectional. As with initiation, the process is controlled by the DNA sequence in the termination region.

Chapter 1

BACTERIAL CHROMATIN (A CRITICAL REVIEW OF STRUCTURE-FUNCTION RELATIONSHIPS)

E. Kellenberger

TABLE OF CONTENTS

I. INTRODUCTION: THE SPECIAL FEATURES OF PROKARYOTIC CHROMATIN

Bacterial chromatin is of interest because rapidly growing bacteria are the cells in which the largest proportion of DNA is metabolically active. In the "well-trained" laboratory strains of *Escherichia coli,* the cell mass doubles in less than 30 min. Macromolecular syntheses are continuous and that of DNA shows a periodic arrest only in cells that grow slowly in poor media.[1,2] When growing in rich media, DNA replication occurs with an estimated number of 2 to 3 replication forks per genome and with 2 to 3 nucleoids per cell. In rapidly growing cells, the number of transcriptional forks was calculated to be some 2300 per cell.[2a]

Another special aspect of prokaryotes is that transcription and translation occur at the same time: a newly produced messenger immediately interacts with ribosomes to form polysomes.[3] This is correlated with the absence of a nuclear membrane.

Prokaryotic DNA is metabolically in a continuous interphase. The calculated concentration (or packing density) of DNA is estimated to be on the order of 20 to 60 mg/ml.[4,5] The packing density of prokaryotic DNA plasms is practically invariant; it does not undergo the cycles of condensation-decondensation which are characteristic of eukaryotic cells. The DNA concentration of an interphase nucleus of hepatocytes is about the same as that of a bacterial nucleoid. In this estimate, differences between hetero- and euchromatin are not considered.[6]

Bacteriophage T4 is interesting for comparison. After infection, the host DNA is degraded and replaced by the replicating phage DNA, the "pool of vegetative phage". The concentration of this pool is again 20 to 50 mg/ml and, hence, nearly identical with the DNA of uninfected bacteria. In contrast to these, bacteriophage DNA condenses about 30 times when it becomes packaged into bacteriophage heads, where it reaches a concentration of 800 mg/ml with phage T4.[4] For T7, the estimate is only 260 mg/ml.[5] Such calculations are very error prone because of the precision in determining the volume of the virus, which is in general calculated from width measurements on electron micrographs of air-dried specimens. (For this topic, compare References 7 and 8.) A new procedure is now available for measuring concentrations *in situ* on unstained thin sections by STEM.[9]

The DNA of a bacterial genome of about 4400 kbase pairs is some 1.3 mm long, continuous as a single chromosome, and closed to form a circle.[10] The vegetative forks, either of replication or transcription, are necessarily associated with a local separation of the two DNA strands, which, in turn, requires a rotation of the double strand. Rotation of one segment relative to another around the helical axis cannot occur freely in a closed ring. Nature has solved these problems of DNA strand separation by the use of DNA topoisomerases. They have been extensively reviewed by Wang,[11,12] Vosberg,[13] and Gellert.[14] Type II topoisomerases, the gyrases in bacteria, are able to transiently break both strands of DNA and, thus, are able to introduce (or remove) supercoiling, with energy (ATP) consumption. They are also involved in decatenation. Type I topoisomerases transiently break only one strand, without energy consumption, and, hence, are particularly suited to relax unrestrained supercoiling.

In eukaryotes, the DNA is coiled around the nucleosomal core; by binding to the protein, the torsional stress becomes restrained. In eubacteria, the supercoil is negative and, for the majority of the supercoil, apparently unrestrained.[15-17] Negative torsional stress can be measured by an increased rate of photobinding of psoralen. The results of experiments of this type, complemented by observations using electron microscopy and other techniques, suggested that the bacterial genome is organized in about 50 domains — possibly as loops similar to the eukaryotic chromosome[18,19] — which behave with respect to coiling as if they were individual rings (reviewed in References 20 to 22); for further discussion, see Section IV.

    In many of the ideas and models summarized above and reported in the excellent reviews that are mentioned, too much weight was occasionally placed on electron microscopy observations. The micrographs were oftern interpreted too "literally" without considering the physical influences exerted onto the biological material during specimen preparation. Very often the cautions presented by the electron microscopist were set aside by adhering to the philosophy "what I see I believe". One should always carefully investigate if the selected structural constellation or arrangement could not have been the consequence of completely different, unrelated causes. Often, very trivial physical phenomena can be found and proven to produce it.

    The purpose of this chapter will be to complement the available, excellent reviews by critical considerations of electron microscopy. The bacterial nucleoid can be studied by two main approaches: (1) by the thin sectioning or freeze fracturing of bacterial cells or (2) by observing what are called "isolated nucleoids". Both approaches have limitations which, fortunately, are not the same. The problems associated with the structure of the nucleoid in the intact cell have been reviewed by Woldringh and Nanninga,[5] where critical emphasis is given not only to thin sections, but also to the freeze fractures. We will not review the latter technique which, for the study of nucleoids, is hampered by the fact that in unfixed cells, nucleoids are not distinct from the cytoplasm. They are eventually visible after fixation because chemical crosslinking alters the physical properties of matter and, thus, the behavior in cleavage (for the physics of cleavage, see Reference 23).

    Since 1985, the techniques of cryofixation and cryosubstitution have allowed some decisive results[24] to be obtained. The present review will therefore concentrate on these recently introduced procedures.

    The structure of the isolated nucleoid is excellently summarized by Pettijohn and Sinden.[20] When working with isolated nucleoids, one has the advantage of being able to observe a complete genome; the disadvantage is that, during isolation, we are necessarily departing from the original intracellular conditions. We know only partially the composition and concentration of small molecular components of the cellular sap. Therefore, we are not yet able to design an isolation buffer which simulates the intracellular ionic conditions. In addition, by the necessarily occurring dilution of the macromolecular components, we also completely modify the chemical equilibria of the association-dissociation of the concerned components. We know at least that the potassium concentration in *E. coli* cells is between 150 to 650 mM.[25] Within strain and species-specific limits, the growing cell is able to increase the intracellular potassium concentration in response to the osmolality of the growth medium[25-28] Apparently, no, or nearly no, intracellular sodium is present. We know also that most of the potassium is present as small molecular salts in solution in the cellular sap, while nearly all the divalent Mg^{2+} (90 to 180 mM) and Ca^{2+} (30 to 40 mM) are bound to the macromolecules.[27,29,30] We still do not know the counterions of potassium: one suspects organic and inorganic phosphates, glutamate, betaine, and possibly a few other organics,[30-33] while Cl^- is believed to be either absent or present only in small amounts. Not knowing more, it is very difficult to make any predictions about the ionic strength of the intracellular sap. This knowledge would be primordial since it affects very strongly the binding of proteins to DNA. To avoid loss of bound proteins, one usually performs isolations at low ionic strength. This raises then, in reverse, the question about an *a posterior;* binding of basic proteins.

    Until recently, the observations by thin sections of bacteria suffered a similar problem: all the conventionally used fixatives permeabilize very rapidly the plasma membrane. Potassium and magnesium leak out of the cell (Table 1; References 27 and 34) while sodium enters, and the pH becomes altered to that of the medium. The reduced intracellular concentration of potassium, together with a suspected entry of Cl^-, are part of the causes for profound topological rearrangments.

TABLE 1
Induced Cellular Leakages in *Escherichia coli* B[a]

Treatment			Potassium[b]	Magnesium[b]
OsO₄[c]	0.1%	10 min	+ + + +	+ + +
Glutaraldehyde	4%	10 min	+ + + +	0
	0.1%	10 min	+ +	0
Formaldehyde[d]	3.3%	10 min	+ + + +	0
Acetaldehyde	2.5%	10 min	+ + + +	0
Benzaldehyde	2.5%	10 min	+ + + +	+ + + +
Acrolein	2.5%	10 min	+ + + +	+
KCN	10 mM	10 min	+ + +	0
4-h culture			+ +	+
16-h culture (overnight)			+ + +	+ +
Chloramphenicol	200 μg/ml	1 h	+[e]	0
Polymixin	10⁻⁶ M	6 min	+ + + +	+ + + +
T4 phage ghosts	m.o.i. = 5	Few min	+ + + +	0

[a]   Aerated cultures in bacto-tryptone medium (1%) with 0.1 M NaCl. The cells were harvested by filtration onto membrane filters. After washing, the deposit was dissolved in HNO_3 and the ions determined by atomic absorption. During deposit and washes, care was taken not to induce any accidental leakage through lack of oxygen or change of temperature or osmolality.

[b]   Legend: 90—100% = + + + +; 70—90% = + + +; 40—70% = + +; 15—40% = +; 10% or less = 0.

[c]   With OsO_4, spermidine and putrescine were also found to be completely lost; for aldehydes, it could not be measured because the staining reaction involved aldehydes.

[d]   Similar results are obtained with solubilized paraformaldehyde.

[e]   Newer data[28] show no leakage during the first 2 h.

From Moncany, M. L. J., Thèse d'Etat, Université de Paris VII, Paris, 1982.

Recently, improved methods of cryofixation, which in most cases are combined with cryosubstitution, have overcome these problems.[24,35] By very rapid freezing, native structures are immobilized. During substitution of the ice with organic solvents, the possibilities for macromolecular rearrangements are very small. We will discuss these new results in some detail. They are of particular interest when combined with immunocytochemistry, which is able to bring morphology nearer to biochemistry. We will discuss the results obtained on the location of the most abundant histone-like protein of *E. coli*, which is not found to be associated with the bulk DNA, but, rather, in regions where transcription occurs, together with RNA polymerase and topoisomerase I.

The remaining limitations of these new techniques for *in situ* observations are twofold: (1) thin sections never allow for high resolution and (2) the concentrations of the macromolecular structures per cell have to be above some limits to become detectable by immunocytochemistry. The lowest concentration of well-dispersed antigens needed for detection by immunolabel was explored experimentally and found to be in the range of 10 to 100 μM or 10⁴ to 10⁵ molecules per cell of *E. coli*.[36] These numbers lead to an observable label. It is, however, hopeless to investigate replication: out of the 6 to 8 replication forks of an *E. coli* cell, only one section out of ten contains a fork. In addition, for immunolabeling, the antigen has to be at the surface, which reduces the chance by a further factor[36] of about ten, resulting in a probability of 0.01. This is the probability for a single antigen per fork. It increases about proportionally with the number of identical proteins per fork, but in the most favorable cases the probability can obviously never exceed 0.1, given by the presence of a fork in a section.

Due to lack of contrast, it is not yet possible to resolve individual chromatin filaments

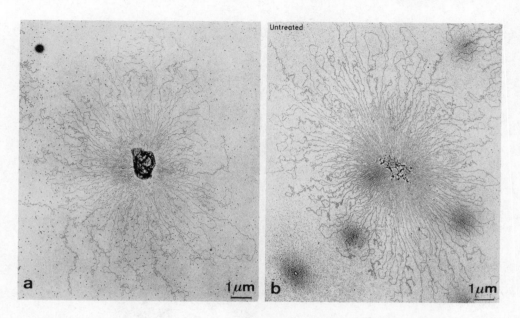

FIGURE 1.   Isolated nucleoids of *Escherichia coli*. (a) The nucleoid when still attached to the membrane; (b) prepared with detergents such that the membrane is detached. In both preparations, the salt concentration was 0.25 *M* for (a) and 0.4 *M* for (b). They were spread with the aid of cytochrome C. Note the loops, originating from the center, of which some show supercoils. (From Kavenoff, R. and Ryder, O., *Chromosoma*, 55, 13, 1976; Kavenoff, R. and Bowen, B., *Chromosoma*, 59, 89, 1976. With permission.)

in thin sections and, hence, it is also not possible to observe a suspected attachment point of the genome to the plasma membrane.[37] Its observability is further restricted by the presumed small number per cell. Because of these problems, only the observation of isolated nucleoids can be of use (Figure 1), particularly when studied in combination with immunolabel[40] or in situ hybridization. For producing specific antibodies, the biochemical characterization of attachment sites has to reach definitive results. Large efforts had been devoted by many laboratories to this problem, hitherto, unfortunately without consistent results[38,39].

So far, we have learned that the major difference of prokaryotic and eukaryotic chromatin is that the latter is constrained by the "classical" nucleosome, while the former is so only partially or not at all. The packing density (or local concentration) of DNA is the same as in eukaryotic interphase. An additional major difference will be discussed in Section III. It consists of a very different reaction of the two chromatins when dehydrated in organic solvents: while one is easily protected by various chemical fixations ("aggregation insensitive"), the other is highly sensitive and is protected only by a few, special procedures.[6]

II. THE NUCLEOID IN THIN SECTIONS

A "history of the nucleoid" is in preparation.[41] It will start from the 1940s and 1950s when it was most important to prove that bacteria also contain DNA, as do the nuclei of higher cells (Figure 2). This was demonstrated particularly well with specific staining methods by Piekarski[42] and Robinow,[43] and by differential enzymatic digestion by Tulasne and Vendrely.[44] As soon as electron microscopes became available, it was obviously valuable to make use of their higher resolving power for observing thin sections. When those fixation procedures of electron microscopic cytology were applied to bacteria, "nuclear vacuoles" were discovered which contained irregular black bodies (Figure 3),[45] later interpreted as either chromosomes or artificially occuring aggregates. Ryter and Kellenberger[46] introduced

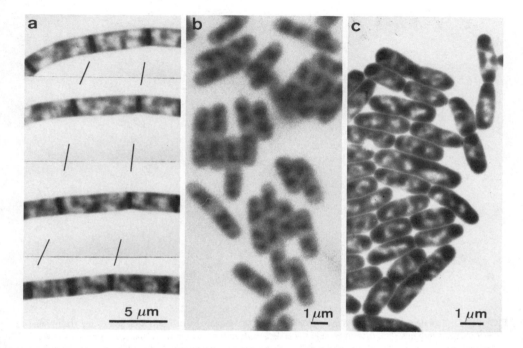

FIGURE 2.   Nonsectioned bacteria in the light and electron microscope. (a) Dividing cells of a bacterial sp. "Medusa" observed by phase contrast (Courtesy C. Robinow, unpublished.); (b) *Escherichia coli* cells of an exponential culture, stained with Giemsa (author's micrograph, 1954, Archive No. 9274); (c) *E. coli* cells of an exponential culture prepared by agar filtration and fixed in vapors of OsO_4 observed in the electron microscope.[48,63] The areas with DNA-containing plasms are less electron scattering (author's micrograph, 1954, Archive No. 12816).

a particular OsO_4 fixation under the so-called RK conditions, i.e., in the presence of amino acids and/or small peptides and of divalent cations, and buffered at a pH around 6. When fixed under these conditions, the "nuclear vacuole" was filled with a homogeneously spread, more-or-less fibrillar DNA-containing plasm (Figure 5). These authors and others[47] also found that a posttreatment with uranyl acetate not only stained, but was also able to compensate for deficient RK conditions (e.g., those caused by the presence of phosphate buffers which chelate divalent cations).

All researchers in the field soon agreed that no nuclear membrane could be detected (see, for example, Reference 48). The name "nucleoid" was thus proposed as it is also used today for membrane-free eukaryotic nuclei. The shape of the nucleoid had been thoroughly investigated, mainly in the hope of understanding more about nuclear replication and division. For more details, the reader is referred to the broader review contained in Reference 41. To summarize, it was soon discovered that the morphology of the nucleoid was highly variable, not only as a consequence of the conditions of growth or of metabolic disturbances, but also according to the techniques used and even to the laboratory which applied them. It is noteworthy that, with very few exceptions, the researchers involved agreed about the limitations imposed by this variability and did not lose time in overinterpretation and related sterile controversies. When fixed at low salt, the nucleoids are dispersed into the cytoplasm showing a cleft or lobed outline. Since the local packing density (or concentration) of the DNA is not modified, we propose to call the nucleoid to be confined when it is in the least dispersed form. The frequently used misnomer of "condensed" should here be avoided and stay reserved for those cases where the packing density is increased (metaphase eukaryotic chromosomes, mature phages). At that time, it was already known that the intracellular osmotic pressure of the bacterial cell appeared to be higher than that

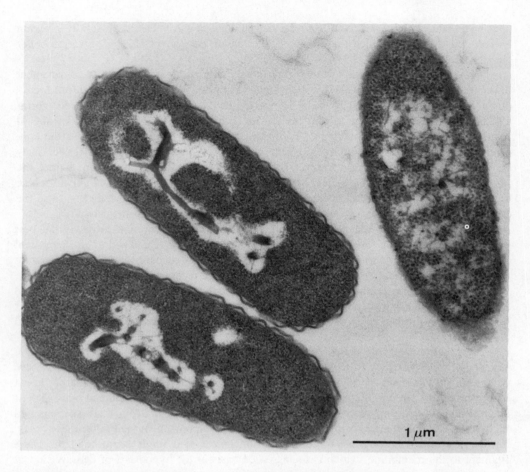

FIGURE 3.   Comparison of sections of *E. coli* fixed by glutaraldehyde and OsO_4. The cells of a growing culture were fixed by adding 1% glutaraldehyde or 1% OsO_4 and then, after mixing, processed for embedding in Epon. The thin sections (40 to 60 nm) were stained by treatment in uranyl acetate, 6%, 15 min, and lead acetate according to Millonig,[122] 3 min. The nucleoid is in both cases visible as a "vacuole" or a ribosome-free space. The content, the DNA-containing plasm, is in both cases aggregated. According to the "size" of the ribosome-free spaces, the aggregates are large in the case of OsO_4 fixation and smaller in the case of glutaraldehyde fixation. Aggregation occurs during dehydration because the DNA plasm is not gelled. For more details, see text. (Embedding and electron microscopy in the author's laboratory by Renate Gyalog.)

of the outside medium. The permeability of the cell envelope was already at that time considered to be strictly controlled. The possibility that the fixative could perturb permeability and thus cause these morphological changes in the nucleoid was already recognized.[49,50] Later, when it was established that the cell compensates for an increased outside osmolality mainly by pumping potassium,[25,26] the idea became attractive that the nucleoid really varies its shape as a consequence of the internal ionic strength.

With the generally increased use of aldehyde fixation introduced by Sabatini et al.,[51] it was soon observed that with aldehydes the nucleoid was always in the more dispersed, lobed, or very cleft form (Figure 3), independent of the osmolality of the growth medium.[27,52-54] Knowing that aldehydes also induce rapid potassium leakage (Table 1) and that after aldehyde fixation the DNA-containing plasm aggregates (Figure 3), it was not possible to make definitive conclusions about the native shape of the nucleoid. The results with cryofixation and direct observation in the frozen-vitrified state,[35] but mainly with cryofixation followed by cryosubstitution (Figure 4; Reference 24), settled the question in favor of a dispersed state being representative of the morphology of the nucleoid of rapidly growing *E. coli* cells. This confirmed the previously expressed view about aldehyde fixation better preserving the native state.[53-55]

Danoe-Moore and Higgins[53,55] investigated several possible causes for the confinement of nucleoids when fixed with OsO_4. They studied the possibility of an OsO_4-induced breakdown of mRNA and, indeed, found a 5 to 30% breakdown with both osmium and formaldehyde fixation, but not with glutaraldehyde. They also investigated the efficiency of fixation by monitoring the binding of radioactively labeled lysine to proteins. At the concentrations studied, they found again much more cross-linking with glutaraldehyde. They concluded that glutaraldehyde fixation is faster and thus gels the cytoplasm before the ribosomes can rearrange themselves. More recently, it was shown that formaldehyde is able to cross-link as well as glutaraldehyde, but higher concentrations are needed because the cross-links are likely to be performed by isomers and polymers of formaldehyde.[56]

Provisionally, we conclude that cryofixation with cryosubstitution provides the best preserved nucleoids. This method is, however, time consuming. The result of an experiment is available only after 2 to 4 weeks. The usual methods of processing at room temperature (or at 0°C) are obviously much faster. For this case, we recommend the use of aldehydes, combined with a postfixation in aqueous uranyl acetate. An initial fixation in OsO_4 has to be avoided. Whether a second fixation in OsO_4 still introduces structural modifications is not yet known. Further comparative studies should be made to strengthen these provisional statements.

The predominantly dispersed form of the nucleoid (Figure 4), which is representative of metabolically active cells, means, however, that it is completely futile to speak any more of a defined "nucleoid morphology". Thin sections are about 40 nm thick, that is, about one twentieth of the cell diameter. Hence, nuclear division cannot be studied by means of single thin sections: only reconstruction from serial sections[57] or from thick sections and high-voltage electron microscopy could provide a valid approach. In neither of these cases, however, would it be possible to detect the attachment of a single segment of the DNA filament to the membrane, which is the assumed basis of the hypothesis of cell division of Jacob et al.[37]

The dispersed, highly lobed form of the nucleoid is more satisfactory for explaining the consequences of the high metabolic activity (i.e., transcription) of the bacterial genome, together with the known simultaneity of transcription and translation.[3] It is no longer necessary to assume loops of DNA emerging from a confined bulk DNA, loops which would reach far into the ribosomal areas. Immunocytochemistry of dsDNA does not reveal a substantial number of such loops, although ssDNA is found in the border zone about 60 nm adjacent to the lobes of the nucleoid[58] (see Section V). The lobes of very cleft nucleoids seem to be sufficient for promoting large contact areas with the ribosomes.

This observation of a highly dispersed, cleft nucleoid should by no means be overinterpreted as a more or less homogeneous random mixture of DNA with the ribosomes. One has to remember that by phase-contrast (Figure 2) or confocal light microscopy, the nucleoid is distinctly visible in live and rapidly dividing cells.[54,59-61]

Independent of the type of fixation, the nucleoid assumes a very specific and regular, nearly spherical shape when the metabolically active cell has its protein synthesis specifically inhibited (Figures 6 and 7). This is achieved either by growing an amino acid-requiring strain in the absence of this amino acid[62] or after the action of aureomycin,[48,63] chloramphenicol,[64,65] or puromycin.[66] This form is not identical and probably not even related to the variable shape of nucleoids in stationary phase cells: the latter have become permeable for K^+, Na^+, and H^+ while it was shown for the action of chloramphenicol that homeostasis is maintained and no leakage is observed[27,67] (Table 1).

The exact interpretation of the spherical form is unsettled (compare, for instance, with

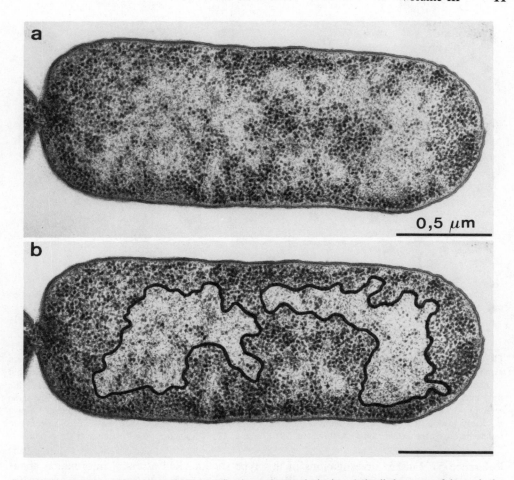

FIGURE 4.   Thin section of *E. coli* after cryofixation and cryosubstitution. A detailed account of the method is found in Reference 24. Exponentially growing cells are rapidly harvested onto a very thin, adsorbent paper, which is then rapidly placed on a polished copper block at 4°K. The frozen cells are then treated at 190 K in acetone with 2.5% OsO$_4$. Further processing is as usual for embedding in Epon. In the lower figure, the ribosome-free space is crudely outlined as a visual aid. By immunocytochemistry, the DNA has been located in the ribosome-free space.[58] (From Hobot, J. A., Villiger, W., Escaig, J., Maeder, M., Ryter, A., and Kellenberger, E., *J. Bacteriol.*, 162, 960, 1985. With permission.)

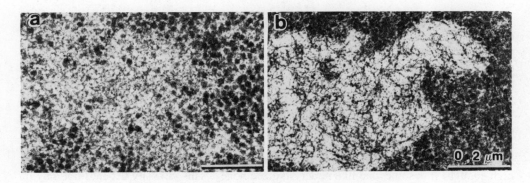

FIGURE 5.   Fine structure of the DNA plasm of *Escherichia coli*. The fibrillar state (a) is obtained when fixed under RK conditions. The DNA plasm has a much more grainy aspect when processed by cryofixation-cryosubstitution, as in Figure 4. (From Hobot, J. A., Villiger, W., Escaig, J., Maeder, M., Ryter, A., and Kellenberger, E., *J. Bacteriol.*, 162, 960, 1985. With permission.)

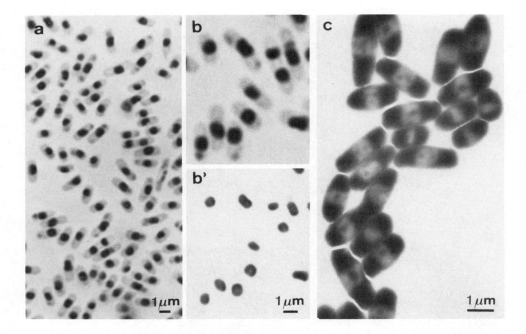

FIGURE 6.   The spherically shaped nucleoid. These cells were obtained after specific inhibition of protein synthesis, as explained in the text. This particular, approximately spherical, shape differs from that of a nucleoid of stationary phase cells in that it contains a central core, which is visible even with stained preparations in the light microscope (a). To show the structures clearly by print requires a lower exposure, as is shown for the same region in (b) and (b'). An electron micrograph of unsectioned whole cells is shown in (c); the "annular" aspect is clearly visible. The two figures were obtained from *Escherichia coli* treated with 4 μg/ml of aureomycin (micrographs by the author, 1954, Archive No. 5970 (c) and, 17284, (a and b); see Reference 48).

Reference 5). Further investigations of this type should yield essential information about the structure of the genome, as we will discuss further below. It is expected that specific morphologies should occur when either the translation, as in the case mentioned above, or the transcription is disturbed and/or when the coiling of DNA is affected by influencing the equilibria between the two essential types of DNA topoisomerases, I and II (see Section I).

III. THE AGGREGATION-SENSITIVITY OF THE DNA PLASM OF NUCLEOIDS

The DNA-containing plasm of nucleoids, independently of how their shape appears, can be more or less aggregated (Figure 3). This is true for all aggregation-sensitive DNA plasms, for example, the pool of replicating, transcribing, and recombining DNA of the T even phages (Figure 8). According to the fixation procedure used, the DNA plasm is or is not sufficiently protected against aggregation in the subsequent transfer into organic liquids, particularly during dehydration. Cryofixed bacteria, when observed while still in the frozen-vitrified state, do not show any coarse structure of the DNA plasm.[35] This behavior is correlated with that observed macroscopically with DNA and with commercially available nucleohistone and nucleoprotamine.[68] Fixations which produce fine-stranded DNA plasms in sections are those which produce a gel, provided the viscosity of the starting solution is high enough. It is somewhat like noncrystallized honey. This viscosity depends not only on the concentration, but also very much on the chain length of the DNA or DNP considered. Eukaryotic chromatin, isolated under conditions such that histones are not lost, is already nearly a gel and cannot be tested for gelation by the usual procedures. Instead, eukaryotic,

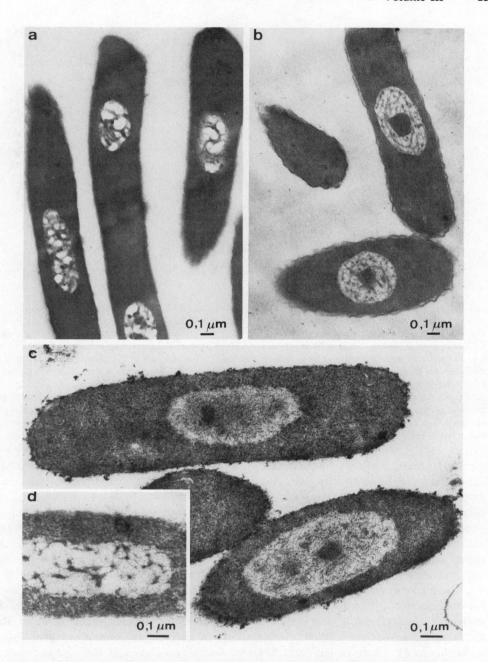

FIGURE 7.   The spherical nucleoid in electron micrographs of thin sections. These spherical nucleoids were obtained by treating growing *Escherichia coli* with 4 μg/ml of aureomycin (a, b) or 100 μg/ml of chloramphenicol (c). Sections (a) and (b) were obtained after conventional OsO_4 fixation[48] and (c) when using OsO_4 under RK conditions and uranyl acetate postfixation.[64] For (d), the postfixation was replaced by a treatment with EDTA, which abolishes in some unknown manner the proper cross-linking which renders the DNA plasm aggregation insensitive. (a) and (b) are selected to show the variable outcome of aggregation when the RK conditions are not fulfilled. The ribbons of (b) are very tempting to interpret as "real" chromosomes. (Micrographs a and b by the author with A. Ryter, 1954, Archive No. 17268 and 18897, respectively; see also, Reference 48. Figures (c) and (d) from Kellenberger, E., Ryter, A., and Séchaud, J., *J. Biophys. Biochem. Cytol.*, 4, 671, 1958. With permission.)

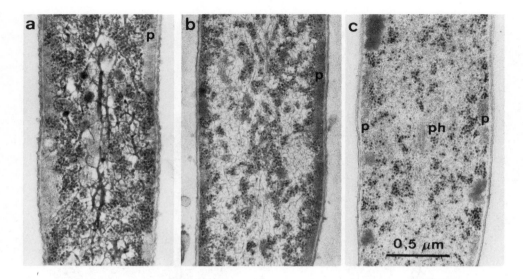

FIGURE 8.   Aggregation sensitivity of the pool of vegetative phage T4 DNA. The figure shows *E. coli* cells after infection with a mutant (21⁻) of bacteriophage T4 that is arrested at the prohead (p) stage. Polyheads (ph) are also visible in some cells. The replicating phage DNA (vegetative pool) is seen as strands which appear coarse, in panel a, finer in panel b, and so fine as to be nearly undetectable in panel c. The same magnification is used for panels a to c. Individual preparation of the three samples differed. (a) Fixation was in 2% glutaraldehyde for 60 min; thereafter, the sample was "classically" treated at room temperature by dehydration in ethanol and embedding in Epon. The sections were stained with uranyl acetate and lead citrate. (b) In this case, after the same fixation, the sample was dehydrated by "progressive lowering of the temperature" (PLT) and then embedded in Lowicryl K4M. Sections are stained as in panel a. (c) The sample was not fixed, but rapidly frozen by the procedure outlined in the text at liquid helium temperature, substituted into acetone containing 2.5% OsO₄ at 190°K over 64 h, and then embedded in Epon and stained as in panels (a) and (b).

membrane-free "nucleoids" were isolated from mouse mastacytoma cells. They show Mg-dependent swelling and contraction. Any fixative used inhibited this response.[68] Since eukaryotic interphase nuclei have about the same concentration of DNA as the bacterial nucleoids (see section I), we conclude that the generally observed aggregation insensitivity of these higher eukaryotic nuclei[6] cannot be explained by a difference in DNA concentration, but must be related to higher relative amounts of its protein partner and/or to other fundamental structural differences. This conclusion correlates very well with the observation that torsional stress is constrained in eukaryotes, while it is much less so in prokaryotes.[15,20]

It is noteworthy that many other aggregation-sensitive DNA plasms are found besides bacterial nucleoids. It is not surprising that the pool of vegetative phage (Figure 8) and the nucleoids of mitochondria[69] and plastids[70] behave the same. It is unexpected, however, that the chromosomes of the eukaryotic dinoflagellates are also very aggregation sensitive (Figure 9; Reference 6). Those of the algae *Euglena,* which at the level of the light microscope have features very similar to dinoflagellates, are not sensitive. In both species, the chromosomes appear to stay individualized throughout the mitotic cycle, including interphase. This difference in aggregation sensitivity is nicely correlated with the presence of normal histones in *Euglena*[7] and the presence of only "histone-like proteins" in much smaller relative amounts in the dinoflagellates.[72,73]

IV. FINE STRUCTURE OF THE DNA-CONTAINING PLASM OF THE NUCLEOID: FACTS AND FANTASY

As we have just seen, aggregation insensitivity is a special aspect of the higher eukaryotes and, as far as can be known from the restricted amount of data, of those which possess the

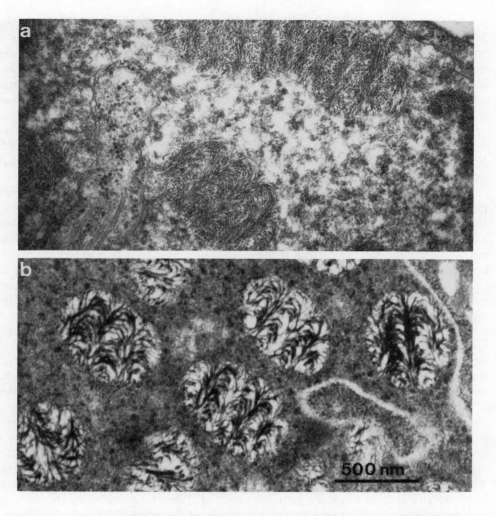

FIGURE 9.   Aggregation sensitivity of dinoflagellate chromosomes. Chromosomes of the dinoflagellate *Crypthecodenium cohnii* are aggregation sensitive. In panel (a), they are shown with a minimum of aggregation, as obtained after fixation in mixtures of aldehydes followed by OsO_4 and uranyl acetate; in panel (b), they are shown with extensive aggregation, as obtained after aldehyde fixation alone. The sections were stained with uranyl acetate and lead, as in Figure 3. The details of the procedures are as follows: (a) Glutaraldehyde 1.25%, paraformaldehyde 2%, OsO_4 2%, uranyl acetate 0.5% in phosphate buffer. Dehydration in ethanol at room temperature. Embedding in Epon with thermal polymerization. (b) Glutaraldehyde 0.5%, paraformaldehyde 3%. Dehydration in ethanol by progressive lowering of the temperature (PLT). Embedding in Lowicryl HM20. Polymerized at $-35°C$. (Cultures by G. Vernet and M. Maeder; specimen preparation and electron microscopy by M. Maeder.)

normal set of five histones. Structural information about the fine structure of these two types of chromatin is therefore urgently needed.

As already mentioned, electron microscopy of thin sections is not able to provide high-resolution data. We are, nevertheless, able to discern small differences in the structure of the DNA plasm according to the fixation used (Figure 5). While the DNA plasm of bacterial nucleoids fixed with OsO_4 under RK conditions is typically fibrillar (Figure 5b), it is much more granular after cryofixation and cryosubstitution (Figure 5a). This speaks for a globular manifestation of the supercoil, which would either be lost upon Os fixation or gained in the methanol or acetone treatment during cryosubstitution.

From the observation of bacterial chromatin, either extruded from cells (see Reference 74) or in isolated nucleoids (for references, see Reference 20), we also do not obtain an unambiguous answer. Griffith[74] observed a beaded structure outside of cells which were made leaky for DNA by treatment with detergents. He found that these beads, which were considered to be prokaryotic nucleosomes, are very labile. This partly explains why other people could not repeat the observation.[75] Another difference in treatment should be considered: while most researchers use ethanol for rapidly drying the specimen and for a better adherence of the filaments to the supporting film, others do not. It is well known through the work of Eickbush and Moudrianakis[76] that even pure, protein-free DNA can easily be induced by ethanol to form beaded structures which look at least as good as the nicest micrographs of eukaryotic nucleosomes.

These putative "labile prokaryotic nucleosomes" show other fundamental differences from the "classical" nucleosomes of eukaryotes. By digestion of the DNA, eukaryotic chromatin produces, upon gel electrophoresis, the characteristic ladder with repeats of about 150 to 200 bp.[77] This is not the case with bacterial chromatin. Only one group[78] found some repeats at 120 bp. A smaller repeat is found when the digestion is pushed further. Broyles and Pettijohn[17] found a repeat of 8.5 bp for *E. coli,* which is smaller than that of eukaryotes (10.5). This repeat is thought to be due to the bending of DNA when it is coiled around the nucleosome core. We have found no data which would preclude the assumption that these repeats could not be the consequence of any other type of supercoiling.

As reviewed by Pettijohn and Sinden,[20] the electron microscopy of isolated nucleoids revealed supercoiled loops (Figure 1), which are compatible with the subdivision of the genome into some 50 domains.[79,110] Similar to the loops of the postulated structure of the eukaryotic chromosome,[18,19,80] these domains would correspond to segments of the DNA within which the supercoiling and its relaxation would be confined and would be independent of that of the other segments. Sinden et al.[16] provided elegant evidence by introducing single-strand nicks by irradiation with γ-rays and found that with few nicks induced, only a small part of the genome had its supercoils released. The *in situ* morphology of the nucleoid in thin sections (see section III) or immunocytochemistry (see Section V) provide no new features which can be correlated with these domains, except the spherical CAP nucleoid. We might assume, by analogy to the loop model of eukaryotic chromatin, that the bacterial loops are also anchored in some particular structure. When these anchoring structures are gathered together, they would form the central core of the spherical nucleoid. The gyrases are putative candidates for being located in this core, as are the topoisomerase II molecules in the eukaryotic chromosome scaffold.[81] It is interesting to note that topiosomerase II was also associated with the nuclear-pore-lamina complex.[82]

The electron microscopy evidence for a bacterial loop model is based on the observation that isolated nucleoids sometimes result in micrographs where supercoiled loops emerge symmetrically from a center (Figure 1; References 38 and 39). Beautiful micrographs as those of Kavenoff (Figure 1) are unfortunately not reproducibly obtained. The published and unpublished micrographs of other authors do show only very few or no individually supercoiled loops. Instead the DNA forms a random "network" frequently with localized smaller structures arranged like multileaf clover. Similar "center-symmetric" arrangements are also very frequently obtained with DNA which is *in situ* released from bacteriophages, although everything known about bacteriophage DNA packaging disagrees with such a model.[83,84] Supercoiling is distinctly visible on the loops of isolated nucleoids, but not at all on those of bacteriophage DNA. This is in agreement with the negative supercoiling of bacterial DNA which is also obvious from the experiments with psoralen-photobinding.[16] The amount of supercoiling which remains visible on the micrographs of isolated nucleoids is much too small to be derived from a chromatin-like nucleosomic structure with its high linking number. If in the native state so much was supercoiled, then we have to assume that

TABLE 2
Isoelectric Points and Molar Ratios in S100 Extracts
of Purified Nucleoid Proteins

Molecular mass (kDa)	Isoelectric point	Molar ratio	Corresponding protein	Ref.
28	7.3	7	—	
27	6.4	1	H protein	86
26	8.0	1	22-kDa protein	87
24	9.8	1	—	
22	7.0	2	—	
17	9.6	4	HLPI	88
			Protein 1	78
			BH1	78
17	7.3	4	—	
14	Not determined	3	—	
9	9.8	100	HLPII	88
			HU	89
			BH2	78

From Yamazaki, K., Nagata, A., Kano, Y., and Imamoto, F., *Mol. Gen. Genet.*, 196, 217, 1984. With permission.

during specimen preparation relaxation of the supercoils must have occurred, which would explain also the lack of reproducibility in the electron microscopy of isolated nucleoids.

We may conclude this section by conceding that we know very little about how supercoiling in the bacterial nucleoid is structurally expressed. Is it in the form of beads and, if so, are these beads comparable to nucleosomes of the eukaryotic type, with a stoichiometrically defined protein core around which the DNA is coiled? Are the beads, which are observed by electron microscopy, native structures or an artifact induced by the procedures involved in specimen preparation? Or, in reverse, could the events involved in specimen preparation induce a relaxation of the existing torsional stress? In section VI we discuss the possibility that OsO_4 fixation could "catalyze" the action of a "nickase" as it activates the lysozyme in cells infected with bacteriophage T4.

In the next section, we will see that protein HU, as the most likely candidate for being a structural protein of a prokaryotic nucleosome core, is not located where normal nucleosomes are presumed to be.

V. THE IMMUNOCYTOCHEMICAL LOCALIZATION OF THE HISTONE-LIKE PROTEIN HU TOGETHER WITH SOME TRANSCRIPTION-RELATED ENZYMES

The histone-like proteins of *E. coli* have recently been reviewed.[6,22] We summarize in Table 2 the findings of Yamasaki et al.,[85] who provide the relative amounts for the known histone-like proteins of *E. coli*. It is obvious that HU is by far the most abundant among them. Previous determinations showed about one dimer of HU (HU_1 and HU_2, which are slightly different isomers) per 200 nucleotides.[90] HU has a molecular weight of about 9.5 kDa, or 19 kDa per 200 bp, which is considerably less than the 110 kDa of the core of a eukaryotic nucleosome. Hence, to produce a compact prokaryotic nucleosome, one would need other, additional proteins. The presently determined amounts for the other known histone-like proteins are too small to account for these. Either the experimental determinations are underestimates or other proteins have yet to be discovered. The other alternatives are

TABLE 3
Localization of the Immunolabel of HU, RNA Polymerase, and
Topoisomerase

Serum	Background per μm^2	Gold over[a]		
		Bulk DNA	Border zone	Cytoplasm
Anti HU	1.4 ± 0.18	0.5	23.6	10.7
Anti RNA polymerase	<0.1	0.6	18.7	9.1
Anti topoisomerase	<0.1	0.05	6.8	0.7

[a] The numbers are expressed as average numbers of gold particles per bacterial section.

From Dürrenberger, M., Bjornsti, M. A., Uetz, T., Hobot, J. A., and Kellenberger, E., *J. Bacteriol.*, 170, 4757, 1988.

either (1) the role of HU is not that of producing nucleosomes of the classical type associated with resting DNA or (2) pseudonucleosomes (''compactosomes'') exist of a type which is not comparable with the eukaryotic ones. The latter possibility of a particular, labile nucleosome was proposed by Broyles and Pettijohn[17] and Drlica and Rouviere-Yaniv[22] in confirmation of the views of Griffith.[74] As an additional, important argument, they used the fact that beaded structures are observed in the electron microscope when DNA is reacted 1:1 with protein HU *in vitro,*[91] although *in vivo* the relative amount of HU is about ten times smaller. One should also not forget, as mentioned in the previous section, that ethanol treatment per se is able to produce beaded structures.[76] Some preliminary experiments with immunolabel of HU on isolated nucleoids did not reveal HU to be equally distributed along the genome, but, rather, to be clustered.[40] For these experiments, it cannot be excluded that HU, whose nativity was present also elsewhere, could have been washed away during specimen preparation.

Immunocytochemistry of cryofixed thin sections of growing cells revealed that HU is located along the periphery of the lobed nucleoid, over the ribosomes in an area covering a width of about 60 nm. The areas of the bulk DNA of the nucleoid were virtually devoid of any label.[92] The necessary controls were made to exclude the possibility that the epitope was covered and thus hidden by the DNA. The location over the ribosomes would also agree with the finding that the two isomers of HU are identical with the two isomers of protein NS which bind to the native 30S ribosomal subunit of *E. coli.*[93] This finding immediately suggested that HU — alias NS — could be involved in the coupled transcription-translation of prokaryotes.[6,92] This hypothesis is supported by the observation that the DNA-dependent RNA polymerase and the single-stranded DNA are also located over the ribosomes adjacent to the nucleoids, which is the same region in which HU is located (Table 3; Reference 92). Surprisingly, topoisomerase I was also found in the same area (Table 3; Reference 92). In the giant chromosomes of *Drosophila,* topoisomerase I was also found associated with transcriptionally active regions of the genome, as are, for example, the puffs.[94] From the experimental results with *E. coli,* one has to conclude that the bulk DNA, if organized as ''compactosomes'', does not involve HU. Rather, this histone-like protein seems to be involved in transcription-translation, together with topoisomerase I and RNA polymerase.

The resolution of immunolabel on thin sections does not yet allow one to determine more precise topological relations — for example, those needed to decide if HU is part of a ''transcriptosome''. It is well known from many systems that negative supercoiling is essential for transcription.[11,14,95] It seems that gyrases (type II topoisomerase) are particularly important.[14] Unfortunately, the results of immunocytochemistry of bacteria are inconclusive, because of the low cellular concentration.[125].

It is noteworthy that with respect to the pool of vegetative DNA of phage T4, the location of HU is identical to that of the nucleoid.[92] The rates of phage DNA and mRNA syntheses stay about the same after phage infection, although those of the host are arrested. No ribosomal RNA being made, the amount of mRNA must be correspondingly higher; since the host polymerase, after modification, is used, the number of transcriptional forks stays the same as before infection, i.e., 2300.[2a] The DNA is concatenated, probably into a single DNA filament; accordingly, when assuming unchanged rates of macromolecular synthesis, we need not assume more replication forks than with rapidly growing bacteria. The number of recombination events, however, is much higher and should become detectable by immunolabel. Indeed, the DNA of every mature T4 phage has undergone an estimated 10 to 20 recombinations before it is packaged. The T4 system would be extremely suitable for further immunocytochemical studies because nearly all genes and their products have been characterized.[96] Of particular interest is the topoisomerase II encoded by T4 genes 39, 52, and 60, which provide its subunits.[97-99] The products of gene 39 are claimed to be located in the membrane,[100,101] an observation worthy of immunocytochemical confirmation. The requirement for topoisomerase in the packaging of phage T4 has been reported.[102]

Also of interest for further immunocytochemical studies is the spherical CAP nucleoid discussed in the previous section. With immunolabeling, HU was not found to be concentrated in its core; an increased amount of topoisomerase I is not excluded.[92]

The pool of vegetative T4 DNA is not transformed into a spherical structure with a central core when protein synthesis is arrested. When protein synthesis is inhibited some 8 min after infection, DNA synthesis proceeds unaltered[103] until a huge pool accumulates because no phage can mature in the absence of late protein synthesis.[64]

Polyclonal antibodies raised against bacterial HU were found to crossreact with a protein of about 9 kDa of *Crypthecodinium cohnii*. On sections of this dinoflagellate, most of the immunolabel was found in the nucleoplasm and little to nothing in the chromosomes, except at their periphery. In another species, Amphidinium carteri, the chloroplasts were labeled, in addition to the nucleoplasm.[104]

VI. CONCLUSIONS AND PERSPECTIVES

As usual, the biochemical data have accumulated far ahead of the correlated structural information, as is seen in the excellent review of Drlica.[21] The new techniques of immunocytochemistry (for references, see References 105 and 106) have also opened new avenues for the study of bacterial nucleoids. With these techniques, it became possible to locate the histone-like protein HU together with RNA polymerase and topoisomerase I in the same region, over the ribosomes in a 60-nm-wide zone adjacent to the lobes of the nucleoid.[92] HU is therefore not part of the bulk of the resting DNA. It does not form ''real'' nucleosomes in their role of ''compactosomes'', which restrain the negative supercoiling. Rather, HU is involved in processes related to transcription-translation, as is also suggested by its dual ability to bind to nucleic acids and to 30S native ribosomes.

Primarily on the basis of electron microscopy,[38,39,107-109] 50 independent domains of the bacterial genome were postulated.[79,110] They account perfectly for the contingencies related to the negative supercoiling and the needs of strand separation in transcription and replication (for references, see Reference 21). This hypothesis is particularly attractive because of its similarity with the loop model of the eukaryotic chromosome (for a recent discussion, see Reference 19), where the loops appear to be attached to a protein scaffold of which topoisomerase II is an essential constituent.[81] The prediction that the bacterial analog, the gyrase, should be situated in the center of the typical spherical nucleoid obtained after inhibition of protein synthesis has not become confirmed by experiment.[125]

Thin sections of cryofixed and cryosubstituted cells seem to be the best method to preserve topological relationships. Chemical fixation induces premature leakage long before

TABLE 4
Similarities and Differences of Chromatin

Characteristic	Prokaryotic	Eukaryotic
Aggregation sensitivity	Very Sensitive	Insensitive
Loops or domains	Good evidence	Strongly suggested
Bulk nucleosomes	Not yet found	Present and well defined
Labile pseudonucleosomes	As beads in E.M.	—
Supercoiling	Negative, mostly unrestrained	Negative, restrained
Location of topoisomerase I	In the area of transcription	With transcription
Location of topoisomerase II (gyrase)	Unknown	In the chromosomal scaffold
Location of phage topo II (gp 39)	In the membrane	—
Protein: DNA	1:10	10:10

the cytoplasm or the DNA plasm eventually becomes fixed. Although it was shown that, upon fixation, all macromolecular syntheses are immediately stopped,[55] a rearrangement due to altered ionic conditions (Table 1) is easily envisaged and has, indeed, been demonstrated.[24] A similar situation is encountered when nucleoids are isolated. Although we know fairly well the quality and quantities of the intracellular *cations* of the sap[27,30] and the intracellular pH of 7.5 to 8,[111,112] we still lack precise data on the corresponding small molecular *anions*. As long as we do not know them, we also do not know their dissociation constants and, consequently, the intracellular ionic strength. Such knowledge is indispensible for designing a correct buffer for the isolation of nucleoids in order to prevent either the dissociation of proteins from the DNA or, conversely, the artificial binding of any basic protein. In this context, one has also to be aware that DNA is most likely present in the form of a salt with Mg, spermidine, and putrescine.[27,113] Thus, it might be possible that nonbasic proteins bind specifically to already neutralized DNA instead of basic proteins using a salt that binds in competition with other cations. It is this Mg salt of DNA that forms an excellent substrate for nucleases. Therefore, a common practice in experiments involving DNA is to inhibit nucleolytic digestion by adding EDTA. This practice might have very deleterious effects on the structure.

The possibility of distinguishing clearly two types of DNA-containing plasms by virtue of their aggregation sensitivity adds one more parameter to the differences between eukaryotic and prokaryotic chromatin (Table 4). In the cases we have compared, the DNA concentrations (packing densities) are approximately the same. Hence, it is very likely that this fundamental difference in aggregation sensitivity is due to different relative amounts of proteins. The bulk of aggregation-sensitive DNA would contain less protein and, thus, in these DNA plasms the existence of classical nucleosomes with the protein core having about the same weight as the DNA itself is very doubtful. One therefore has to envisage other types of expression of the negative, unconstrained supercoiling of prokaryotic DNA. That some globular compaction exists ("compactosomes") is suggested by the rather granular appearance of the prokaryotic DNA plasm in thin sections. It would be interesting to know if the fibrillar aspect which is observed after Os fixation is due to an OsO_4-induced relaxation of the supercoiling. Some binding of OsO_4 to DNA, when it possesses particular secondary structures, has been reported.[114] A sensitivity of such regions toward nicking enzymes can be envisaged, similar to the OsO_4-induced activity of lysozymes observed with T4-infected bacteria.[123] It seems to us that a nonnucleosomal type of beaded supercoiling should not be excluded *a priori* if we consider the production of beads from protein-free DNA by the action of ethanol. It is important to reach an understanding of the physical basis of such beads — for instance, to know if the removal of the hydration shell[115] leads to a lower melting point of DNA. It is well known to electron microscopists that ssDNA and RNA have a strong tendency to fold into globular structures. Their preparation for electron microscopy in an extended form needs special procedures.[116,117]

In all these considerations, we should not forget that high packing densities need not be achieved solely by a torsional compaction. In the head of bacteriophage T4, the DNA, with a packing density of 800 mg/ml, is arranged in some sort of liquid crystal, forming domains of parallel packing of the DNA strands.[84] It is also relevant to mention here that in the dinoflagellate chromosomes, the chromatin filaments form characteristic, arc-shaped structures. Two to three rows of arcs had previously been observed in Gram-positive bacteria after RK-osmium fixation. This similarity corroborated a strong relatedness between the DNA plasms of prokaryotes and dinoflagellates,[118,119] a relationship which is also strongly suggested by showing the same type of aggregation sensitivity. Dinoflagellates, like prokaryotes, also show much less basic protein staining with the optical microscope.

The observations available until now are fully consistent with the existence of only two classes of chromatin, distinguished by the amount and quality of protein associated with the bulk DNA (or resting DNA), which is not involved in transcription. This difference might vanish for metabolically active DNA, where experiments demonstrated for eukaryotes an absence of nucleosomes and even histones, for instance with the rDNA in the nucleolus. Experiments with other material suggested that transcription occurs in the presence of nucleosomes (for references, see Reference 6 and 120). Recently, Lorch and La Pointe[121] have shown that nucleosomes inhibit initiation of transcription *in vitro,* while elongation of the transcript proceeds with transient histone displacement. Despite these different possibilities,[6,120] it will be very interesting to ascertain whether, throughout the living world, metabolically active vegetative DNA are similarly organized.

REFERENCES

1. **Ingraham, J. L., Maaløe, O., and Neidhardt, F. C.,** *Growth of the Bacterial Cell,* Sinauer Associates, Sunderland, MA, 1983.
2. **Schaechter, M., Neidhardt, F. C., Ingraham, J. L., and Kjeldgaard, N. O.,** *The Molecular Biology of Bacterial Growth,* Jones and Bartlett, Boston, 1985.
2a. **Pato, M.-L. and von Meyenburg, K.,** Residual RNA-synthesis in *E. coli* after inhibition of transcription by rifampicin, *Cold Spring Harbor Symp. Quant. Biol.,* 35, 497, 1970.
3. **Miller, O. L., Jr., Hamkalo, B. A., and Thomas, C. A., Jr.,** Visualization of bacterial genes in action, *Science,* 169, 392, 1970.
4. **Kellenberger, E., Carlemalm, E., Séchaud, J., Ryter, A., and de Haller, G.,** Considerations on the condensation and the degree of compactness in non-eukaryotic DNA-containing plasmas, in *Bacterial Chromatin,* Gualerzi, C. and Pon, C. L., Eds., Springer-Verlag, Heidelberg, 1986, 11.
5. **Woldringh, C. L. and Nanninga, N.,** Structure of nucleoid and cytoplasm in the intact cell, in *Molecular Cytology of E. coli,* Academic Press, Cambridge, 1985, 161.
6. **Kellenberger, E.,** About the organisation of condensed and decondensed non-eukaryotic DNA and the concept of vegetative DNA, *Biophys. Chem.,* 29, 51, 1988.
7. **Earnshaw, W. C., King, J., and Eiserling, F.,** The size of the bacteriophage T4 head in solution with comments about the dimension of virus particles as visualized by electron microscopy, *J. Mol. Biol.,* 122, 247, 1978.
8. **Baschong, W., Aebi, U., Baschong-Prescianotto, C., Dubochet, J., Landmann, L., Kellenberger, E., and Wurtz, M.,** Head structure of bacteriophage T2 and T4, *J. Ultrastruct. Mol. Struct. Res.,* 99, 189, 1988.
9. **Reichelt, R., Carlemalm, E., Villiger, W., and Engel, A.,** Concentration determination of embedded biological matter by STEM, *Ultramicroscopy,* 16, 69, 1985.
10. **Cairns, J.,** The bacterial chromosome and its manner of replication as seen by autoradiography, *J. Mol. Biol.,* 6, 208, 1963.
11. **Wang, J. C.,** DNA-topoisomerases, *Annu. Rev. Biochem.,* 54, 665, 1985.
12. **Wang, J. C.,** Recent studies of DNA-topoisomerases (a review), *Biochem. Biophys. Acta,* 909, 1, 1987.

13. **Vosberg, H. P.,** DNA topoisomerases: enzymes that control DNA conformation, *Curr. Top. Microbiol. Immunol.,* 114, 19, 1985.
14. **Gellert, M.,** DNA topoisomerases, *Annu. Rev. Biochem.,* 50, 879, 1981.
15. **Pettijohn, D. E. and Pfenninger, O.,** Supercoils in prokaryotic DNA restrained *in vivo, Proc. Natl. Acad. Sci. U.S.A.,* 77, 1331, 1980.
16. **Sinden, R., Carlson, J., and Pettijohn, D.,** Torsional tension in the double helix measured with trimethylpsoralen in living *E. coli* cells: analogous measurements in insect and human cells, *Cell,* 21, 773, 1980.
17. **Broyles, S. S. and Pettijohn, D. E.,** Interaction of the *E. coli* HU protein with DNA. Evidence for formation of nucleosome-like structures with altered DNA helical pitch, *J. Mol. Biol.,* 187, 47, 1986.
18. **Benyajati, E. and Worcel, A.,** Isolation, characterization and structure of the folded interphase genome of *Drosophila melanogaster, Cell,* 9, 393, 1976.
19. **Gasser, S. M. and Laemmli, U. K.,** A glimpse at chromosomal order, *TIG,* 3, 16, 1987.
20. **Pettijohn, D. E. and Sinden, R.,** Structure of the isolated nucleoid, in *Molecular Cytology of E. coli,* Nanninga, N., Ed., Academic Press, Cambridge, 1985, 199.
21. **Drlica, K.,** The nucleoid, in *Escherichia coli and Salmonella typhimurium,* Neidhardt, F. C., Ed., American Society for Microbiology, Washington, D. C., 1987, 91.
22. **Drlica, K. and Rouviere-Yaniv, J.,** Histonelike proteins of bacteria, *Microbiol. Rev.,* 51, 301, 1987.
23. **Acetarin, J. D., Carlemalm, E., Kellenberger, E., and Villiger, W.,** Correlation of some mechanical properties of embedding resins with their behavior in microtomy, *J. Electron Microsc. Tech.,* 6, 63, 1987.
24. **Hobot, J. A., Villiger, W., Escaig, J., Maeder, M., Ryter, A., and Kellenberger, E.,** Shape and fine structure of nucleoids observed on sections of ultrarapidly frozen and cryosubstituted bacteria, *J. Bacteriol.,* 162, 960, 1985.
25. **Epstein, W. and Schultz, S. G.,** Cation transport in *E. coli.* V. Regulation of cation content, *J. Gen. Physiol.,* 49, 221, 1965.
26. **Epstein, W.,** Osmoregulation by potassium transport in *E. coli, FEMS Microbiol. Rev.,* 39, 73, 1986.
27. **Moncany, M. L. J.,** Determination des conditions intracellulaires chez *E. coli.* Conséquences biologiques de leur modification, (sous la direction de E. Kellenberger), These d'Etat, Université de Paris VII, Paris, 1982.
28. **Alatoassava, T., Stauffer, E., and Seiler, H.,** Osmotic adaptation of T4 infected *E. coli, FEMS Microbiol. Lett.,* 43, 177, 1987.
29. **Chang, C. F., Shuman, H., and Somlyo, A. P.,** Electron probe analysis, X-ray mapping, and electron energy-loss spectroscopy of calcium, magnesium, and monovalent ions in log-phase and in dividing *E. coli* B cells, *J. Bacteriol.,* 167, 935, 1986.
30. **Richey, B., Cayley, D. S., Mossing, M. C., Kolka, C., Anderson, C. F., Farrar, T. C., and Record, M. T., Jr.,** Variability of the intracellular ionic environment of *E. coli, J. Biol. Chem.,* 262, 7157, 1987.
31. **Levrino, S., Harrison, C., Cayley, D. S., Burgess, R. R., and Record, M. T., Jr.,** Replacement of potassium chloride by potassium glutamate dramatically enhances protein-DNA interactions *in vitro, Biochemistry,* 20, 2095, 1987.
32. **Perroud, B. and Le Rudulier, D.,** Glycine betaine transport in *E. coli:* osmotic modulation, *J. Bacteriol.,* 161, 393, 1985.
33. **Chambers, S. T., Kunin, C. M., Miller, D., and Hamada, A.,** Dimethylthetin can substitute for glycine betaine as an osmoprotectant molecule for *E. coli, J. Bacteriol.,* 169, 4845, 1987.
34. **Woldringh, C. L.,** Effect of cations on the organisation of the nucleoplasm in *E. coli* prefixed with osmium tetroxide or glutaraldehyde, *Cytology,* 8, 97, 1973.
35. **Dubochet, J., McDowall, A. W., Menge, B., Schmid, E. N., and Lickfeld, K. G.,** Electron microscopy of frozen-hydrated bacteria, *J. Bacteriol.,* 155, 381, 1983.
36. **Kellenberger, E., Dürrenberger, M., Villiger, W., Carlemalm, E., and Wurtz, M.,** The efficiency of immunolabel on Lowicryl sections compared to theoretical predictions, *J. Histochem. Cytochem.,* 35, 959, 1987.
37. **Jacob, F., Brenner, S., and Cuzin, F.,** On the regulation of DNA replication in bacteria, *Cold Spring Harbor Symp. Quant. Biol.,* 28, 329, 1963.
38. **Kavenoff, R. and Ryder, O.,** Electron microscopy of membrane-associated folded chromosomes of *E. coli, Chromosoma,* 55, 13, 1976.
39. **Kavenoff, R. and Bowen, B.,** Electron microscopy of membrane-free folded chromosomes from *E. coli, Chromosoma,* 59, 89, 1976.
40. **Meyer, J. and Arnold-Schulz-Gahmen, B.,** Electron microscopic approaches to the study of bacterial DNA organization, *Microbiol. Sci.,* 5, 68, 1988.
41. **Robinow, C. F. and Kellenberger, E.,** The bacterial nucleoid, in preparation.
42. **Piekarski, G.,** Cytologische Untersuchungen an Paratyphus und Coli Bact., *Arch. Mikrobiol.,* 8, 428, 1937.
43. **Robinow, C. F.,** Cytological observations on bact. coli, proteus vulgaris and various aerobic spore-forming bacteria with special reference to the nuclear structures, *J. Hyg.,* 43, 413, 1944.

44. **Tulasne, R. and Vendrely, R.,** Demonstration of bacterial nuclei with ribonuclease, *Nature (London),* 160, 225, 1947.

45. **Birch-Andersen, A., Maaløe, D., and Sjostrand, F. S.,** High-resolution electron micrographs of sections of *E. coli, Biochim. Biophys. Acta,* 12, 395, 1953.

46. **Ryter, A., Kellenberger, E., Birch-Andersen, A., and Maaloe, O.,** Etude au microscope electronique de plasmas contenant de l'acide desoxyribunucleique, *Z. Naturforsch.,* 9, 598, 1958.

47. **Schreil, W. H.,** Studies on the fixation of artificial and bacterial DNA plasms for the electron microscopy of thin sections, *J. Cell Biol.,* 22, 1, 1964.

48. **Kellenberger, E. and Ryter, A.,** Contribution a l'etude du noyau bacterien, *Schweiz. Z. Allg. Pathol. Bakteriol.,* 18, 1122, 1955.

49. **Whitfield, J. F. and Murray, R. G. E.,** The effects of the ionic environment on the chromatin structures of bacteria, *Can. J. Microbiol.,* 2, 245, 1956.

50. **Kellenberger, E.,** The physical state of the bacterial nucleus, *Symp. Soc. Gen. Microbiol.,* 10, 39, 1960.

51. **Sabatini, D. O., Bensch, K., and Barrnett, R. J.,** The preservation of cellular ultrastructure and enzymatic activity by aldehyde fixation, *J. Cell Biol.,* 17, 19, 1963.

52. **Séchaud, J. and Kellenberger, E.,** Electron microscopy of DNA-containing plasms. IV. Glutaraldehyde-uranyl acetate fixation of virus infected bacteria for their sectioning, *J. Ultrastruct. Res.,* 39, 598, 1972.

53. **Daneo-Moore, L., Dicker, D., and Higgins, M. L.,** Structure of the nucleoid in cells of *Streptococcus faecalis, J. Bacteriol.,* 141, 928, 1980.

54. **Woldringh, C. L. and Nanninga, N.,** Organization of the nucleoplasm in *E. coli* cells visualized by phase-contrast light microscopy, freeze fracturing and thin sectioning, *J. Bacteriol.,* 127, 1455, 1976.

55. **Daneo-Moore, L. and Higgins, M. L.,** Morphokinetic reaction of *Streptococcus faecalis* (ATCC 9790) cells to the specific inhibition of macromolecular synthesis: nucleoid condensation on the inhibition of protein synthesis, *J. Bacteriol.,* 109, 1210, 1972.

56. **Baschong, W., Baschong-Prescianotto, C., Wurtz, M., Carlemalm, E., Kellenberger, C., and Kellenberger, E.,** Preservation of protein structures for electron microscopy by fixation with aldehydes and/ or OsO₄, *Eur. J. Cell Biol.,* 35, 21, 1984.

57. **Ryter, A.,** Etude au microscope electronique du transformation nucleaires de *E. coli* K12S et K12S(λ26) apres irradiation aux rayons ultraviolets et aux rayons X, *J. Biophys. Biochem. Cytol.,* 8, 399, 1960.

58. **Hobot, J. A., Bjornsti, M. A., and Kellenberger, E.,** The distribution of bacterial DNA investigated by on-section immunolabelling and cryo-substitution, *J. Bacteriol.,* 169, 2055, 1987.

59. **Mason, D. J. and Powelson, D. M.,** Nuclear division as observed in live bacteria by a new technique, *J. Bacteriol.,* 71, 474, 1956.

60. **Robinow, C. F.,** The chromatin bodies of bacteria, *Bacteriol. Rev.,* 20, 207, 1956.

61. **Valkenburg, J. A. C., Woldringh, C. L., Brakenhoff, G. J., van der Voost, H. T. M., and Nanninga, N.,** Confocal scanning light microscopy of the *E. coli* nucleoid: comparison with phase contrast and electron microscopy images, *J. Bacteriol.,* 161, 478, 1985.

62. **Chai, N. C. and Lark, K. G.,** Segregation of deoxyribonucleic acid in bacteria: association of the segregating unit with the cell envelope, *J. Bacteriol.,* 94, 415, 1967.

63. **Kellenberger, E.,** Les formes caracteristiques des nucleoides de *E. coli* et leurs transformations dues a l'action d'agent mutagenes-inducteurs et de bacteriophages, in *Symp. Citologia Batterica* (Roma), Supplemento Rendiconti Istituto Superiore di Sanita, 1953, 45.

64. **Kellenberger, E., Ryter, A., and Séchaud, J.,** Electron microscopy study of DNA-containing plasms. II. Vegetative and mature phage DNA as compared with normal bacterial nucleoids in different physiological states, *J. Biophys. Biochem. Cytol.,* 4, 671, 1958.

65. **Zusman, D. R., Carbonell, A., and Haga, J. Y.,** Nucleoid condensation and cell division in *E. coli* MX74T2 ts52 after inhibition of protein synthesis, *J. Bacteriol.,* 155, 1167, 1973.

66. **Kellenberger, E.,** unpublished data.

67. **Kellenberger, E., Alatossava, T., Stauffer, E., Jütte, H., and Seiler, H.,** Is the constitutive KfP-K⁺ pumping system of *E. coli* able to adapt to osmotic shifts without new protein synthesized?, in *Ion Pumps, Structure, Function and Regulation,* Stein, W., Ed., Alan R. Liss, New York, in press.

68. **Kellenberger, E., Carlemalm, E., Stauffer, E., and Wunderli, H.,** *In vitro* studies of the fixation of DNA, nucleoprotamine, nucleohistone and proteins, *Eur. J. Cell. Biol.,* 25, 1, 1981.

69. **Nass, M. K. and Nass, S.,** Intramitochondrial fibers with DNA characteristics. I. Fixation and electron staining reaction, *J. Cell Biol.,* 19, 593, 1963.

70. **Ris, H. and Plant, W.,** Ultrastructure of DNA-containing areas in the chloroplast of *Chlamydomonas, J. Cell Biol.,* 13, 383, 1962.

71. **Jardine, N. J. and Leaver, J. L.,** The fractionation of histones isolated from *Euglena gracilis, Biochem. J.,* 169, 103, 1978.

72. **Rizzo, P. J.,** Comparative aspects of basic chromatin proteins in dinoflagellates, *BioSystems,* 14, 433, 1981.

73. **Rizzo, P. J., and Burghardt, R. C.,** Histone-like protein and chromatin structure in the wall-less dino-flagellate *Gymnodinium nelsoni, BioSystems,* 15, 27, 1982.
74. **Griffith, J. D.,** Visualization of prokaryotic DNA in a regularly condensed chromatin-like fiber, *Proc. Natl. Acad. Sci. U.S.A.,* 73, 563, 1976.
75. **van Iterson, W. and Aten, J. A.,** Nuclear and cell division in *Bacillus subtilis, J. Ultrastruct. Res.,* 54, 135, 1976.
76. **Eickbush, T. H. and Moudrianakis, E. N.,** The compaction of DNA helices into either continuous supercoils or folded fiber rods and toroids, *Cell,* 13, 295, 1987.
77. **Noll, M. and Kornberg, R. G.,** Action of micrococcal nuclease on chromatin and the location of histone Hl, *J. Mol. Biol.,* 109, 393, 1977.
78. **Varshavski, A. J., Bakayev, V. V., Nedospasov, S. A., and Georgiev, G. P.,** On the structure of eukaryotic, prokaryotic and viral chromatin, *Cold Spring Harbor Symp. Quant. Biol.,* 42, 457, 1977.
79. **Pettijohn, D. E. and Hecht, R.,** RNA molecules bound to the folded bacterial genome stabilize DNA folds and segregate domains of supercoiling, *Cold Spring Harbor Symp. Quant. Biol.,* 38, 31, 1973.
80. **Paulson, J. R. and Laemmli, U. K.,** The structure of histone-depleted metaphase chromosomes, *Cell,* 12, 817, 1977.
81. **Earnshaw, W. C., Halligan, B., Cooke, C. A., and Liu, L. F.,** Topoisomerase II is a structural component of mitotic chromosome scaffolds, *J. Cell Biol.,* 100, 1701, 1985.
82. **Berrios, M., Osheroff, N., and Fisher, P. A.,** *In situ* localization of DNA-topoisomerase II, a major polypeptide component of the *Drosophila* nuclear matrix fraction, *Proc. Natl. Acad. Sci. U.S.A.,* 82, 4142, 1985.
83. **Casjens, S.,** Nucleic acid packaging by viruses, in *Virus Structure and Assembly,* Casjens, S., Ed., Jones and Bartlett, Portola Valley, CA, 1985, 75.
84. **Lepault, J., Dubochet, J., Baschong, W., and Kellenberger, E.,** Organization of double-stranded DNA in bacteriophages: a study by cryo-electron microscopes of vitrified samples, *EMBO J.,* 6, 1507, 1987.
85. **Yamazaki, K., Nagata, A., Kano, Y., and Imamoto, F.,** Isolation and characterization of nucleoid proteins from *E. coli, Mol. Gen. Genet.,* 196, 217, 1984.
86. **Hübscher, U., Lutz, H., and Kornberg, A.,** Novel histone H2A-like protein of *E. coli, Proc. Natl. Acad. Sci. U.S.A.,* 77, 5097, 1980.
87. **Kishi, F., Ebina, Y., Micki, T., Nakazawa, T., and Nakazawa, A.,** Purification and characterization of a protein from *E. coli* which forms complexes with superhelical and single-stranded DNA, *J. Biochem.,*
88. **Pettijohn, D. E.,** Structure and properties of the bacterial nucleoid, *Cell,* 30, 669, 1982.
89. **Rouviere-Yaniv, J. and Gros, F.,** Characterization of a novel, low molecular-weight DNA-binding protein from *E. coli, Proc. Natl. Acad. Sci. U.S.A.,* 72, 3428, 1971.
90. **Rouviere-Yaniv, J.,** Localization of the HU protein on the *E. coli* nucleoid, *Cold Spring Harbor Symp. Quant. Biol.,* 42, 439, 1977.
91. **Rouviere-Yaniv, J., Yaniv, M., and Germond, J. E.,** *E. coli* DNA binding protein HU forms nucleosome-like structures with circular double-stranded DNA, *Cell,* 17, 265, 1979.
92. **Dürrenberger, M., Bjornsti, M. A., Uetz, T., Hobot, J. A., and Kellenberger, E.,** The intracellular location of the histone-like protein HU in *E. coli, J. Bacteriol.,* 170, 4757, 1988.
93. **Suryanarayana, T. and Subramanian, A. R.,** Specific association of two homologous DNA-binding proteins to the native 30-S ribosomal subunits of *E. coli, Biochim. Biophys. Acta,* 520, 342, 1978.
94. **Fleischmann, G., Pflugfelder, G., Steiner, E. K., Javaherian, K., Howard, G. C., Wang, J. C., and Elgin, S. C. R.,** *Drosophila* DNA topoisomerase I is associated with transcriptionally active regions of the genome, *Proc. Natl. Acad. Sci. U.S.A.,* 81, 6958, 1984.
95. **Yang, H. L., Heller, K., Gellert, M., and Zubay, G.,** Differential sensitivity of gene expression *in vitro* to inhibitors of DNA gyrase, *Proc. Natl. Acad. Sci. U.S.A.,* 76, 3304, 1979.
96. **Mosig, G.,** T4 genes and gene products, in *Bacteriophage T4,* Mathews, C. K., Kutter, E. M., Mosig, G., and Berget, P. B., Eds., American Society for Microbiology, Washington, D.C. 1983, 362.
97. **Kreuzer, K. N. and Jongeneel, C. V.,** *E. coli* phage topoisomerase, *Methods Enzymol.,* 100, 144, 1983.
98. **Kreuzer, K. N. and Alberts, B. M.,** Site-specific recognition of bacteriophage T4 DNA by T4 type II DNA topoisomerase and *E. coli* DNA gyrase, *J. Biol. Chem.,* 259, 5339, 1984.
99. **Seasholtz, A. F. and Greenberg, G. R.,** Identification of bacteriophage T4 gene 60 product and a role for this protein in DNA topoisomerase, *J. Biol. Chem.,* 258, 1221, 1983.
100. **Huang, W. M.,** Membrane-associated proteins of T4-infected *E. coli, Virology,* 66, 508, 1975.
101. **Takacs, B. J. and Rosenbusch, J. P.,** Modification of *Escherichia coli* membranes in the prereplicative phase of phage T4 infection, *J. Biol. Chem.,* 250, 2339, 1975.
102. **Zachary, A. and Black, L. W.,** Topoisomerase II and other DNA-delay and DNA-arrest mutations impair bacteriophage T4 DNA-packaging *in vivo* and *in vitro, J. Virol.,* 60, 97, 1986.
103. **Hershey, A. D. and Melechen, N. E.,** Synthesis of phage-precursor nucleic acid in the presence of chloramphenicol, *Virology,* 3, 207, 1957.

104. **Vernet, G., Maeder, M., and Kellenberger, E.,** to be published.
105. **Verkleij, A. J. and Leunissen, J. L. M.,** *Immuno-gold Labeling Methods,* CRC Press, Boca Raton, FL in press.
106. **Bullock, G. R. and Petrusz, P.,** *Techniques in Immuno-cytochemistry,* Vols. 1 to 3, Academic Press, London, 1982, 1983, 1985.
107. **Worcel, A. and Burgi, E.,** On the structure of the folded chromosome of *Escherichia coli, J. Mol. Biol.,* 71, 127, 1972.
108. **Worcel, A. and Burgi, E.,** Properties of a membrane-attached form of the folded chromosome of *E. coli, J. Mol. Biol.,* 82, 91, 1974.
109. **Delius, H. and Worcel, A.,** Electron microscopic visualization of the folded chromosome of *E. coli, J. Mol. Biol.,* 82, 107, 1974.
110. **Sinden, R. S. and Pettijohn, D. E.,** Chromosomes in living *E. coli* cells are segregated into domains of supercoiling, *Proc. Natl. Acad. Sci. U.S.A.,* 78, 224, 1981.
111. **Repaske, D. R. and Adler, J.,** Change in intracellular pH of *E. coli* indicates the chemotactic response to certain attractants and repellents, *J. Bacteriol.,* 145, 1196, 1981.
112. **Slonczewski, J. L., Rosen, B. P., Alger, J. R., and Mamab, R. M.,** pH homeostasis in *E. coli:* measurement by ^{31}P nuclear magnetic resonance of methylphosphonate and phosphate, *Proc. Natl. Acad. Sci. U.S.A.,* 78, 6271, 1981.
113. **Bjornsti, M. A., Hobot, J. A., Kelus, A. S., Villiger, W., and Kellenberger, E.,** New electron microscopic data on the structure of the nucleoid and their functional consequences, in *Bacterial Chromatin,* Gualerzi, C. O. and Pon, C. L., Eds., Springer-Verlag, Berlin, 1986, 64.
114. **Lilley, D. M. J. and Palecek, E.,** The supercoil-stabilised cruciform of ColEl is hyper-reactive to osmium tetroxide, *EMBO J.,* 3, 1187, 1984.
115. **Hearst, J. E. and Vinograd, J.,** The net hydration of T4 bacteriophage deoxyribonucleic acid and the effect of hydration on buoyant behavior in a density gradient of equilibrium in the ultra-centrifuge, *Proc. Natl. Acad. Sci. U.S.A.,* 47, 1005, 1961.
116. **Brack, C.,** DNA electron microscopy, *CRC Crit. Rev. Biochem.,* 10, 113, 1981.
117. **Meyer, J.,** Electron microscopy of viral RNA, in *Current Topics in Microbiology and Immunology* 94/95, Henle, W., Hofschneider, P. H., Koprowski, H., Maaloe, O., Melchers, F., Rott, R., Schweizer, H. G., and Vogt, P. K., Eds., Springer-Verlag, Berlin, 1981, 209.
118. **Giesbrecht, P.,** Uber das "supercoiling" System der Chromosomen von Bakterien und Flagellaten und seine Beziehungen zu Nucleolus und Kerngrundsubstanz, *Zentralbl. Bakteriol. Parasitenkd. Infectionskr. Hyg. Abt. Orig. Reibe A,* 183, 1, 1961.
119. **Drews, G. and Giesbrecht, P.,** Die Bauelemente der Bakterien und Blaualgen, in *Die Zelle, Struktur und Funktion,* Melzner, H., Ed., Wissenschaftliche Verlagsgesellschaft, Stuttgart, 1971, 407.
120. **Kellenberger, E.,** The compactness of cellular plasmas; in particular chromatin compactness in relation to function, *TIBS,* 12, 105, 1987.
121. **Lorch, J. and La Pointe, J.,** Nucleosomes inhibit the initiation of transcription but allow chain elongation with the displacement of histones, *Cell,* 49, 203, 1987.
122. **Millonig, C.,** A modified procedure for lead staining of thin sections. *J. Biophys. Biochem. Cytol.,* 11, 736, 1961.
123. **Kellenberger, E.,** unpublished observations.
124. **Han, M., Chang, M., Kim U.-J., and Grunstein, M.,** Histone H2B repression causes cell-cycle-specific arrest in yeast: effects on chromosomal segregation replication and transcription, *Cell,* 48, 589, 1987.
125. **Dürrenberger, M., Yang, Y., Ferro-Luzzi Ames, G., Gellert, M., and Kellenberger, E.,** Immuno-cytological localization of gyrase in E. Coli, unpublished.

Chapter 2

THE CHROMOSOMAL DNA REPLICATION ORIGIN, *ORIC*, IN BACTERIA

Douglas W. Smith and Judith W. Zyskind

TABLE OF CONTENTS

I. INTRODUCTION

The bacterial chromosome is a single, circular, double-stranded DNA molecule of molecular weight 500×10^6 (*Mycoplasma hominus*)[1] to about 3000×10^6 Da. In *Escherichia coli,* this large molecule (4700 kb)[2] is present as a highly compact structure, enclosed within a space of about 0.5 μm.[3] The DNA concentration is thus very high (13 to 26 mg/ml packing density),[3] yet this chromosome functions rapidly and efficiently in gene expression and in DNA recombination, repair, and replication. Very little information is yet available concerning these topological considerations.

Within the cell, the nuclear region or nucleoid is not enclosed within a membrane (see Kellenberger, Chapter 1, for electron microscopy data), and the chromosome can be isolated as a highly compact, nonviscous structure from *E. coli* lysed by nonionic detergents plus 1 *M* NaCl,[4] a structure that contains the chromosome as an intact DNA molecule.[5] When lysed at room temperature, the folded chromosome or nucleoid contains little cellular membrane material and the primary protein is RNA polymerase.[6] These structures possess a superhelical density similar to other supercoiled DNA molecules, but many nicks are needed to completely relax the DNA.[6,7] These data plus electron microscopy indicate that a single folded chromosome possesses about 40 to 50 domains of superhelicity. RNase and rifampicin experiments[5-7] indicate a structural role for nascent RNA molecules in the integrity of these superhelical domains, although protein may also be involved. When isolated at low temperature, more protein species and membrane material are found associated with the folded chromosomes,[8,9] including the histone-like protein HU.[10] The function of the supercoiled domains, and RNA and proteins, in the structure and function of the nucleoid *in vivo* remains ill understood (for recent reviews, see Drlica,[11] Pettijohn and Sinden,[12] and Kellenberger, this volume).

Replication of the bacterial chromosome proceeds sequentially and in an overall semiconservative manner from a unique heritable origin, *oriC,* bidirectionally to a terminus region roughly halfway around the circular chromosome. Events involved in beginning a round of DNA replication at *oriC* are termed the initiation process, those at the replication forks in the subsequent DNA synthesis events are termed the elongation process, and the terminus region events resulting in completion of the round of DNA replication and resolution of daughter chromosomes constitute the termination process. The time between an initiation event and termination of a round of replication, the C period, and that between this termination and subsequent cell division, the D period, are roughly constant (values of about 40 and 20 min, respectively) at rapid growth rates (doubling times of 60 min or less), but become longer at slower growth rates (see Leonard and Helmstetter, this volume). Under balanced growth conditions, cells exactly double their DNA content each generation time. The constancy of the C period, then, implies that temporal regulation of DNA synthesis (amount of DNA synthesized per unit time) is controlled at the level of the initiation (or termination) process. That is, to synthesize more DNA per unit time, cells initiate more often rather than increase the rate of chain elongation. Understanding these control mechanisms is a primary objective of studies of the initiation process, and recent reviews are available on this topic.[13-15] The structure of the origin, *oriC,* found in bacteria and the events, catalytic and regulatory, which occur in initiation at *oriC* are the subjects of this review.

II. STRUCTURAL ANALYSIS OF THE ENTEROBACTERIAL ORIGIN

In *E. coli,* marker transfer experiments using synchronized cells showed that the origin mapped near the *ilv* operon;[16,17] however, precise mapping awaited restriction enzyme mapping approaches. Using gradients of ³H found in restriction fragments isolated from *E. coli*

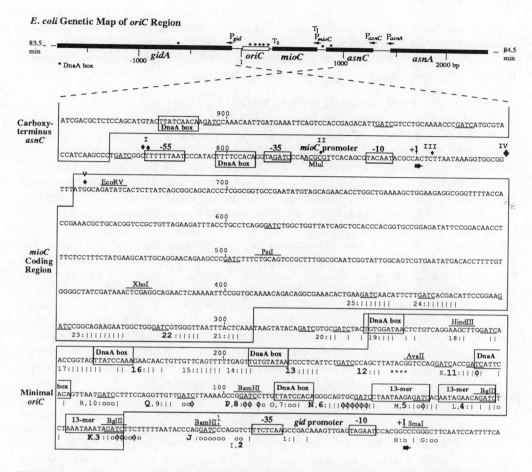

FIGURE 1.   Section of *E. coli* genetic map containing the origin of replication, *oriC,* and nucleotide sequence of region associated with *oriC* function. The coding regions are based on nucleotide sequence[21,226] and maxicell analysis of labeled proteins encoded by λ$_{asn}$ transducing phage.[227] Diamonds in the *mioC* promoter region indicate the 3′ ends of *asnC* transcripts, with frequency of termination designated by size of diamond.[84] The −10 and −35 promoter regions[228,250] are boxed, as are the 9-mer (DnaA box)[34,35] and 13-mer[50] DnaA recognition sequences. *Vertical lines* indicate sites of RNA polymerase pausing *in vitro* and 3′ termini of *in vivo* transcripts originating from the *mioC* promoter.[113,114,121] The RNA:DNA transition clusters are designated by *letters* G-S and the sites where transitions occur are indicated (o);[112] sites corresponding to both RNA:DNA transitions and 3′ termini of *in vivo* transcripts as well as *in vitro* pause sites are indicated by φ. Bold letters and numbers refer to clusters present in higher frequency. GATC sites are *underlined* and relevant restriction sites are indicated.

dnaC and *dnaA* cells, Marsh and Worcel[18] showed that the origin was contained within a 9-kb *Eco*RI fragment adjacent to the *E. coli asn* and *tna* genes. This led von Meyenburg and co-workers[19] and others[20] to isolate and localize the *oriC* region using lambda transducing phage; without using recombinant DNA techniques, this yielded a small plasmid, pCM959, of size 4012 bp,[21] containing only *E. coli* chromosomal DNA.[22] Simultaneously, the *oriC* region was cloned by conversion of a nonreplicating DNA restriction fragment containing a gene conferring antibiotic resistance into a self-replicating plasmid,[23] and this rapidly led to the DNA sequence determination of *oriC*.[24,25] Deletion derivative analyses of these *oriC* plasmids, coupled with complementation of *E. coli* mutants, positioned *oriC* between the *asnA* and the *gidA* genes[19,26] at 84 min on the revised *E. coli* genetic map (see Figure 1) and yielded a minimal region required for autonomous replication of only 238 to 245 bp.[27,28] This minimal origin contains no apparent open reading frames (ORFs) or genes and is unusually AT-rich (59%).

Two complementary approaches have been used to determine the relative importance of specific nucleotides within *oriC* to its function in initiation: sequence comparison of the *E. coli* origin with other naturally occurring bacterial origins of replication that are functional in *E. coli* and mutational analysis of the *E. coli* origin.

A. SEQUENCE COMPARISON OF ENTEROBACTERIAL ORIGINS

In this approach, chromosomal origins from bacteria "closely related" to *E. coli* are cloned and analyzed in *E. coli*. If they function in *E. coli* using *E. coli trans*-acting initiation factors, they can reasonably be viewed as "multiply mutated" *E. coli* origins and the resulting DNA sequences can be meaningfully compared. Bacterial origins from four other members of the Enterobacteriaceae and one from Vibrionaceae were cloned and each yielded functional replicons in *E. coli*. These were analyzed genetically and physically and their DNA sequences were determined. Comparisons of these sequences yielded an unambiguous alignment, shown in Figure 2.

Genetically, the *oriC* region from each of these bacteria resembles that of *E. coli*.[29] The *Eco*RI chromosomal fragment containing *oriC* from *Salmonella typhimurium* contains genes that complement *E. coli* mutations in *asnA* and in *atpB (uncB)*,[30] as do the *Sal*I *oriC* fragments from *Enterobacter aerogenes* and from *Klebsiella pneumoniae*.[31] The *Sal*I *oriC* fragment cloned from the plant pathogen *Erwinia carotovora* complements *E. coli asnA* mutants,[32] and the *Sal*I *oriC* fragment cloned from the marine bacterium *Vibrio harveyi* contains sequences on either side of *oriC* homologous to those found in the *gidA* and *mioC* genes of *E. coli*, although sequences more distant from the *V. harveyi* origin do not complement *E. coli asnA* mutants and express proteins of sizes different from those of *E. coli*.[33] This genetic analysis also provides convincing evidence that these functional origins are indeed the bacterial origins rather than origins from a plasmid or cryptic prophage.

The most striking feature apparent from the consensus sequence (Figure 2) is that regions of complete identity (e.g., positions 60 to 71 and 80 to 88) are separated by regions of fixed numbers of nucleotides, but where nearly any nucleotide can be used (e.g., positions 72 to 79). This is unlike a typical protein coding region where sequence changes occur most often every third base, reflecting redundancy in the genetic code; it is, however, typical of regulatory regions such as promoters. Three sets of repeat structures are found among the conserved regions. The first are the "R sites": R1 and R4 are perfect inverted repeats, R3 and R5 (positions 135 to 143) are direct repeats of R1, and R2 is a direct repeat of R4 (Figure 2). These sites are essential for the binding of DnaA protein,[34,35] a protein required positively for initiation at *oriC* both *in vivo*[36] and *in vitro*,[37,38] as discussed below, and the function associated with the binding of DnaA protein to these repeats appears to be dependent on the orientation of the repeat.[39] The consensus R5 site is not conserved in the *V. harveyi* sequence (Figure 2); however, there is a potential DnaA 9-bp site at positions 121 to 129 in this sequence which may be the R5 of the *V. harveyi* origin. Note also that the *V. harveyi* origin contains two GATC sites spaced 3 bp apart at positions 107 to 117 which may be the functional equivalents to the two GATC sites found at positions 142 to 152 in all the other origins.

Similar 9-bp DnaA protein binding sites are found in the origin regions of several *E. coli* bacteriophage and plasmids, and in the promoter regions and coding regions of many genes, particularly those involved in DNA metabolism (see Figure 3). DnaA protein has been shown to bind to some of these sites,[34] and functions in control of expression of some of these genes, including the *dnaA* gene itself[40-43] (see below and Figure 7A), *dam*,[44] *mioC*,[41,43,45,46] (see below and Figure 1), and *uvrB*.[47] Similar sites are found in the origin regions from the bacteria *Pseudomonas aeruginosa*,[48] *P. putida*,[48] and *Bacillus subtilis*[49] (see below and Figures 10 and 11). The primary DnaA sites from these bacterial origins, plus those from some *E. coli* bacteriophage, transposons, and genes are compared in Figure

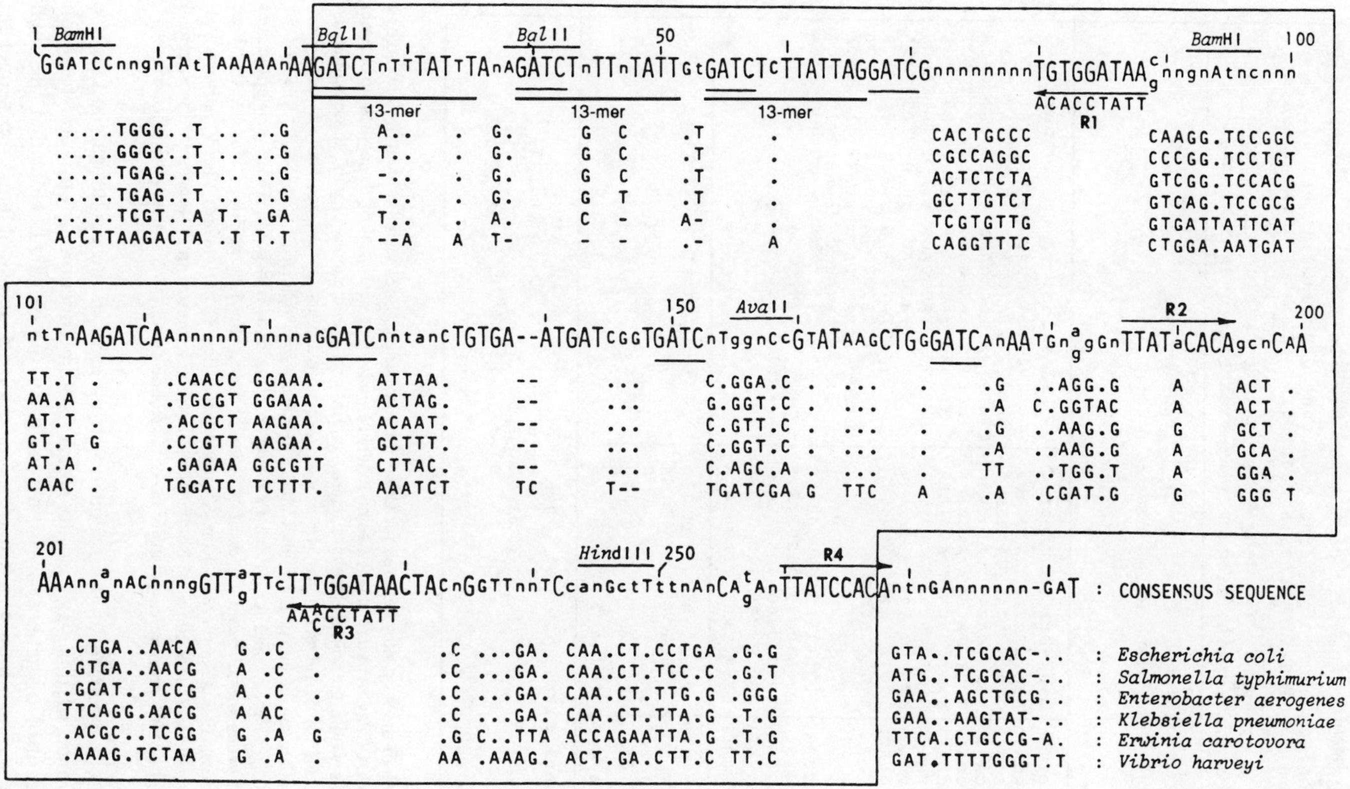
BamHI
BglII
BglII
50
13-mer
13-mer
13-mer
BamHI
100
R1
101
150
AvaII
R2
200
201
HindIII 250
R4
R3
CONSENSUS SEQUENCE
Escherichia coli
Salmonella typhimurium
Enterobacter aerogenes
Klebsiella pneumoniae
Erwinia carotovora
Vibrio harveyi

FIGURE 2. Consensus sequence of the minimal origin of the enterobacterial chromosome. The consensus sequence is derived from six bacterial origin sequences, all of which function in *E. coli*[19,20,24-26,30-33,151,229] A large capital letter means that the same nucleotide is found in all six origins; a small capital letter indicates the nucleotide is present in five of the six sequences; a lower-case letter is used when that nucleotide is present in three or four of six bacterial origins, but only two different nucleotides are found at that site; and where three or four of the four possible nucleotides, or two different nucleotides plus a deletion, are found at a site, the letter n is used. In the individual origin sequences, − means a deletion relative to the consensus site is present and a dot indicates that the nucleotide in the bacterial sequence is the same as the nucleotide in the consensus sequence. GATC sites are underlined, as are the 13-mers.[50] The minimal origin of *E. coli* is enclosed within the box.[27] The numbering of the nucleotide positions is that used for *E. coli,* and the 5′ end is at the upper left. The four related 9-bp DnaA boxes, R1, R2, R3, and R4, are indicated by arrows, with the 5′→3′ consensus sequence listed below the arrows for those DnaA boxes in the opposite orientation. The fifth DnaA box, R5 (not indicated), is found between *E. coli* nucleotide positions 135 and 143;[35] a potential R5 in *V. harveyi* is at positions 121 to 129 (see text and Figure 3).

```
          Eco-oriC  Sty-oriC  Eae-oriC  Kpn-oriC  Eca-oriC  Vha-oriC
R1:       TTATCCACA TTATCCACA TTATCCACA TTATCCACA TTATCCACA TTATCCACA
R2:       TTATACACA TTATACACA TTATGCACA TTATACACA TTATACACA TTATGCACA
R3:       TTATCCAAA TTATCCAAA TTATCCAAA TTATCCAAA TTATCCCAA TTATCCAAA
R4:       TTATCCACA TTATCCACA TTATCCACA TTATCCACA TTATCCACA TTATCCACA
R5:       TCATTCACA TCATTCACA TCATTCACA TCATTCACA TCATTCACA TGATCCAAA

G:        000000000 000000000 000010000 000000000 000000000 010010000
A:        005010515 005010515 005000515 005010515 005010415 005000525
T:        540510000 540510000 540510000 540510000 540510000 540500000
C:        010035040 010035040 010035040 010035040 010035140 000045030
Con Seq:  TTATcCACA TTATcCACA TTATcCACA TTATcTACA TTATcCACA TTATCCAcA

      oriC DnaA Sites:    G:  0  1  0  0  2  0  0  0  0
                          A:  0  0 30  0  4  0 29  7 30
                          T: 30 24  0 30  5  0  0  0  0
                          C:  0  5  0  0 19 30  1 23  0

Net Consensus Sequence:       T  T  A  T  c  C  A  C  A
```

```
Pae-oriC  Ppu-oriC | Bsu1-oriC Bsu2-oriC |    Plasmids              Others
TTATGCACA TTATCCACA| ATATCCACA TTATTCACA | ColE1:TTATCCACA    Tn5:TTATACACA
TTATCCACA TTATCCACA| GTATCCACA TCATACACA | pBR322:TTATCCACA   M13:TTATACAAT
TTTTCCACA TTATCAACA| TTATCGACA CTATCCACA | CloDF13:TTACCCACA  P1:TTATCCACA
TTATCCACA TTATCCACA| TTGTCCACA TTATCCACA | CloDF13:TTATACACG  P1:TTATCCACT
TTCTCGACA TCATCCACA| TTATCCACA TTATAAAAA | p15A:TTAACCAAA     mioC:TTTTCCACA
TTATCCACA TTATCCACT| TTATCCACA           | RSF1030:TTATCCACA  asnC:TTATCAACA
          TTATTCACA| TTCTACACA           | NTP1:TTATCCACA     dnaA:TTATCCACA
          TTATCCACA| TTGTCCACA           | F:TTATCCACG        dam:TTCTCCACA
                   |                     | F:TTATCCACT        polA:TTATCCACA
                   |                     | pSC101:TTATACACA   uvrB:TTATCCACA
                   |                     | RK2:TTATCCCCA      rpoH:TTATCCACA
G:000011000 000000000 102001000 000000000| RK2:TTGTCCACA      nrdA:TTATCCACA
A:00.4000606 008001807 106010909 005021515| R751:TTATCCACA
T:661600000 870810001 790900000 440510000| R751:TTATCCACA
C:001055060 010077080 001088090 110024040|
  TTaTCCACA TTATCCACA  tTaTCCACA TTATnTACA

G: 0  0  0  0  1  1  0  0  0    1  0  2  0  0  1  0  0  0    G: 0  0  1  0  0  0  0  0  2
A: 0  0 12  0  0  1 14  0 13    1  0 11  0  3  1 14  1 14    A: 0  0 23  1  4  1 25  2 21
T:14 13  1 14  1  0  0  0  1   11 13  0 14  1  0  0  0  0    T:26 26  1 24  0  0  0  0  3
C: 0  1  1  0 12 12  0 14  0    1  1  1  0 10 12  0 13  0    C: 0  0  1  1 22 25  1 24  0

   T  T  A  T  C  C  A  C  A    t  T  a  T  c  c  A  C  A       T  T  A  T  C  C  A  C  A

     Total DnaA Sites:     G:  1  0  3  0  3  2  0  0  2
                           A:  1  0 76  1 11  3 82 10 78
                           T: 81 76  2 82  7  0  0  0  4
                           C:  1  6  3  1 63 79  2 74  0

Final Consensus Sequence:      T  T  A  T  c/a C  A  C/a A
```

FIGURE 3.   Comparison of the 9-bp sites required for DnaA protein binding. Sites with 8 or 9 out of 9 homology to the sequence TTAT(C/A)CA(C/A)A are shown; some evidence exists that nearly all of these sites do interact with DnaA protein. Sites within the minimal enterobacterial origins and pseudomonas origins, those within the two potential origin regions for *B. subtilis*, plus 9-bp sequences from plasmids, phage, and some *E. coli* genes relevant to DNA metabolism are included: plasmids ColE1,[230] pBR322,[188,230] CloDF13,[230] p15A,[230] RSF1030,[230] NTP1,[231] F,[232,233] pSC101,[234] RK2,[235] and R751,[236] a *P. aeruginosa* plasmid, transposon Tn5,[237] bacteriophages M13[34] and P1,[232,238] and genes *mioC*,[113] *asnC*,[113] *dnaA*,[40-43] *dam*,[44] *polA*,[47] *uvrB*,[47] *rpoH*,[240,249] and *nrdA*.[241] Eco: *E. coli*; Sty: *S. typhimurium*; Eae: *E. aerogenes*; Kpn: *K. pneumoniae*; Eca: *E. carotovora*; Vha: *V. harveyi*; Pae: *P. aeruginosa*; Ppu: *P. putida*; Bsu: *B. subtilis*. Bold capitals: nucleotide present in 95% or more cases; capital letters: nucleotide present in 80% or more cases; lower case: nucleotide present in less than 80% of cases.

3. Although similar, the consensus 9-bp sequence from each of these sources shows some variation; the relevance of this variation to DnaA protein function in initiation for each bacterium and in control of gene expression remains to be clarified. Note that the final consensus shows a strong preference for Cs at positions 5 and 8 of the 9-bp sequence. Although Figure 3 is a useful summary of the 9-bp DnaA binding site, it has two limitations: (1) some sites of less homology are known to be DnaA interaction sites, for example, the weak binding site TTAAACCAA found in the origin region of plasmid RK2,[235] and (2) the

nucleotide distribution flanking the 9-bp sites is not random and probably influences DnaA interaction.

In addition, between 9 and 14 GATC sites are found within each minimal origin and 8 of these are conserved in position; possible function(s) of these sites in initiation is discussed below. Furthermore, three 13-bp direct repeats of consensus sequence GATCTnTTnTTTT are found beginning at positions 23, 39, and 54, as well as a fourth found just outside the minimal origin, beginning at position 2.[50] These sequences appear to be recognized by DnaA protein after it is bound to *oriC,* with a melting of this origin region for binding of DnaB protein, as discussed below. The degree of conservation for each of the 13-bp repeats *between* the six enterobacterial origins is greater than between each of the three 13-bp repeats *within* a given origin (Figure 4), suggesting that nucleotide differences between each of the three 13-bp repeats may be significant for the initiation process. Further, the first 13-bp repeat is the one least required for prepriming complex formation and for *in vitro* replication[50] (see discussion below), yet is an absolute requirement for initiation *in vivo.* This repeat may thus perform a function in initiation beyond that determined to date.

Each of the enterobacterial origin regions is AT-rich, ranging from 59 to 66% A + T, and the abundance of stop codons precludes this region encoding any proteins. The few relative insertions and deletions occur only in the region containing the 13-bp direct repeats (see Figure 2).

The DNA sequence from the *oriC-mioC* region from E. coli B/r has been determined and shown to differ at 11 positions from that of *E. coli* K12, two of which are in the minimal origin.[127] The two changes are an A to C at position 213 and a T to C at position 221. Both of these are "new" changes not present in other enterobacterial origins (Figure 2) and not found by mutagenesis of the *E. coli* K12 origin (Figure 5). The T at position 221, part of the DnaA site R3, is completely conserved in all other enterobacterial origins. The B/r *oriC* sequence thus extends the list of nucleotides within *oriC* that can be altered and yield a functional origin, although the B/r region on a plasmid has not been shown to function in a normal *E. coli* K12 background.[127]

B. MUTAGENESIS OF *ESCHERICHIA COLI ORIC*

Initial mutational analysis[27,51] centered on extensions of the deletion derivative approach used to define the minimal origin and on introduction of small insertions or deletions at restriction sites within *oriC.* These derivatives all lack initiation function (Ori⁻ phenotype), with the exception of a 12-bp deletion at the *Hind*III site and a 4-bp GATC duplication at the *Bgl*II site at position 38 (Figure 5).

In contrast, isolation of point substitution mutants within *oriC* yielded mutants with both Ori⁺ and Ori⁻ phenotypes.[28,52,53] These Ori⁺ point mutations are found in the same regions as the Ori⁻ deletion/insertion mutations, suggesting that *oriC* is composed of binding regions for *trans*-acting factors whose separation distance must be precisely maintained (spacer regions).[28] This is as expected for a complex prokaryotic regulatory region, exemplified, for example, by the prokaryotic promoter with its 16- to 18-bp spacer region separating the −10 and −35 regions. Many of the point mutations so obtained, however, map to other regions of *oriC* and there is a wealth of potentially useful information here for correlation of structure with function.

Interpretation of the mutants is, however, complicated by two factors: (1) many mutants so obtained contain more than one base change and (2) assays for Ori phenotypes do not always yield strict plus/minus results. Regarding the first factor, effects of one base change can be negated or accentuated (particularly true of the Ori⁻ mutants) by the presence of one or more other mutations; sorting out the effects is complex. With regard to the second factor, none of the single base substitution Ori⁻ mutants show a complete loss of origin function. The assays used, either *in vitro* incorporation assays or ability to function *in vivo* as an

FIGURE 4.

Enterobacterial-type Origins

13mers:	1st	2nd	3rd
E. coli:	GATCTATTTATTT	GATCTGTCTATT	GATCTCTTATTAG
S. typhimurium:	GATCTTTTATTT	GATCTGTCTATT	GATCTCTTATTAG
E. aerogenes:	GATCTTTTATTTA	GATCTGTTCTATT	GATCTCTTATTAG
K. pneumoniae:	GATCTTTTATTTA	GATCTGTTTATT	GATCTCTTATTAG
E. carotovora:	GATCTTTTATTT	GATCTCTTTATTT	GATCTCTTATTAG
V. harveyi:	GATCTATATATAT	GATCTTTTATTGA	GATCTATTATTAG

	1st	2nd	3rd
G:	6000000000000	6000040000010	6000000000006
A:	0600020124012	0600000011401	0600010006060
T:	0060646542654	0060616625255	0060606606600
C:	0006000000000	0006010030000	0006050000000
Each Con Seq:	GATCTtTTtaTTt	GATCTgTTnTaTT	**GATCTCTTATTAG**

All 13mers:

G:	**18**	0	0	0	0	4	0	0	0	0	0	1	6
A:	0	**18**	0	0	0	3	0	1	9	5	4	7	3
T:	0	0	**18**	0	**18**	5	**18**	17	6	13	14	10	9
C:	0	0	0	**18**	0	6	0	0	3	0	0	0	0

Net Con Seq: G A T C T n T T n T T n n

Pseudomonad-type Origins

13mers:	1st	2nd	3rd
P. aeruginosa:	GATGACGTAATAG	GAAGTCATTTTC	GAAACGGTTAATG
P. putida:	GATCGGGGACAAC	GAAGAGACATATA	GAAGAACTTATAA

	1st	2nd	3rd
G:	2001112100001	2002010000000	2001011000001
A:	0200100021120	0220102010101	0221110002111
T:	0020000100100	0000100112120	0000000220110
C:	0001010001001	0000010100001	0000101000000
Each Con Seq:	GATnnnGnAnnAn	GAAGnnAnnTnTn	GAAnnnnTTAnnn

All 13mers:

G:	**6**	0	0	4	1	3	3	1	0	0	0	0	2
A:	0	**6**	4	1	3	1	2	0	3	3	3	3	2
T:	0	0	2	0	1	0	0	4	3	2	4	3	0
C:	0	0	0	1	1	2	1	1	0	1	0	0	2

Net Con Seq: G A A G a g g T n n n n n

FIGURE 4.   Comparison of the 13-bp direct repeats found in the enterobacterial and pseudomonad origins. The 13-bp repeats from *E. coli* are those identified by Bramhill and Kornberg[50] as functional in initiation at *oriC in vitro*. Those from the other five enterobacterial-type origins are identified from sequence homology. The pseudomonad 13-bp repeats are based on sequence homology to those of the enterobacterial origins and on their position conservation in the pseudomonas origins (see Figure 10). Bold capitals: nucleotide present in 90% or more cases; normal capitals: nucleotide present in 70% or more cases; lower case: nucleotide present in 60% or more cases; n: nucleotide present in less than 60% of cases.

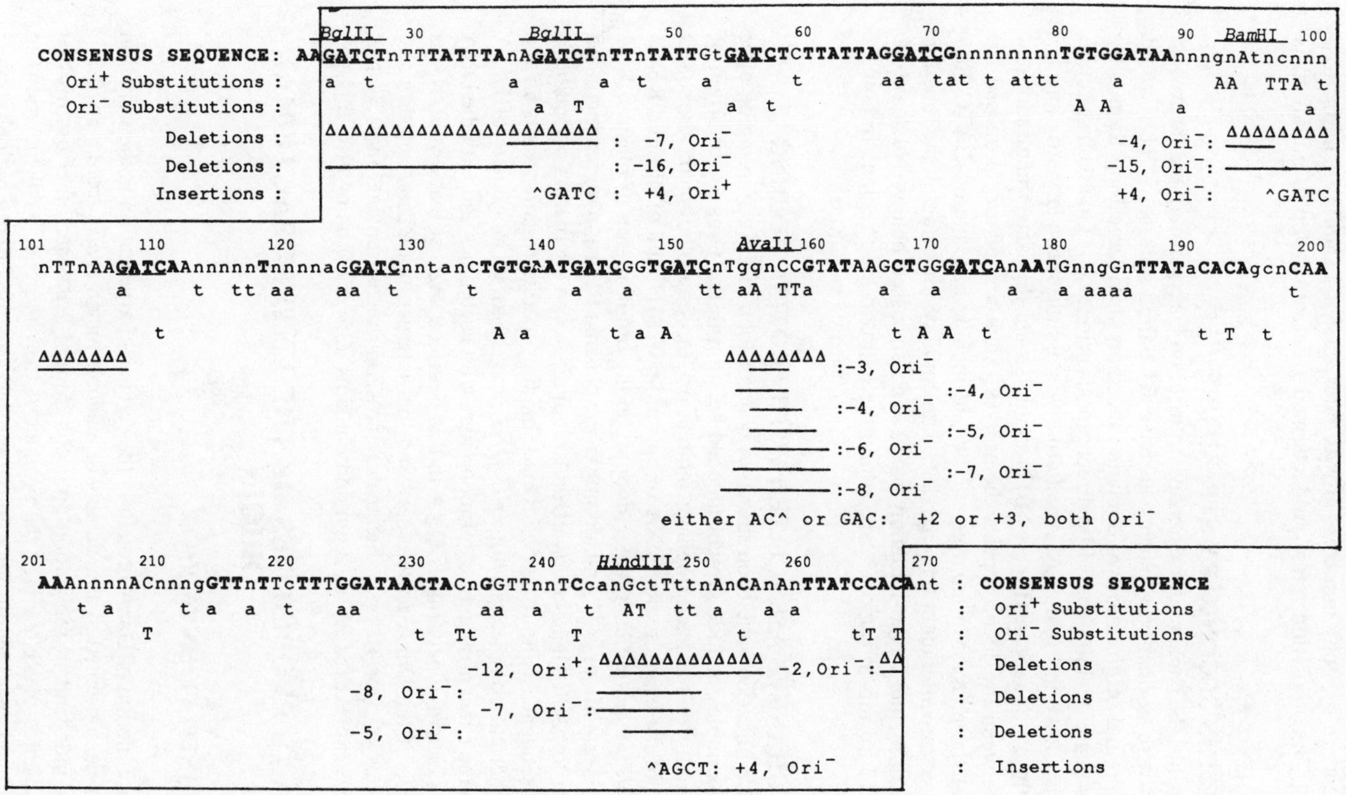

FIGURE 5.   Mutagenesis of the *E. coli* bacterial origin region. Mutated *E. coli oriC* sequence changes are shown below the enterobacterial consensus sequence. In the consensus sequence, bold capital letters: same nucleotide is found in all six origins; normal capital letter: nucleotide is found in five of the six origins; lower-case letter: nucleotide is present in three or four of the six origins, but only two different nucleotides are found at that site; n: three or four different nucleotides, or two different nucleotides plus a deletion, are found at that site. In Ori⁺ substitutions, capital letter: mutation at this site only; lower case: mutation present in an origin with multiple base changes. In Ori⁻ substitutions, capital letter: mutation at this site only; lower case: mutation present in an origin with multiple base changes, but not found in any Ori⁺ multiple mutants. In insertion and deletion derivatives, ΔΔΔΔΔ: region of *E. coli* deletion derivatives; ——: specific *E. coli* deletion derivatives; Λ: point of insertion for insertion derivatives. GATC sites are underlined in the consensus sequence and representative restriction sites are shown. The sequence within the box is the minimal bacterial origin and the nucleotide numbering scheme is the same as in Figure 1.

autonomous replicon, often show intermediate phenotypes: less total incorporation or at a slower rate, or decreased transformation frequencies with subsequent slow growth rates or altered plasmid copy numbers or stabilities. Qualitative assignment of these intermediate phenotypes to "classes"[53] does not address the defect present; quantitative information, both from *in vivo* experiments (e.g., copy numbers, growth rates, plasmid stability) and from *in vitro* experiments (e.g., incorporation rates, products made, defective step in initiation) is needed.

C. POTENTIAL ORIGIN SECONDARY STRUCTURE

Each of the origins has the capacity for extensive intrastrand base pairing, and the lowest free energy structures with normal base pairing all have R1 paired with R4.[54] This would effectively split the minimal origin into two domains, the short stem-large loop formed by the R1-R4 domain and that containing the 13-bp direct repeats prior to R1. One such structure for the *E. coli* origin is shown in Figure 6A. An alternative possible secondary structure for the region containing the 13-bp repeats (Figure 6B) is an interstrand slippage structure formed from the homology between the 13-bp repeats themselves. Such a structure, proposed for the promoter region of the phage Mu *mom* gene,[55] would include base-paired GATC sites of different methylation state after an initiation event; such could be important in the function of the GATC methylation state in the initiation process, as discussed below. Nuclease P1 cleavage sites in the 13-bp repeat regions[50] are within the exposed loop regions in both of these structures.

D. STRUCTURAL CHANGES ASSOCIATED WITH PROTEIN BINDING

Several origins of replication have been shown to bend or loop in the origin region after binding to proteins. Examples include protein-induced bending in the virus SV40 origin of replication,[193] the bending of the bacteriophage lambda origin upon binding of lambda O protein,[56] the bending of the origin region DNA of plasmid pSC101[57] and of plasmid R6K[58] upon binding of *E. coli* integration host factor, looping in the origin region of plasmid R6K mediated by its initiator protein, pi,[59,239] and looping in the plasmid P1 origin region mediated by its initiator protein, RepA.[60] There are five DnaA binding sites in alternating orientations in the minimal *oriC* sequence (Figure 2).[34,35] DnaA binding to this region results in the formation of large spherical complexes with *oriC* DNA and 20 to 40 monomers of DnaA protein.[34,61] Although the exact arrangement of monomers with each other and with the DNA is unknown, DNAse I cleavage patterns of DNA in this complex suggest that *oriC* DNA is wrapped on the outside of the DnaA protein core and that interaction of DnaA protein and DNA occurs at intervals of 5 or 10 bp.[34] Electron microscopy measurements indicate that the diameter of these complexes is 21 ± 2 nm and contains 225 ± 50 bp of *oriC* DNA.[61]

III. FUNCTIONAL PROPERTIES OF THE ENTEROBACTERIAL ORIGIN

A. FEATURES OF INITIATION *IN VIVO*
1. DnaA Protein

The only essential function of DnaA protein in the cell is its participation in the initiation of DNA replication, as demonstrated by the viability of a double mutant, *rnh⁻ dnaA*::Tn*10*.[62,63] In this mutant, DnaA-independent replication proceeds from at least four other origins that are activated in the absence of RNase H,[64] the product of the *rnh* gene.[65] All mutations in the *dnaA* gene yet isolated specifically block initiation, but have no effect on elongation.[66] Phenotypes of different *dnaA* mutants include reversibility of temperature inactivation of the DnaA protein, cold sensitivity, and suppressibility by *rpoB* mutations, and these phenotypes appear to correlate with clusters of mutations within the gene.[66] All *dnaA*(Ts) mutations can

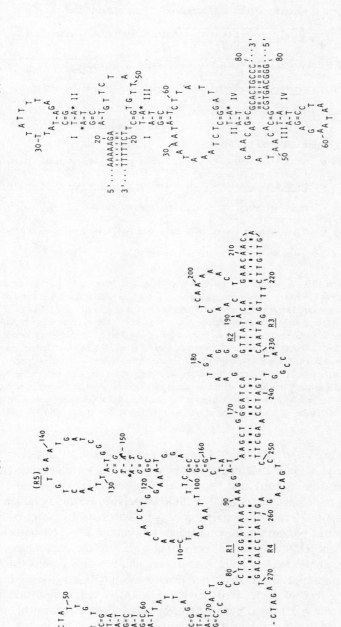

FIGURE 6.   Possible secondary structures within the enterobacterial origin. (A) Low free-energy structure generated by the Ooi and Takanami DNASEQ2B secondary structure computer program[242] with minimal helix length of 4 bp and a maximal loop size of 300 bp. The DnaA binding sites are shown as the R sites. Italicized GATC sites: sites which remain methylated in *E. coli dam-13* cells[54]; *: the methylated adenines in these two GATC sites. (From Smith, D. W., Garland, A. M., Herman, G., Enns, R. E., Baker, T. A., and Zyskind, J. W., *EMBO J.*, 4, 1319, 1985. With permission.) (B) Interstrand "slippage" structure in the region of the 13-bp repeats, as patterned after the structure proposed for the phage mu *mom* gene promoter region.[55] Structure shown is that just after an initiation event; *: methylated adenine residues in GATC sites on the parental DNA strand. I, II, III, IV: the GATC sites located at positions 23, 39, 54, and 67, respectively, in *oriC* (see Figure 2). Nucleotide numbering is as in Figure 1.

A. *E. coli dnaA* promoter region

```
220
AGCACGGAAGCCGTGAGAACGGTTGCGCTTCAGTACAGACGGTTGAAAAGTGCGTTTCATGGCGATTTCTACCTAAACTTGAATAAATTCAATGGCTTTA
AlaArgPheGlyHisSerArgAsnArgLysLeuValSerProGlnPheThrArgLysMet
                                              Amino terminus, RpmH
320
TTGGATATCCGCCGAAAAATGAAACGATGGACACCGAAGCCATGGGTGATTAAAGAGGCCGG|ATTGTAATAATTGTACACTCCGGAGTCAA|TTCTCTTTC
                                                                                   rpmH2p
420
CTTATTTACCGCGCTTTTCCGCACCTTTTCGCAGGGAAAATGTACGACCTCACACCAGTGGAAACCAGCATGGCGCGCCGGGTGGAGG|ATTATACGGGCT
                                                                                    ◄▬
520
|GATGGGTAAAGCGCAA|GGATC|GTCCTGG|GATC|TTT|ATTAGATCGATTAAGCCAATTTTTGTCTA|TGGTCATTAAATTTTCCAATATGCGGCGTAAAT|GTG
 rpmH1p                 ◄                   (rpmH3p)
620
CCCGCCTCGCGGCAGG|GATC|GTTTACACT|TAGCGAGTTCTGGAAAGTC|GTGTGGATAA|ATCGGGAAAATCTGTGAGAAACA|AAGATC|TCTTGCGCAGTTT
            ,dnaA1p                 ▬►        DnaA box                              dnaA2p
720
AGGCTATGAT|CCGCGGTCCCGATC|GTTTTGCAGGATC|TTGATCGGGCATATAACCGCAGACAGCGGTTCGTGCGTCACCCTCAAGCAGGGTCTTTTCGAC
           ▬►
820
GTACGTCAACAATCATGAATGTTTCAGCCTTAGTCATTATCGACTTTTGTTCGAGTGGAGTCCGCCGTGTCACTTTCGCTTTGGCAGCAGTGTCTTGCC
                                              MetSerLeuSerLeuTrpGlnGlnCysLeuAla
                                              Amino terminus, DnaA
```

B. *E. coli* DnaA protein: 467 amino acids; 52,551 kDa

```
                        50                                        v          ^     100
MSLSLWQQCL ARLQDELPAT EFSMWIRPLQ AELSDNTLAL YAPNRFVLDW VRDKYLNNIN GLLtsFCGaD APqLRFEVGt KPvTQTpqaa VtSnvaaPAQ
                                    |helix?| turn  150helix  E(dnaA167)         V(dnaA46)
                                                                 ↑              ↑           200
VaqtQPQRaA PstRsgWDNv pApaEptYRS NVNVv|HTFDN FV|EGKSN|QLA RAAARQVAD|N PGGAYNPLF|L YGGTGLGKTH LLHAVGNGIM ARKpNAKVVY
                                                              putative ATP binding domain
                        250                                                              300
MHSERFVQDM VKALQNNAIE EFKRYYRSVD ALLIDDIQFF ANKERSQEEF FHTFNALLEG NQQIILTSDR YPKEINGVED RLKSRFGWGL TVAIEPPELE
   |helix|    |turn|  |helix|    350
|TRVAILMKKA DEND|I|RLPG|E VAFFIAKRL|R SNVRELEGAL NRVIANANFT GRAITIDFVR EALRDLLALQ EKLVTIDNIQ KTVAEYYKIK VADLLSKRRS
                        450
RSVARPRQMA MALAKELTNH SLPEIGDAFG GRDHTTVLHA CRKIEQLREE SHDIKEDFSN LIRTLSS
```

FIGURE 7. Nucleotide sequence of the *E. coli dnaA* promoter region[74] (A) and amino acid sequence of the *E. coli* DnaA protein[172] (B). (A) Arrows and their sizes represent initiation sites, direction, and level of transcription from the different promoters. Promoters, DnaA box, and GATC sites are boxed or underlined.[74] (B) Amino acid sequence of the *E. coli* DnaA protein as translated from the nucleotide sequence is listed. Capital letters indicate *E. coli* sequence identity with the *S. typhimurium* and *S. marcescens* DnaA sequences,[173] while lower-case letters indicate that a different amino acid is found in these latter proteins. A caret signifies a deletion and an inverted caret indicates an insertion is found in either the *S. typhimurium* or *S. marcescens* sequences[173] when compared to the *E. coli* sequence. Two helix-turn-helix motifs predicted from secondary structural analysis (Figure 8) are indicated, as are the putative ATP-binding domain[179] and the single amino acid substitutions found in two temperature-sensitive *dnaA* mutants.[172] The four amino acids changes in the *dnaA*(Cs) mutant that lead to overinitiation at 32°C are the *dnaA46* change at residue 184, Gln to Leu at residue 156, His to Tyr at 252, and Tyr to His at 271.[174]

be suppressed by at least one *rpoB* mutation;[67] however, certain amber mutations in the *dnaA* gene are not suppressed by *rpoB* mutations.[68] Presence of the DnaA protein apparently is required for *rpoB* suppression. Schaus et al.[68] propose three possibilities to account for this observation: (1) DnaA protein interacts directly with RNA polymerase during initiation, (2) *rpoB* mutations increase the level of transcription of the *dnaA* mutant gene providing enough partially active mutant protein to allow initiation to occur, or (3) *rpoB* mutations lead to synthesis of a mutant RNA polymerase, permitting bypass of the requirement for one function of a multifunctional DnaA protein. A fourth possibility would be that there is increased synthesis of a transcript in the RNA polymerase mutants that activates *oriC*, overcoming the inhibitory effects of the reduced activity of DnaA protein. Reinitiation in a thermoreversible *dnaA* mutant is inhibited by rifampin, but not by chloramphenicol, suggesting that DnaA protein acts before or during the synthesis of an RNA required for initiation and that RNA polymerase synthesizes this RNA.[69]

The *dnaA* gene is the first gene in an operon that also contains the *dnaN* gene,[70,71] which encodes the β subunit of DNA polymerase III holoenzyme (Figure 11).[72] The *recF* gene is adjacent to *dnaN* and its expression is primarily from promoters within the *dnaN* gene.[73] S1 nuclease protection studies have located two promoters for the *dnaA* operon, *dnaA1p* and *dnaA2p* (Figure 7A),[74] with 73 to 88% of the transcripts originating from *dnaA2p*.[40-42] Both promoters are controlled by DnaA protein, which acts as a repressor of transcription apparently by binding to the DnaA box located between the two *dnaA* promoters.[40-43] The *dnaA2p* promoter efficiency is reduced when the promoter region is unmethylated at GATC sites in the promoter region (see Figure 7A); transcription from the *dnaA2p* promoter is reduced 2.7- to 4.5-fold in *dam* mutants.[42,75] If the *dnaA* promoter in the hemimethylated state also has reduced activity compared to when it is fully methylated, this may provide reduced DnaA expression for a period of time after initiation from *oriC* because the *dnaA* gene, which is 2 min away from *oriC,* would be replicated soon after initiation. The *dnaA2p* is stringently regulated[248] so the rate of expression of DnaA protein is coupled to the rate of protein synthesis.

The observation by Donachie[76] that the cell mass per origin ratio is invariant with growth rate led him to coin the term "initiation mass", defined as the mass of a cell at which time initiation from *oriC* occurs. The interpretation of the constancy of this ratio is that there is a critical protein whose concentration determines when initiation events occur in the cell cycle. The properties of DnaA protein suggest that it may be this critical protein as well as both the initiator and autorepressor proposed in the "autorepressor" model[77] for the control of DNA replication. To test this, Atlung et al.[78] examined whether overproduction of the DnaA protein increases the frequency of DNA initiation. When a temperature shift was used to increase DnaA protein concentration expressed from the lambda P_L promoter, DNA initiation frequency increased, as would be expected if DnaA protein is the initiator protein; however, the new initiation events were subsequently aborted. A component(s) of replication may have been present at insufficient concentrations to accommodate the additional replication forks.[78] In these experiments, initiation was assayed by DNA/DNA hybridization with probes containing genes proximal to *oriC* and from different locations around the chromosome. That the additional replication forks induced by increased DnaA synthesis stalled at some short distance from the origin suggests why experiments measuring the DNA to mass ratio[79-81] failed to correlate changes in this ratio with changes in DnaA concentration. Increasing the intracellular DnaA protein concentration via induction from the *tac* promoter can also lead to inhibition of the rate of DNA replication.[82] The experiments of Atlung et al.[78] and others[251] support the possibility that DnaA protein could be the initiator protein whose critical concentration is one of the determinants of initiation timing. Two observations suggest that the concentration of DnaA protein may not be the *only* determining factor in initiation. First, in cells where the *dnaA* gene is cloned in overexpression vectors, the uninduced levels of DnaA protein are increased several fold,[83,84] yet the cells appear normal. Second, the amount of DnaA protein per origin increases with growth rate rather than remaining constant.[84] Transcription from the *mioC* promoter as well as the state of supercoiling and methylation may also be factors controlling the initiation frequency, as discussed below.

Protein synthesis inhibition studies imply that synthesis of a protein must occur before a new round of DNA replication can begin.[85,86] If this protein is the initiator protein, the role proposed for DnaA protein, then this protein synthesis requirement suggests that an initiation event is dependent on initiator concentration. If the concentration of DnaA in the cell is a critical and rate-limiting factor in initiation, then initiation of DNA replication from *oriC* would cease if the synthesis of DnaA were halted. Blockage of further *de novo* synthesis of DnaA protein in *dnaA* amber mutants by heat inactivation of a temperature-sensitive suppressor tRNA led to the rapid cessation of DNA accumulation[87] as well as to a decrease in the rate of DNA synthesis,[88] supporting this prediction.

Mutations in the *dnaA* gene confer a conditional lethal phenotype; however, some of the mutants show replication timing defects even at the permissive temperature. The use of flow cytometry has been exploited to examine synchrony of initiation in different *E. coli* strains.[89-94] Cells are stained with the DNA-specific dye, mithramycin, and ethidium bromide, and excitation and fluorescence wave lengths are such that fluorescence is a measure of DNA and not RNA content. A three-dimensional plot of cell number versus fluorescence (DNA content) versus light scattering (cell mass) after treatment of wild-type cells with rifampicin or chloramphenicol yields 2n (n = 1, 2, 3) number of chromosomes per cell. Cells defective in synchrony of timing would contain numbers of chromosomes different from 2n. A high degree of asynchrony at the permissive temperature was found with *dnaA* mutations mapping in the central part of the *dna* gene, while mutations mapping in either of the distal parts of the gene yielded a low degree of asynchrony.[90]

2. DNA Gyrase and Topoisomerase I

Gyrase catalyzes a number of reactions, including the ATP-dependent introduction of negative supercoils into DNA. It contains two subunits, α and β, which are products of the *gyrA* and *gyrB* genes. *In vivo* and *in vitro* studies suggest that DNA gyrase plays an important role in all three stages of replication: initiation, elongation, and termination. That gyrase is an important component of the initiation machinery is demonstrated by the effects of gyrase mutants on initiation and its requirement in the *in vitro oriC* replication system. There is a strong gyrase binding site within the minimal *oriC* sequence (Figure 1) covering the *Hind*III site, which is protected from *Hind*III digestion by gyrase binding. Binding to this site is suppressed both by oxolinic acid, which is an inhibitor of the GyrA subunit, and by a nonhydrolyzable ATP analog.[95]

Conditional lethal strains, defective in *gyrB* activity at the nonpermissive temperature, are blocked at the initiation stage of chromosome synthesis.[96,97] A mutation in the *rpoB* gene has been described that suppresses the temperature sensitivity as well as the replication and transcription deficiency at the nonpermissive temperature of *gyrB*-ts mutants.[98] These observations suggest that the primary effect of gyrase is to alter promoter activity within or near *oriC* or to provide the necessary superhelicity required for initiation, or both. With respect to the effect of superhelicity on initiation, the absence of topoisomerase I in a Δ *(topA)* strain suppresses the temperature sensitivity of a *dnaA46* mutant, and the *dnaA46* gene product apparently continues to be required at the high temperature since a *dnaA*::Tn*10* allele could not be introduced into the Δ *(topA)* strain.[99] This might be caused by either an increased expression of the *dnaA46* gene in the Δ *(topA)* strain or a change in supercoiling in this strain that favors DnaA binding.[99] Because of the requirement for the mutant DnaA46 protein and the fact that DnaA protein is not required for replication in the absence of RNase H (stable replication), the suppression of *dnaA46ts* by Δ *(topA)* does not involve activation of other origins besides *oriC*.

3. DnaB and DnaC Proteins

Genetic and biochemical studies suggest that the DnaB and DnaC proteins are involved in events at the replication fork as well as at the origin of replication. These two proteins form part of the primosome responsible for priming ϕX174 in a purified *in vitro* replication system.[100] DnaB and DnaC proteins form a stable complex in the presence of rNTPs,[101-103] and the primary function of the DnaC protein in the replication of *oriC* plasmids may be to deliver DnaB protein to the *E. coli* replication origin after DnaA protein has formed a complex with *oriC*.[61] The DnaB protein has helicase activity[104,105] and appears to unwind DNA at *oriC* so that priming by primase can occur in the *oriC in vitro* replication system.[105] It is doubtful that either protein contributes to the control of timing events at *oriC* during steady-state growth; however, there are only 10 to 20 molecules of DnaB protein per cell, limiting the number of replication forks possible.[14]

4. RNA Polymerase

Rifampicin, a specific inhibitor of the β subunit of RNA polymerase, inhibits a new cycle of replication at a time when protein synthesis is no longer required for initiation.[69,106,107]

Thus, RNA polymerase-catalyzed synthesis of one or more RNA species appears to be required for initiation. The role of this RNA, be it as a primer for DNA synthesis or as a regulator or activator of initiation, is unknown. So far, three different types of mutations in RNA polymerase subunits have been isolated that affect initiation of DNA replication from *oriC*. One type suppresses *gyrB*-ts mutants, as mentioned above. Another type of mutation causes an increase in the mass/origin ratio correlated with initiation events and is temperature sensitive for growth; these mutations map in either *rpoB* or *rpoC* genes.[108,109] A third type of mutation, mapping in the *rpoB* gene, causes allele-specific suppression of *dnaA*(ts) mutations.[67,110] Some of these *rpoB* mutations also show an enhancement of termination at the *trp* attenuator, while *dnaA* mutations have the opposite effect,[68,111] suggesting that a transcription termination event may be required either for initiation of replication from *oriC* or for controlling the expression of *dnaA*.

5. RNA-DNA Junction Sites and *mioC* Transcript 3′ Ends

RNA-linked DNA molecules have been isolated from *E. coli dnaC*(ts) cells that were synchronized and then allowed to initiate a new round of chromosomal replication.[112] The nucleotides at the 5′ ends of the DNA moiety of RNA-linked DNA molecules were identified[112] for both replication directions and are indicated as circles in Figure 1. Within the minimal origin sequence, the positions of junctions from primer RNA to DNA for the leading strand in the counterclockwise-moving fork are clustered in nine regions in the left half of *oriC* (Figure 1). RNA to DNA transitions on the opposite strand were not detected within the minimal origin, suggesting that priming of the leading strand in the clockwise-moving fork occurs outside of *oriC*. We[113,121] and others[114] have shown by S1 nuclease protection analysis that leftward-moving transcripts enter and terminate throughout the *oriC* region at several clusters, nine of which correspond closely in location to these RNA-DNA transitions. The 3′ ends of these transcripts are indicated as vertical lines in Figure 1. The length and instability (half life = 1.51 min)[84] of the *mioC* transcript may have contributed to the difficulty in determining whether it acts as a primer for replication or whether its role is to activate *oriC* in a manner similar to the replication of λ *dv*.[115] The source of a primer, whether from RNA polymerase or primase, can be deduced if the 5′ end of the RNA is identified. The original 5′ end of a primer can be labeled with (α-^{32}P)GTP in the vaccinia guanylyltransferase capping reaction. Substrates include 5′-terminal tri- or diphosphate, but not monophosphate, ends, so only intact primer molecules are labeled with this enzyme. This approach was used successfully to define the length and consensus 5′ dinucleotide of the *E. coli* RNA primer for Okazaki fragments.[116] So far, attempts to similarly label *oriC* primer RNA with vaccinia capping enzyme have been unsuccessful,[244] suggesting that this RNA is processed prior to or simultaneously with its use as a primer. Kohara et al.[112] have proposed that priming of the leading strand in the counterclockwise-moving fork is an early and unique event accomplished by either RNA polymerase or primase because replication *in vivo* appears to be initiated first in the counterclockwise direction.[117,118] After formation of the counterclockwise-moving fork, priming at an Okazaki start site to the left of *oriC* could form the primer for the leading strand in the clockwise direction.

6. Stringent Control, Growth Rate Regulation, *dnaA* Gene Expression *mioC* Transcription, and Initiation

An interesting facet of the effects of chloramphenicol (Cm) treatment and amino acid starvation on initiation is that Cm induces a burst of initiation from the *oriC* region in a *dnaA46* mutant growing at the partially restrictive temperature of 36.5°C.[119] One explanation for this observation is that Cm, which causes an increase in the rate of stringently controlled RNA synthesis, presumably by reducing the intracellular level of ppGpp,[120] overcomes the decreased DnaA activity by stimulating an increased synthesis of a stable RNA required for initiation. This RNA may either be rate limiting for initiation or increased transcription of this RNA may activate the origin such that initiation occurs at decreased DnaA activity. A likely candidate for such an RNA is the *mioC* transcript, which has been shown to be stringently controlled in that expression is strongly reduced *in vivo* by amino acid starvation and stimulated by chloramphenicol.[113] Also, in an *in vitro* transcription system, the presence of ppGpp or pppGpp reduced the level of transcription from the *mioC* promoter by a factor of 10.[121]

Several lines of evidence suggest that the region to the right of *oriC*, which contains the *mioC* gene encoding a 16-kDa protein of unknown function, is required for correct replication from *oriC in vivo*. Deletion of this region from an *oriC* plasmid leads to loss of bidirectional replication,[122] decreased *oriC* plasmid copy number, and increased segregational instability.[123-127] Most of these properties are determined by the *mioC* promoter, which is required for maintaining copy number and stability of *oriC* plasmids,[125] and neither the *asnC* promoter[125] nor the induced *lac* promoter[123] can replace it.

The fact that the *mioC* promoter (P$_2$, Figure 1), when assayed with reporter genes,[43,45,82,126] is inhibited by increased concentrations of DnaA protein is puzzling if both DnaA protein and the *mioC* transcript are required for initiation. The chromosomally located *mioC* promoter is not very sensitive to increased DnaA concentrations, however. No decrease in the *mioC* transcript concentration as assayed by two methods, primer extension and S1 nuclease protection, was found in cells containing a tenfold increase in DnaA protein,[84] suggesting that the *mioC* promoter structure is different when in the chromosome than when in the plasmid, perhaps caused by differences in supercoiling or that the promoter on the chromosome is sequestered in some way.

In *B. subtilis*, Séror-Laurent and co-workers[128,129] identified the RNA in an RNA-DNA copolymer to contain ribosomal RNA sequences. The synthesis of this RNA is correlated with initiation of replication[130] and the stringent response appears to regulate the transcription step involved in initiation in *B. subtilis*.[131]

The significance to replication of the stringent regulation of a transcript required for initiation is that the stringently regulated promoter, P1, of rRNA promoters is also under growth-rate-dependent control.[132-134] Growth-rate-dependent regulation of the *mioC* transcript could provide a mechanism whereby the rate of chromosomal duplication would be coupled to the rate of protein synthesis. We recently examined whether the *mioC* promoter was subject to growth-rate-dependent regulation.[84] The original observation describing growth control of stable RNA synthesis was that the rate of stable RNA synthesis relative to the total rate of RNA synthesis, r_s/r_t, displays an inverse correlation with growth rate.[135] Experiments correlating the rate of stable RNA synthesis and levels of ppGpp suggest that both stringent control and growth control reflect the same ppGpp-dependent mechanism.[136,137] The *mioC* transcript concentration in cells grown at six different growth rates was determined using S1 nuclease protection analysis, and the amount of *mioC* RNA relative to the amount of total RNA per cell, RNA$_{mioC}$/RNA$_{total}$, was found to be inversely correlated with growth rate, increasing sevenfold as the generation time decreased from 109 to 30 min.[84] Because it is stringently controlled, inhibited by ppGpp, and growth-rate regulated, the *mioC* promoter appears to be uniquely subject to the same mechanisms that control expression from stable RNA promoters. Control of *mioC* transcription may be one mechanism by which the cell couples the rate of DNA replication to the protein synthesis rate.

Our recent finding that DnaA protein expression is stringently controlled[248] suggests a major mechanism by which the cell couples DNA replication to growth rate. The *dnaA2p* promoter activity is reduced considerably after valine or serine hydroxamate addition, whereas it increases in activity after chloramphenicol treatment or in a *relA1* mutant after treatment with serine hydroxamate.

7. Role of GATC sites within *oriC*

The *E. coli dam* gene product, a DNA adenine methylase, specifically recognizes GATC sites, catalyzing methylation of both adenine residues to yield 6-methyl-adenine.[138] Mutations in the *dam* gene have been obtained,[139-141] including insertion[142] and deletion mutants.[143,144] The *dam* insertion and point mutants, however, exhibit some residual adenine methylation,[54] based on *Dpn*I, *Mbo*I, and *Sau*3A digestion patterns, although this amount is low. These mutants are highly sensitive to growth in the presence of 2-aminopurine and exhibit a variety of phenotypes consistent with involvement of the Dam methylase in mismatch repair. These properties include increased rates of spontaneous and induced mutagenesis, high spontaneous rate of induction of prophage lambda, required induction of SOS repair functions, hyper-recombination ability, and inviability when combined with *recA, lexA, recB, recC, recJ,* or *polA* mutations. Many of these phenotypic properties are also exhibited by cells over-producing the Dam methylase.[145] The proposal of Wagner and Meselson[146] that the methylated parental DNA strand is recognized as the "correct" strand for mismatch repair has largely been substantiated using genetically marked hemimethylated DNA substrates.[147,148] Several of the phenotypic properties of *dam* mutants are suppressed by *recL, uvrE, mutH, mutL,* and *mutS* mutations, genes that code for mismatch repair enzymes[148] (for recent reviews on Dam methylation, see Sternberg[149] and Marinus[150]). One possible function of the high number of GATC sites found in *oriC* is to conserve the DNA sequence of this region by serving as a "sink" for the mismatch repair system,[151] which itself may even be coupled to DNA replication, the coupling being established at the time of initiation.

Several lines of evidence indicate that the Dam methylase and the state of methylation of GATC sites are directly involved in the initiation process. First, plasmids having only *oriC* as an origin transform *E. coli dam* mutants some 30-fold less efficiently than they do isogenic *dam*+ strains.[54,152,153] This decrease in transformation frequency is less pronounced (only some tenfold) in plasmids containing *oriC* but using a ColE1-type origin. Thus, *oriC* must not only be present, but must also be functioning as an origin to see the maximal decrease in transformation frequency. Second, this reduced transformation frequency is not seen in *E. coli dcm* mutants defective in the *E. coli* cytosine methylase, demonstrating specificity for the Dam methylase. Third, the reduced transformation frequency is also observed in double mutants, where the *dam* mutation is suppressed by a mismatch repair mutation, e.g., a *mutH* mutation. This argues that the role of the GATC sites and the Dam methylase in *oriC* function is at least in part different from that in mismatch repair. Fourth, *oriC* plasmid DNA isolated from *dam* cells again transforms *dam* mutants at low frequency,[54] arguing against a modification/restriction system type of function for the Dam methylase. Fifth, M13-*oriC* RFI DNA isolated from *dam* mutants functions two- to fourfold less well in the DnaA-dependent, *oriC*-specific *in vitro* initiation system,[54] using either the Fraction II system[37] or a purified component system,[181] which are results similar to those found using unmethylated *oriC* plasmid DNA.[152,154] This decreased template activity was restored two- to threefold if the DNA from the *dam* mutants was first methylated with purified Dam methylase.[54] These results argue that *oriC* DNA must be methylated at one or more of its GATC sites to serve as a functional template in these *oriC*-specific *in vitro* initiation systems.

Dam methylation of GATC sites obviously changes the covalent structure of the DNA. This change provides the information needed by the mismatch repair system in its decision as to which strand to repair.[146] Likewise, this could be used to distinguish a parental origin from the two daughter origins formed immediately after an initiation event (note that the two daughter origins are different from each other in this respect due to the nonpalindromic structure of *oriC*). Thus, Dam methylation could reasonably play a role in any aspect of initiation whereby daughter origins need to be distinguished from parental origins. Two such aspects are the timing between rounds of replication and the process of segregation of daughter origins (or chromosomes). Some evidence exists for a role of Dam methylation in each of these aspects of the initiation process.

Russell and Zinder, [153] using substrates constructed *in vitro* via renaturation of denatured, unmethylated and methylated, linearized chimeric plasmid DNA molecules and purification of circular products, showed that hemimethylated *oriC* plasmids transform *dam* mutants at low frequency. Further, when *dam* mutants were transformed with methylated plasmid DNA, hemimethylated molecules accumulated in these cells, suggesting that these hemimethylated molecules, presumed products of one round of replication, were unable to replicate. These results suggest that one function of GATC methylation in initiation at *oriC* is involvement in the timing between rounds of DNA replication. A period of about 20 min is needed after an initiation event before a second initiation event can occur, even with cells that have excess "initiation potential".[155,156] Conversion of daughter origin DNA from the hemimethylated state to the fully methylated state, if required for the next initiation event, would contribute to this observed time interval. In support of this hypothesis, Messer et al.[152] observed a dramatic decrease in the 20-min time interval using *dnaA* cells harboring a chimeric plasmid which could overproduce the Dam methylase. More recently, flow cytometry measurements of rifampicin-treated cells[246] and density transfer experiments using exponentially growing cells[157] indicate that timing of initiation in *dam* mutants occurs essentially at random throughout the cell cycle. Further, flow cytometry of *E. coli dam* mutants containing a plasmid with the *dam* gene expressed from the bacteriophage lambda *pL* promoter and the *c*I857 gene showed a strong temperature-sensitive pattern of synchrony of timing: a high degree of asynchrony was observed at both high and low temperatures.[247] This strongly suggests that synchrony of initiation timing requires a precise intracellular concentration of Dam methylase; asynchrony occurs when the concentration is either too low or too high.

B. ROLE OF THE CELL ENVELOPE IN *oriC* FUNCTION

Interaction of the bacterial chromosome with the cell envelope was proposed as part of the "replicon hypothesis"[158] to account for the precise segregation of daughter chromosomes into daughter cells during cell division. Such interactions have been sought by many workers (see Ogden and Schaechter[159] for a recent review), but no essential role for such interactions has yet been demonstrated. A protein termed B', isolated from a membrane fraction, shows binding to *oriC* DNA at two sites, one within the minimal origin and one within the *mioC* gene.[160] No function has been demonstrated for this protein. Binding of the *B. subtilis* origin region, and that of the *B. subtilis* plasmid pUB110,[211] to the *B. subtilis* cell surface has been shown,[202-206] and this association requires the product of the *B. subtilis dnaB* gene,[212] a gene which as yet has no known homolog in *E. coli*.

More recently, Schaechter and co-workers, using a simplified isolation procedure of a previously described *E. coli* replication origin complex,[161] have shown that (1) this complex contains predominantly outer membrane material from the *E. coli* cell surface,[162] (2) the complex contains predominantly origin region DNA,[163] and (3) binding occurs to origin DNA only when it is in a hemimethylated state.[164] Based on the latter observation, Schaechter et al.[164] propose that this outer membrane-hemimethylated *oriC* DNA complex is the equivalent of a prokaryotic centromere. This binding is visualized to occur *in vivo* immediately after an initiation event and to exist for the 8 to 10 min before hemimethylated GATC sites are methylated by Dam methylase.[164] The cell surface site may be equivalent to the "periseptal annulus" previously described.[165] During this time, cell surface growth separates the two, bound daughter origins into cell surface regions destined for each of the daughter cells resulting from the subsequent cell division event, thereby effecting segregation of the daughter chromosomes. This model thus would implicate both Dam methylation and cell surface interactions in the process of segregation.

No cell envelope structures or components are yet required for initiation *in vitro* at *oriC*; however, diphosphatidylglycerol (cardiolipin), a diacidic membrane phospholipid, is capable of dissociating the ADP bound to DnaA protein and, in the presence of ATP and the components of the replicative system, it is able to restore the inactive ADP form to full activity.[166] Further, approximately half of the DnaA protein is recovered in a particulate fraction after heat lysis of *E. coli* cells with lysozyme and EDTA.[83] The principal cellular location of DnaA protein, therefore, may be the cell surface, where it could be easily interconverted from the ADP-bound form to the cardiolipin-bound form from which ATP and ADP are dissociated or to the ATP-bound and active form.[166]

C. NONESSENTIAL BUT IMPORTANT CHARACTERISTICS IN *oriC* FUNCTION

Several characteristics of *oriC* function in the initiation process are nonessential to features of initiation *in vivo*. Such characteristics are either dispensable, although with probable selective advantage, or they can be replaced by alternative mechanisms. DnaA-dependent initiation at *oriC* is itself a nonessential process for cell survival. The *rnh* gene encodes RNase H, a ribonuclease that specifically degrades the RNA moiety of RNA:DNA duplexes.[65] In the absence of this activity, DNA replication can initiate from at least four other sites than *oriC*; these sites have been termed *oriK* sites.[64] This mode of replication is capable of continued DNA replication in the absence of on-going protein synthesis, a phenotype termed stable DNA replication (SDR).[167] Thus, SDR can proceed in cells deleted for *oriC* or with a Tn*10* insertion in the *dnaA* gene,[63] but is dependent on a functional RecA gene product[94] (for review, see Kogoma[168]).

Timing between rounds of DNA replication is a dispensable characteristic of the initiation process. The C and D periods mentioned above for the time of a round of replication and the time between completion of such a round and cell division have very low coefficients of variation from cell to cell, as shown elegantly by flow cytometry measurements. This timing, however, is lost in *E. coli recA* mutants. The viability of these cells argues, then, that timing between rounds of DNA replication is a dispensable function for the initiation process.

A dramatic demonstration of the nonessential nature of timing between rounds of replication involved substitution of the plasmid R1 replication origin region for the 16-bp *Bgl*II fragment present within the minimal *oriC* region (see Figure 2) and required for *oriC* function.[169] The resulting chromosomes initiated asymmetrically in a bidirectional manner from the integrated R1 origin, present in either orientation, and this initiation required the presence of the R1 *repA* gene product and was repressed by the R1 *copA* repressor RNA, all properties typical of plasmid R1 replication (see Nordström et al.[170] for recent review). However, initiation of replication in these cells is essentially random during the cell cycle, as demonstrated using a density shift approach with exponentially growing cells. Using a similar approach, initiation in *dam* mutants also was shown to be essentially random throughout the cell cycle, as mentioned above.[157]This implies that *E. coli* can survive in the absence of proper initiation timing and helps to clarify why *dam* deletion mutants are viable. Thus, timing of replication is a dispensable property of the initiation process.

D. STRUCTURE OF DnaA PROTEIN

The *dnaA* gene of *E. coli* has been sequenced,[171,172] as have the *S. typhimurium* and *Serratia marcescens dnaA* genes,[173] both of which function in *E. coli*. In a comparison of the amino acid sequences of these three DnaA proteins, shown in Figure 7B, 63 amino-terminal amino acids and 333 carboxy-terminal amino acids are identical except for a conservative change at amino acid residue 194 from proline to alanine. The amino acid changes in the DnaA protein for the *dnaA46* and *dnaA167* mutations are indicated. From the nucleotide sequence of the cold-sensitive mutation, *dnaA*(Cs), it was shown to contain three point mutations in addition to the original *dnaA46*(Ts) mutation,[174] the locations of which are indicated in the legend to Figure 7B. The *B. subtilis*[49] and *P. putida*[48,175] DnaA proteins are highly conserved when compared to enterobacterial DnaA amino acid sequences; however, these proteins do not complement *E. coli dnaA* mutants. The predicted secondary structure of the *E. coli* DnaA protein using two different methods[176,177] is shown in Figure 8. Locations of the helix-turn-helix motif frequently seen in prokaryotic DNA-binding proteins[178] in the DnaA protein may suggest possible DNA-binding domains. ATP binds to DnaA protein with high affinity (K_D = 0.03 μM).[179] We used the Pearson FASTP program to compare the amino acid sequence of DnaA protein with ATPases.[245] The F_1 ATPase β-chain peptides were most similar to the DnaA protein, especially in the underlined stretch (Figure 7B) corresponding to the ATP-binding consensus sequence.[179,180] The *dnaA46* amino acid substitution lies in the putative ATP-binding domain, and such a mutant protein is predicted to have reduced affinity for ATP. The ATP-binding domain is in a hydrophobic region and may be organized as a nucleotide-binding fold. Five different types of interactions suggesting different domains of activity have been identified for the DnaA protein: (1) binding to ATP and ADP,[179] (2) binding to DNA containing the 9-mer 5'TTATCCACA3' and similar sequences (Figure 3),[34] (3) binding to cardiolipin at a site which may overlap with the DNA-binding site,[166] (4) binding to itself to form multimers,[34] and (5) possible direct interaction with the 13-mer direct repeats leading to strand opening (Figures 4 and 9).[50]

E. *IN VITRO* REPLICATION OF *oriC* PLASMIDS

The elegant work of Kornberg and co-workers has led to the design of cell-free and purified systems that replicate minichromosomes and are dependent on DnaA protein and the *oriC* minimal sequence.[37,38,50,61,179,181-186] Replication of *oriC* plasmids in these purified systems has been resolved into several stages, all of which require a negatively supercoiled template;[50,184,185] the early stages are depicted in Figure 9. The "open complex" is an *oriC*-DnaA·ATP structure containing 20 to 40 monomers of DnaA protein,[34,61] as mentioned earlier. This complex is very labile at temperatures below 30°C,[50,182] and a high level of ATP (5 m*M*) and a low level of Mg^{2+} are needed to keep the DnaA protein in the complex in an active form.[185] Using the single-strand-specific nuclease, P1, as a probe, Bramhill and Kornberg[50] showed that the duplex becomes P1 sensitive at specific sites in the region containing the three 13-mers upon binding of DnaA protein to *oriC* and that this strand opening by DnaA protein was dependent on the presence of at least one 13-mer. This duplex opening facilitates the formation of the "prepriming complex", which assembles upon addition of DnaB and DnaC proteins at temperatures ≥30°C. The DnaB protein associates with the open complex in the region of the 13-mers,[50,61] and DnaC protein appears to be absent from the resulting prepriming complex.[61] If only single-stranded binding protein (SSB) and gyrase are added to the prepriming complex, the template is extensively unwound by the helicase action of DnaB protein;[187] however, when SSB, gyrase, primase, and DNA polymerase III holoenzyme are added to the prepriming complex, concurrent unwinding, priming, and DNA synthesis result in bidirectional replication from *oriC*.[186] DnaA protein, then, facilitates entry of DnaB protein and primase into the open complex. A similar role for DnaA protein in the *in vitro* replication of pBR322 is suggested by the ability of DnaA protein to replace the primosomal protein, i.[188] The DnaA binding site in pBR322 is required for this activity, as are its location and orientation.[39]

DnaA protein binds both ATP and ADP with high affinity, but only the DnaA·ATP form is active in the *in vitro* replication systems.[179] Conversion of the ATP form to the ADP form may be a process by which the cell marks the DnaA protein for destruction or removes it from the pool of initiation-active ATP-bound molecules, thereby providing an eclipse phase after initiation until new DnaA protein is synthesized or the ADP-bound form is converted to the active ATP-bound form. In examining compounds for the ability to rejuvenate the inactive ADP form, Sekimizu and Kornberg[166] discovered that one of the three

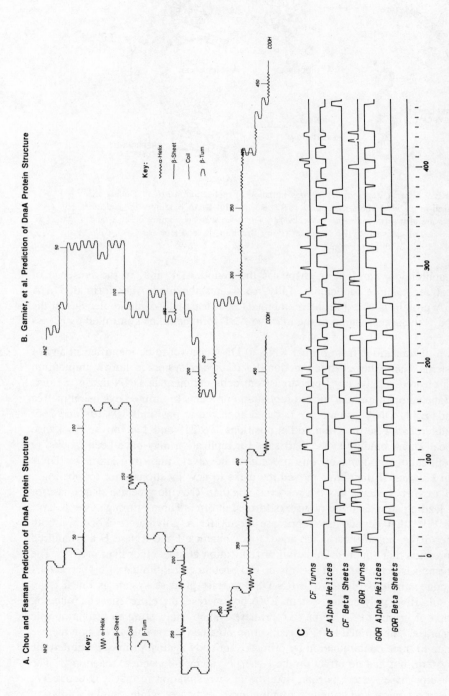

FIGURE 8.   Schematic diagram showing the predicted secondary structure of *E. coli* DnaA protein by two different methods. (A) according to Chou and Fasman;[176] (B) according to Garnier et al.[177] Residues are represented in α-helical, β-sheet, and coil conformations. β-turns are denoted by chain reversals. Only "strong" structures are considered. Consecutive β-turns appearing in the schematic representation do not reflect the actual number of β-turns that may be present, but signify regions with high probability for such a structure. From crystallographic data, it is known that four consecutive amino acid residues are typically involved in each β-turn. The predicted secondary structures were generated with the programs PEPTIDESTRUCTURE and PLOTSTRUCTURE[243] in which the original predictive methods have been slightly modified for deciding final assignments of conformational states. The secondary structure predictions by the two methods used are summarized graphically for comparison in (C).

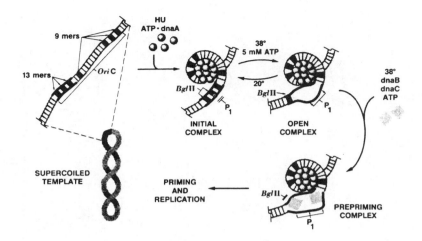

FIGURE 9.   Early stages in the *in vitro* initiation of replication at *oriC*.[50] Details are discussed in the test. The 9-mers are R1, R2, R3, and R4 DnaA binding sites, and the 13-mers are also proposed to be recognized by DnaA protein (see Figures 1, 2, 3, and 4). P_1: sites of cleavage by P1 nuclease. (From Bramhill, D. and Kornberg, A., *Cell,* 52, 743, 1988. With permission.)

membrane phospholipids, cardiolipin, displaces the bound ADP and, in the presence of components that reconstitute replication, fully reactivates the inert ADP form of DnaA protein. If DnaA protein is based in the membrane, as at least half of the molecules in the cell appear to be,[83] then reactivation of the inactive ADP form could be controlled by access to cardiolipin.

The location of transitions from primer RNA to DNA has yet to be identified in any of the purified systems. Using the cell-free Fraction II *oriC in vitro* system (a narrow ammonium sulfate cut of a cleared cell lysate)[37] prepared from cells deficient in DNA ligase, Seufert and Messer[189] labeled open-circular nicked replication products by limited extension of free 3' hydroxyl ends using DNA polymerase I. Label appeared at positions 194/199 and 265/272 for the counterclockwise direction and at positions 209/219 and 254 for the clockwise direction, all near DnaA binding sites. Start sites for replication may have been labeled at nicks in the replication products by this procedure; however, nick translation by DNA polymerase I in Fraction II may have moved the nicks to new locations prior to labeling.

The use of *in vitro* systems has not resolved whether RNA polymerase or primase or both prime the leading strands at the two forks formed at *oriC*. Three priming systems have been described[181,182] that depend upon primase alone, RNA polymerase alone, or both combined. When RNA polymerase is included, topoisomerase I and RNase H are required as specificity proteins that suppress initiation of replication at sites other than *oriC*.[181] The solo primase system is active at lower levels of HU protein and inhibited at higher levels. At the higher concentrations of HU, the RNA polymerase-primase system is much more active than the solo RNA polymerase system. RNA polymerase, therefore, must be relieving the HU inhibition of primase seen in the solo primase system, suggesting an activation role for RNA polymerase. The limited level of replication obtained in the solo RNA polymerase system is not due to trace contamination by primase; antibody against primase inhibited the solo primase system, but had no effect on the RNA polymerase-dependent reaction.[181] The RNA polymerase-primase system is more like the *in vivo* situation in that it is dependent on the presence of a correct and complete minimal *oriC* sequence, while templates missing the DnaA binding site, R4, or the left most 13-mer will replicate in the solo primase system.[50] Ultimately, these *in vitro* systems will provide the definitive biochemical and physiologically relevant characteristics of the components required for assemblage at *oriC* of the two replisomes that will move with the replication forks.

IV. OTHER BACTERIAL ORIGINS

Although bacterial origins have been cloned from three *Hemophilus* strains[190] and some evidence for initiation capability in a DNA cell surface complex obtained from *Streptococcus pneumoniae* has been presented,[191] origin DNA from these bacteria has not been further characterized. Much, however, has been learned about the origin regions from *B. subtilis* and from the pseudomonads, *P. aeruginosa* and *P. putida*. We focus now on the origin regions from these two bacterial genera.

A. THE PSEUDOMONAD BACTERIAL ORIGIN

The Dam-GATC methylation system arose in Gram-negative eubacteria recently in evolutionary history, since the time the pseudomonads diverged from the enterobacteria.[192] The pseudomonads thus have no Dam methylation system. What alternative mechanisms do these bacteria use to substitute for the Dam methylation function in initiation found in the enterobacteria? To address this question, the origin regions from *P. aeruginosa* and from *P. putida* have been cloned in these bacteria and their DNA sequences have been determined.[48] These origins function in both pseudomonas species and neither functions in *E. coli;* enterobacterial origins also do not function in either pseudomonas species. The pseudomonad origins thus constitute a second general class of chromosomal origins from Gram-negative bacteria.

The DNA sequences of these origins show both similarities and differences when compared with the enterobacterial origin. Regarding the differences, first, the pseudomonas origins lose origin activity gradually as their size is decreased; the "minimal" origins, defined by a partial *Hae*III fragment for *P. aeruginosa* and a *Sal*I-*Bam*HI fragment for *P. putida,* are about 500 bp (see Figure 10). Second, the origin regions contain very few GATC sites, and no other 4-bp sequence is present or conserved in position to the extent of the GATC sites in the enterobacterial origin. Third, no sequence homologies to the enterobacterial *mioC* gene are present.

However, these two classes of origins also show distinct similarities. First, each origin region contains between 5 and 7 *E. coli*-type DnaA binding sites, of homology 8 or 9 out of 9 (see Figure 3), and several others at 7 out of 9 homology. Of these, four are conserved in position between the two origins (see Figure 10). Two of these have precisely the same spacing between them as is found between the R1 and R4 sites of the enterobacterial origin (compare Figures 2 and 10). This remarkable structural similarity argues for considerable commonality of structure and function between the Gram-negative chromosomal origins. Although many DnaA binding sites are also found in the likely *B. subtilis* origin region (see below), distances between these do not correlate with the R1-R4 distance of the Gram-negative origins. Second, three 13-bp sequences, showing considerable sequence conservation between themselves and between these origins (see Figure 4), are found at a position in these origins, relative to the R1 and R4 equivalent DnaA binding sites, nearly identical to the positions found in the enterobacterial origins (Figure 10, underlined). Although the spacing between these is greater than that found in the enterobacteria, they may perform a similar function in the initiation process.

Sequence determination coupled with computer analysis of sequences on either side of the minimal origin regions showed a dramatic and striking gene organization. Open reading frames to one side show an amino acid sequence essentially identical to the *E. coli rpmH* protein and strong homology to the *E. coli rnpA* protein, and, in the other direction, to the *E. coli dnaA* protein. Sequences of the *P. putida dnaA* ORF independently determined by

```
                    -10          1         10         20         30         40         50         60      69
CONSENSUS SEQ:     ...TAC...G...G..GA...CG.G...G...G..C.C.T.....G.......GA...TA........A..C..TA..T
P. aeruginosa:     cca    --- ggc cg   aac   t gac cga cg g t gagggt ctgggttg  aag  cgtttcatg tt gg__cc
   P. putida:      tgt    ctg ttt_tc__cga   g ccg gat gc c c ttttaa agaccggc  ttc__gagaaagca gc ca  gg
                HaeIII           SalI                                             XbaI               KpnI

70         80         90        100        110        120        130        140        150        160     169
...TT..C.AC..G...T..CA...A...C.T.T.A.G.GAC.G....T..T..T.A...C..A.GGG..G.T..A.T...A..CA..GAT...G.A..A
ggg  ga g  tt agg cg  gtg ccc g_t_a_a a   c gcga tc ag g aat ga c   ca g ca t tcc ac  gc   gac t at
caa  tc a  ca tct tc  tat gag a g g t -   g tgcc gt_ga_c_gtg cc a   gt c tg - cgg ca  cg   cgg g ca
                             AhaIII                BclI

170        180        190        200        210        220        230        240        250        260     269
.AT..A.A.AA.GAAG..A..T.T..........T.GAA.__.TTA...cTCTT..G..G......TTCTGTG.ATAAC..G.....GCCCA..AATT.
g  ag t ca g     tc tt t cttttaaagga a   acGg    atg     gg ac gcgctt      c      tc atgaa    gc    g
c  ga a aa a     ag ca a aaaaagctttt t   gaac    taa     aa tg ataacc      g      ct cgctg    tg    a
                        AhaIII HindIII

270        280        290        300        310        320        330        340        350        360     369
CG.G.T...C.G......A.AA..TTGT...GA.CC.GTG..G.G..TGTT.C..TC.TGAGCGAA..CTG.GGATGAA.......GTTATCCACAGCGG
  t t tct c gcaggc a  gg     cga   a  g   tc a gc    t ct  c        gc   g      cgagatg
  g g gta a agtttt c  ct     tct   t  c   ct c ct    c aa  g        ag   t      aacacct

370        380        390        400        410        420        430        440        450        460     469
..TT.TC.ACA.G.....G.G  GGG.TGT.C......C..A.....G.TTATCCACA..G...A.....A..G.C.....GCC....TG.T...AT.AA
tt   t  c    c gctgt_c_ca   a   a cccctt aa gcaag g           aa Tcc ggacg cc t cgtcg   tgcc  c ttt  t
ag   a  a    g ctaag g tg   t   g atagcc tc tggtc t           gg ctt ttcac ga g gaaaa   gttt  g cg-  a
                  MstI

470        480        488
.G.CT...TT..C.TGG..   : CONSENSUS SEQ
g t  tga  tg t   gg   : P. aeruginosa
t g  gtt  gt g   tt   : P. putida
               HaeIII
```

FIGURE 10. Comparison of the minimal bacterial origins for *P. aeruginosa* and *Pseudomonas putida*. Capital letters: identical nucleotide in each sequence, defining the consensus sequence; lower-case letters: nucleotide present in each sequence; −: deletion in one sequence relative to the other. Numbering is relative to the *Sal*I site in the *P. putida* sequence; 5′ end is at the upper left. Representative restriction sites are shown and underlined. The three putative 13-bp repeats are shown underlined. In the consensus sequence, the 9-bp DnaA binding site sequences are shown in bold. If the 9-bp DnaA binding site is present only in one individual sequence, it is also shown in bold.

two groups[48,193] are in complete agreement. Homologies are strongest between the two pseudomonas sequences and stronger with the corresponding enterobacterial sequences (Figure 7B) than with those of *B. subtilis*. Although the homology is very strong, distinct differences are present, and the *P. aeruginosa dnaA* ORF does not complement *E. coli dnaA46* or *dnaA508* mutations. A nine-of-nine DnaA binding site is found in the probable promoter region of both pseudomonas *dnaA* ORFs, which possibly functions in autoregulation of expression of the pseudomonas DnaA protein, as found in *E. coli*.[40-43] Thus, the bacterial origins from both of these *Pseudomonas* species are found between the equivalent of the *E. coli rpmH* and *dnaA* genes, a genetic organization found in the Gram-positive bacterium *B. subtilis* (see below and Figure 11), but different from that of *E. coli*. The comparative genetic organization in these regions for these bacteria is shown in Figure 11.

Availability of these origins and probable *dnaA* genes (plus adjacent genes) provides a powerful resource for comparative studies of the Gram-negative bacterial origin and of the initiation process. Although the pseudomonas *dnaA* ORF does not complement *E. coli dnaA* mutants, the 9-bp site required for binding is conserved in sequence and partly in number and position in these origins. What features of DnaA protein are needed for binding and what for other functions in initiation? Direct repeats of 13 bp are present, but are different from those of *E. coli* in sequence and in spacing. What functions equivalent to GATC methylation and to *mioC* function are operational in pseudomonas initiation? The only DNA methylation system yet described for the pseudomonads to our knowledge is the plasmid-encoded *Pae*R7 Type II restriction/modification system in *P. aeruginosa*.[194,195] These are among the many questions that can now be addressed.

B. THE *BACILLUS SUBTILIS* ORIGIN REGION

Replication of the *B. subtilis* chromosome proceeds bidirectionally[196] from a unique region near the *purA* locus.[197] Using germination of spores of a *B. subtilis thy* mutant, Yoshikawa and co-workers have isolated the initially replicated DNA sequences using several approaches;[198-201] these cloned sequences contain the site of initiation *oriC* for chromosome replication. Sequences from this region can also be isolated as membrane DNA complexes.[202-205] Analysis of gradient of label present in restriction fragments isolated from pulse-labeled germinated spores indicates that the site of initiation is within a 3-kb *Sal*I-*Pst*I fragment some 4 to 5 kb from the nearby *rrnO* operon.[201] Other workers, however, using *in vitro* synthesized fragments from a membrane DNA complex[206] or using a temperature-sensitive *B. subtilis dnaB37* initiation mutant[129] find the initiation site to be in the *Bam*HI fragment B7, which includes the 5′ end of the *rrnO* operon. This region also contains the sequences coding for the RNA and DNA found in an RNA-DNA copolymer whose time of synthesis correlates well with initiation timing.[129] This RNA, which hybridizes to the coding strand for ribosomal RNA, is a clear candidate for a priming function in the initiation process. Evidence based on analysis of earliest-labeled restriction fragments using the *dnaB37* initiation mutant has also been presented, suggesting that *B. subtilis* uses two closely linked origins.[128]

To better determine the genetic components within this region and their expression, Yoshikawa and co-workers determined the DNA sequence[49] and transcription patterns[207] of some 10 kb in this region. This powerful and definitive approach immediately identified specific open reading frames (ORFs) with the *gyrA* and *gyrB* genes via restriction map comparisons[208] and limited the *recF* gene to one of two ORFs. In addition, sequence comparisons with *E. coli* sequences showed strong homologies to the *E. coli dnaA*, *dnaN*, and *rpmH* genes, with weaker homology to the *E. coli rnpA* gene.[209] The *dnaN* ORF is in fact a *dna* gene, as shown by complementation of an elongation mutation, *dnaG5*.[210] Remarkably, the chromosome organization of these ORFs and genes is identical to that of the *E. coli rnpA-gyrB* region,[66] except for the presence of small ORFs between some of the *B. subtilis*

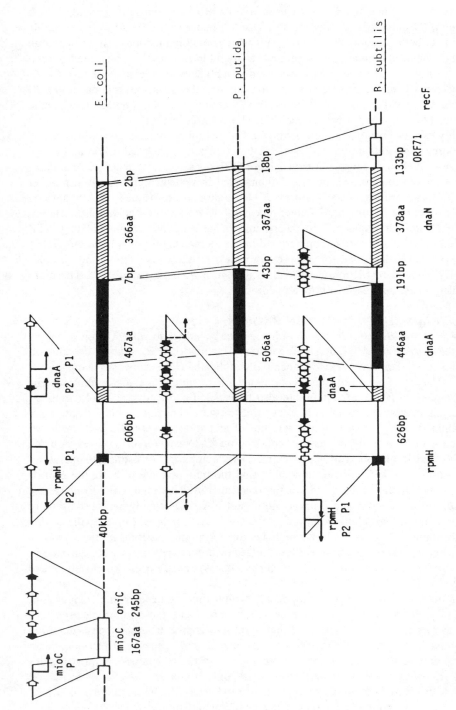

FIGURE 11. Comparison of *E. coli*, *P. putida*, and *B. subtilis* *dnaA* and *oriC* chromosome regions. Black areas in the *rpmH* and *dnaA* genes are highly homologous and shaded areas are moderately homologous. DnaA sites are shown by "arrow boxes"; solid boxes are identical to the consensus sequence; open boxes differ by one or more bases from the consensus. Arrows: transcription initiation sites; those for *P. putida* (dotted arrows) were deduced from the sequence. (From Fujita, M. Q., Yoshikawa, H., and Ogasawara, N., *Mol. Gen. Genet.*, 215, 381, 1989. With permission.)

major ORFs and the presence of the *B. subtilis gyrA* gene adjacent to the *gyrB* gene. Even more remarkably, the region between the *rpmH* and *dnaA* ORFs contains nine copies of the 9-bp *E. coli* DnaA binding site to an identity of eight of nine to the consensus sequence, and that between the *dnaA* and *dnaN* ORFs contains four copies of this sequence (see Figure 3)! Presence of these sequences coupled with the strong homology of the *B. subtilis dnaA* ORF to the *E. coli dnaA* gene argues strongly that one of these two regions (or both) is the *B. subtilis oriC* site(s) of initiation. No putative DnaA binding sites are found in that part of the *Bam*HI fragment B7 encoding the RNA-DNA copolymer found by Séror-Laurent and co-workers.[129] However, demonstration that any one of these three regions has the properties of an autonomously replicating sequence (replication origin) remains lacking. Which of these regions may be the primary one involved in the cell surface interaction via *B. subtilis* DnaB protein[211,212] also is unclear.

More recently, Yoshikawa and co-workers[213] have shown that the presence of the DnaA sites from these three regions renders difficult or impossible the transformation of *B. subtilis* with high- or low-copy-number plasmids. This has prevented the demonstration that this region of the *B. subtilis* chromosome contains the chromosomal origin. In *E. coli*, DNA sequences containing DnaA binding sites have been shown to titrate DnaA protein, causing increased DnaA promoter activity.[214] The rate of synthesis of DnaA protein may be insufficient to provide the DnaA protein concentration necessary for initiation because of titration by the DnaA sites in the *B. subtilis* cloned sequences. This may also be true in *B. subtilis*, and expressing the *B. subtilis dnaA* gene from a strong promoter on a plasmid may facilitate identification of the *B. subtilis oriC* site, the site required in *cis* for autonomous replication.

V. DISCUSSION AND FUTURE DIRECTIONS

Flow cytometry studies of DNA distributions of exponentially growing cells showed that initiation of DNA replication at multiple origins in a single cell appeared synchronous in that initiation occurred at four origins within the same cell within a time period of 10% of the doubling time.[89] Surprisingly, minichromosomes do not perturb the synchrony[90] or timing[216] of chromosomal initiation and they are initiated in apparent synchrony with the chromosome.[217,218] The concentration of unbound DnaA protein is probably normal in the minichromosome-containing cell, even though there is an increased number of *oriC* sequences, because the *dnaA* promoter is autoregulated[40,41] and titration by DnaA boxes on minichromosomes leads to increased synthesis of DnaA protein.[216]

Several models have been proposed to describe theoretical systems for the control of replication. The first model to discuss control of timing of initiation in the cell cycle was the "replicon hypothesis".[158] In this model, the frequency and timing of initiation are determined by interaction between an initiator protein and a fixed activator site on each replicon, and membrane attachment of replicons insures correct segregation into daughter cells. The "inhibitor dilution model"[219,220] is based on changes in the concentration of a repressor-like DNA-binding protein. By coupling synthesis of this protein to initiation such that each initiation event leads to the production of a burst of inhibitor molecules (either because the inhibitor is unstable and constitutively expressed or because it is synthesized only after its gene is replicated), initiation would raise the inhibitor concentration and reduce the probability of further initiations. Growth would reduce inhibitor concentration by dilution and so increase the probability of initiation. The "autorepressor model"[77] proposes that an autorepressor and initiator are cotranscribed, and the synthesis of initiator is regulated by the autorepressor. Initiation occurs when a certain concentration of initiator molecules has accumulated. Applying the latter model to the *oriC* system, DnaA would function as both initiator and autorepressor. It has been suggested by Nordström et al.[170] that any replication control system will contain a replicon-specific negative feedback loop. So far, no inhibitor of *oriC* initiation has been identified, although hemimethylated DNA at *oriC* and perhaps also at the *dnaA* and *mioC* promoters may provide an eclipse phase, preventing initiation until methylation has occurred.[153,164] Also, the fact that the *dnaA* gene is stringently controlled[248] and the probability that a stringently controlled RNA synthesis event is required for initiation[113,121] suggests that ppGpp may act as an inhibitor of initiation, coupling the rate of growth and protein synthesis to initiation.

Our current working model for the conditions in the cell that allow initiation at multiple origins is the following. When four events have occurred at *oriC* — namely, (1) methylation of *oriC* GATC sites, (2) transcription from the *mioC* promoter, (3) sufficient negative superhelicity, and (4) a critical number of DnaA monomers bound to *oriC* — two new replication forks will be formed. One or more of these conditions in slight excess may compensate for absence or deficiency in another. For example, a DnaA concentration that is too low may be compensated for by an increase in transcription from the *mioC* promoter or a decreased rate of transcription may be compensated for by increased negative super-helicity (perhaps explaining the high copy number of *oriC* plasmids). Synchrony of initiation timing within a given cell may be difficult to explain on the basis of these four events occurring simultaneously at every origin. Rather, we favor the notion that DnaA protein (20 to 40 monomers) is released from an origin after initiation and subsequently binds to other ''slightly less prepared'' origins in the same cell, leading to a cascade of initiation events and release of DnaA protein so that all origins in the cell initiate in a short time frame. Recognition of which origin has initiated and which has not would be determined by the methylation state: fully methylated — not yet initiated, hemimethylated — already initiated. Binding of the hemimethylated ''just initiated'' daughter origin to the outer membrane[164] would insure correct segregation of daughter chromosomes into daughter cells. Soon after initiation of replication, the *mioC* and *dnaA* promoters would be hemimethylated and possibly less active, decreasing the rate of *mioC* transcription and DnaA protein synthesis. Topological constraints imposed by movement of the two forks as they move away from the origin may produce a different superhelical density at *oriC* than that facilitating initiation. Modulation of several of these events could lead to changes in the rate of DNA replication so that the cell would be capable of responding to changes in the environment in a variety of ways. The slightly asymmetric nature of the two forks might be explained by collision of the clockwise-moving fork with RNA polymerase as it transcribes from the *asnC* or *mioC* promoters; avoidance of such collisions is apparent in the arrangement of active transcription units on the *E. coli* chromosome.[221] The required level of superhelicity at *oriC* might be achieved by gyrase acting within the *oriC* nucleoid domain or by transcription. Transcription moving toward *oriC* from the *mioC* promoter would presumably introduce local positive supercoiling, followed by local negative supercoiling as RNA polymerase moves through *oriC*.[222]

Many other questions remain to be answered, some of which, due to recent advances addressed in this review, can only now be raised. Given that hemimethylated *oriC* DNA is associated with the outer membrane of the cell surface, is this indeed the basis for chromosome segregation, as proposed by Schaechter et al.[164]? If so, what are the details of this association? How does the *oriC* region of the chromosome pass through the inner membrane to find the sites on the outer membrane? Could the DnaA protein facilitate this passage by providing a pore in the membrane? Using the Finer-Moore and Stroud antipathicity program,[223] DnaA protein was found to have an antipathic region that correlates with a helical structure similar to that of an alpha helix.[224] Antipathic regions are often associated with such pores. Alternatively, is a protein, as yet unidentified in *E. coli*, required that is similar to the *B. subtilis* DnaB protein[212]? Are the stimulatory factors real that are needed for replication in a crude *in vitro* system and reported to be present in the periplasmic space[225] and, if so, are they relevant here? What is the role of the two forms of DnaA protein, with

ATP or ADP bound? Is one the initiator and the other the autorepressor? What is the role of cardiolipin and the cell surface association of DnaA protein in these processes? Concerning the process of initiation itself, what are the primer molecules for the initial DNA synthesis step, and is the enzyme DNA primase or RNA polymerase? Assuming the primer molecule is an RNA species, what other RNA species are important in initiation? What is the relationship of the *trans*-acting molecules involved in initiation to those involved in the establishment and maintenance of the nucleoid domain structure? Is the model whereby a lagging strand Okazaki fragment from the counterclockwise-moving replication fork becomes the leading strand for the clockwise replication fork[112] correct, or is there a separate mechanism for establishment of this second fork in bidirectionally replicating initiation events? Evidence has been reviewed here indicating that the GATC methylation state is involved in both timing of initiation and in segregation of daughter chromosomes. Which GATC sites in *oriC* are important in each of these functions? Are the same GATC sites involved in each of these two functions? The minimum time required between two successive initiation events at the same origin is about 20 min.[155,156] Does the GATC methylation state alone provide the mechanism for this 20-min interval? It could, if one or more specific GATC sites remained hemimethylated for this period. Most GATC sites within *oriC* are converted to a fully methylated state within 10 min after an initiation event.[164] However, because of the close proximity of GATC sites within *oriC,* experiments described to date are not sufficiently precise to demonstrate that *all* GATC sites are so converted. One or two close sites may remain hemimethylated for an extended period of time, perhaps because of a persistent unique *oriC* secondary structure (see Figure 6).

Concerning initiation in other bacteria, how do bacteria that lack the Dam methylation system distinguish daughter origins from parental origins? Is such distinction required? Is timing between rounds of replication precise in bacteria lacking the Dam methylation system? How different are the control mechanisms and the ''actors'' in the initiation process? This review has focused on the three known classes of bacterial origins, and even here the precise origin for *B. subtilis* is not yet known. How many other classes of bacterial origins remain to be discovered, and how diverse will they be? How extensive in the prokaryotic world is the 9-bp DnaA binding site going to be?

In this review, we have attempted to organize the experimental results bearing on the initiation process of DNA replication. Based on considerable genetic, physiological, and biochemical evidence, four conditions — methylation, transcription, supercoiling, and DnaA protein concentration — appear critical in determining when an initiation event will occur. The next few years should provide more definitive insight into the molecular detail of how the cell is able to regulate the DNA replication rate during steady-state growth and in response to environmental changes.

ACKNOWLEDGMENTS

We would like to extend our thanks to all those authors who sent reprints, preprints, and unpublished information and to D. Chakravarti for his assistance in the DnaA protein structural analysis. This work was supported by National Science Foundation Grant DMB-861108 to J. W. Z. and NIH grant GM31839 to D. W. S.

REFERENCES

1. **Bode, H. R. and Morowitz, H. J.,** Size and structure of the *Mycoplasma hominis* H39 chromosome, *J. Mol. Biol.* 23, 191, 1967.
2. **Kohara, Y., Akiyama, K., and Isono, K.,** The physical map of the whole *E. coli* chromosome: application of a new strategy for rapid analysis and sorting of a large genomic library, *Cell,* 50, 495, 1987.
3. **Pettijohn, D.,** personal communication, 1987.
4. **Stonington, O. G. and Pettijohn, D. E.,** The folded genome of *Escherichia coli, Proc. Natl. Acad. Sci. U.S.A.,* 68, 6, 1971.
5. **Kavenoff, R. and Ryder, O.,** Electron microscopy of membrane-associated folded chromosomes of *Escherichia coli, Chromosoma,* 55, 13, 1976.
6. **Pettijohn, D. E. and Hecht, R. M.,** RNA molecules bound to the folded bacterial genome stabilize DNA folds and segregate domains of supercoiling, *Cold Spring Harbor Symp. Quant. Biol.,* 38, 31, 1973.
7. **Worcel, A. and Burgi, E.,** On the structure of the folded chromosome of *Escherichia coli, J. Mol. Biol.,* 72, 127, 1972.
8. **Pettijohn, D. E., Hecht, R. M., Stonington, O. G., and Stomato, T. D.,** Factors stabilizing DNA folding in bacterial chromosomes, in *DNA Synthesis In Vitro,* Wells, R. D. and Inman, R. B., Eds., University Park Press, Baltimore, 1973, 145.
9. **Ryder, O. A. and Smith, D. W.,** Isolation of membrane-associated folded chromosomes from *Escherichia coli:* effect of protein synthesis inhibition, *J. Bacteriol.,* 120, 1356, 1974.
10. **Rouviere-Yaniv, J.,** Localization of the HU protein on the *Escherichia coli* nucleoid, *Cold Spring Harbor Symp. Quant. Biol.,* 42, 439, 1978.
11. **Drlica, K.,** The nucleoid, in *Escherichia coli and Salmonella typhimurium: Cellular and Molecular Biology,* Neidhardt, F. C., Ingraham, J. L., Low, K. B., Magasanik, B., Schaechter, M., and Umbarger, H. E., Eds., American Society for Microbiology, Washington, D. C., 1987, chap. 9.
12. **Pettijohn, D. E. and Sinden, R. R.,** Structure of the nucleoid, in *Molecular Cytology of Escherichia coli,* Nanninga, N., Ed., Academic Press, New York, 1986, chap. 7.
13. **von Meyenburg, K. and Hansen, F. G.,** Regulation of chromosome replication, in *Escherichia coli and Salmonella typhimurium: Cellular and Molecular Biology,* Neidhardt, F. C., Ingraham, J. L., Low, K. B., Magasanik, B., Schaechter, M., and Umbarger, H. E., Eds., American Society for Microbiology, Washington, D. C., 1987, chap. 98.
14. **McMacken, R., Silver, L., and Georgopoulos, C.,** DNA Replication, in *Escherichia coli and Salmonella typhimurium: Cellular and Molecular Biology,* Neidhardt, F. C., Ingraham, J. L., Low, K. B., Magasanik, B., Schaechter, M., and Umbarger, H. E., Eds., American Society for Microbiology, Washington, D. C., 1987, chap. 39.
15. **Messer, W.,** Minireview: initiation of DNA replication in *Escherichia coli, J. Bacteriol.,* 169, 3395, 1987.
16. **Bird, R., Louarn, J., Martuscelli, J., and Caro, L.,** Origin and sequence of chromosome replication in *Escherichia coli, J. Mol Biol.,* 70, 549, 1972.
17. **Louarn, J., Funderburgh, M., and Bird, R. E.,** More precise mapping of the replication origin in *Escherichia coli, J. Bacteriol.,* 120, 1, 1974.
18. **Marsh, R. C. and Worcel, A.,** A DNA fragment containing the origin of replication of the *Escherichia coli* chromosome, *Proc. Natl. Acad. Sci. U.S.A.,* 74, 2720, 1977.
19. **von Meyenburg, K., Hansen, F. G., Nielsen, L. D., and Riise, E.,** Origin of replication, *oriC,* of the *Escherichia coli* chromosome on specialized transducing phages λ $_{asn}$, *Mol. Gen. Genet.,* 160, 287, 1978.
20. **Miki, T., Hiraga, S., Nagata, T., and Yura, T.,** Bacteriophage λ carrying the *Escherichia coli* chromosomal region of the replication origin, *Proc. Natl. Acad. Sci. U.S.A.,* 75, 5099, 1978.
21. **Bühk, H.-J. and Messer, W.,** The replication origin region of *Escherichia coli:* nucleotide sequence and functional units, *Gene,* 24, 265, 1983.
22. **von Meyenburg, K., Hansen, F. G., Riise, E., Bergmans, H. E. N., Meijer, M., and Messer, W.,** Origin of replication, *oriC,* of the *Escherichia coli* K12 chromosome: genetic mapping and minichromosome replication, *Cold Spring Harbor Symp. Quant. Biol.,* 43, 121, 1979.
23. **Yasuda, S. and Hirota, Y.,** Cloning and mapping of the replication origin of *Escherichia coli, Proc. Natl. Acad. Sci. U.S.A.,* 74, 5458, 1978.
24. **Sugimoto, K., Oka, A., Sugisaki, H., Takanami, M., Nishimura, A., Yasuda, S., and Hirota, Y.,** Nucleotide sequence of the replication origin of *Escherichia coli, Proc. Natl. Acad. Sci. U.S.A.,* 76, 575, 1979.
25. **Meijer, M., Beck, E., Hansen, F. G., Bergmans, H. E. N., Messer, W., von Meyenburg, K., and Schaller, H.,** Nucleotide sequence of the origin of replication of the *Escherichia coli* K12 chromosome, *Proc. Natl. Acad. Sci. U.S.A.,* 76, 580, 1979.
26. **Messer, W., Meijer, M., Bergmans, H. E. N., Hansen, F. G., von Meyenburg, K., Beck, E., and Schaller, H.,** Origin of replication, *oriC,* of the *Escherichia coli* K12 chromosome: nucleotide sequence, *Cold Spring Harbor Symp. Quant. Biol.,* 43, 139, 1979.

27. **Oka, A., Sugimoto, K., Takanami, M., and Hirota, Y.,** Replication origin of the *Escherichia coli* K12 chromosome: the size and structure of the minimum DNA segment carrying the information for autonomous replication, *Mol. Gen. Genet.,* 178, 9, 1980.

28. **Hirota, Y., Yamada, M., Nishimura, A., Oka, A., Sugimoto, K., Asada, K., and Takanami, M.,** The DNA replication origin *(ori)* of *Escherichia coli:* structure and function of the *ori*-containing DNA fragment, *Prog. Nucleic Acid Res. Mol. Biol.,* 26, 33, 1981.

29. **Zyskind, J. W., Harding, N. E., Takeda, Y., Cleary, J. M., and Smith, D. W.,** The DNA replication origin region of the enterobacteriaceae, *ICN-UCLA Symp. Mol. Cell. Biol.,* 22, 13, 1981.

30. **Zyskind, J. W., Deen, L. T., and Smith, D. W.,** Isolation and mapping of plasmids containing the *Salmonella typhimurium* origin of DNA replication, *Proc. Natl. Acad. Sci. U.S.A.,* 76, 3097, 1979.

31. **Harding, N. E., Cleary, J. M., Smith, D. W., Michon, J. J., Brusilow, W. S. A., and Zyskind, J. W.,** Chromosomal replication origins (*oriC*) of *Enterobacter aerogenes* and *Klebsiella pneumoniae* are functional in *Escherichia coli, J. Bacteriol.,* 151, 983, 1982.

32. **Takeda, Y., Harding, N. E., Smith, D. W., and Zyskind, J. W.,** The chromosomal origin of replication (*oriC*) of *Erwinia carotovora, Nucleic Acids Res.,* 10, 2639, 1982.

33. **Zyskind, J. W., Cleary, J. M., Brusilow, W. S. A., Harding, N. E., and Smith, D. W.,** Chromosomal replication origin from the marine bacterium *Vibrio harveyi* functions in *Escherichia coli: oriC* consensus sequence, *Proc. Natl. Acad. Sci. U.S.A.,* 80, 1164, 1983.

34. **Fuller, R. S., Funnell, B. E., and Kornberg, A.,** The DnaA complex with the *E. coli* chromosomal replication origin *(oriC)* and other DNA sites, *Cell,* 38, 889, 1984.

35. **Matsui, M., Oka, A., Takanami, M., Yasuda, S., and Hirota, Y.,** Sites of DnaA protein-binding in the replication origin of the *Escherichia coli* K-12 chromosome, *J. Mol. Biol.,* 184, 529, 1985.

36. **Hirota, Y., Mordoh, J., and Jacob, F.,** On the process of cellular division in *Escherichia coli.* III. Thermosensitive mutants of *Escherichia coli* altered in the process of DNA initiation, *J. Mol. Biol.,* 53, 369, 1970.

37. **Fuller, R. S., Kaguni, J. M., and Kornberg, A.,** Enzymatic replication of the origin of the *Escherichia coli* chromosome, *Proc. Natl. Acad. Sci. U.S.A.,* 78, 7370, 1981.

38. **Fuller, R. S. and Kornberg, A.,** Purified DnaA protein in initiation of replication at the *Escherichia coli* chromosomal origin of replication, *Proc. Natl. Acad. Sci. U.S.A.,* 80, 5817, 1983.

39. **Seufert, W., Dobrinski, B., Lurz, R., and Messer, W.,** Functionality of the dnaA protein binding site in DNA replication is orientation-dependent, *J. Biol. Chem.,* 263, 2719, 1988.

40. **Atlung, T., Clausen, E. S., and Hansen, F. G.,** Autorepression of the *dnaA* gene of *Escherichia coli* K12, *Mol. Gen. Genet.,* 200, 442, 1985.

41. **Braun, R. E., O'Day, K., and Wright, A.,** Autoregulation of the DNA replication gene *dnaA* in *E. coli* K12, *Cell,* 40, 159, 1985.

42. **Kücherer, Y., Lother, H., Kölling, R., Schauzu, M., and Messer, W.,** Regulation of transcription of the chromosomal *dnaA* gene of *Escherichia coli, Mol. Gen. Genet.,* 205, 115, 1986.

43. **Wang, Q. and Kaguni, J. M.,** Transcriptional repression of the *dnaA* gene of *Escherichia coli* by dnaA protein, *Mol. Gen. Genet.,* 209, 518, 1987.

44. **Jonczyk, P., Hines, R., and Smith, D. W.,** The *Escherichia dam* gene is expressed as a distal gene of a new operon, *Mol. Gen. Genet.,* 217, 85, 1989.

45. **Lother, H., Kölling, R., Kücherer, C., and Schauzu, M.,** *DnaA* protein-regulated transcription: effects on the *in vitro* replication of *Escherichia coli* minichromosomes, *EMBO J.,* 4, 555, 1985.

46. **Stuitje, A. R., de Wind, N., van der Spek, J. C., Pors, T. H., and Meijer, M.,** Dissection of promoter sequences involved in transcriptional activation of the *Escherichia coli* replication origin, *Nucleic Acids Res.,* 14, 2333, 1986.

47. **van den Berg, E. A., Geerse, R. H., Memelink, J., Bovenberg, R. A. L., Magnée, F. A., and van de Putte, P.,** Analysis of regulatory sequences upstream of the *E. coli uvrB* gene: involvement of the DnaA protein, *Nucleic Acids Res.,* 13, 1829, 1985.

48. **Yee, T. W. and Smith, D. W.,** submitted.

49. **Moriya, S., Ogasawara, N., and Yoshikawa, H.,** Structure and function of the region of the replication origin of the *Bacillus subtilis* chromosome. III. Nucleotide sequence of some 10,000 base pairs in the origin region, *Nucleic Acids Res.,* 13, 2251, 1985.

50. **Bramhill, D. and Kornberg, A.,** Duplex opening by dnaA protein at novel sequences in initiation of replication at the origin of the *E. coli* chromosome, *Cell,* 52, 743, 1988.

51. **Asada, K., Sugimoto, K., Oka, A., Takanami, M., and Hirota, Y.,** Structure of replication origin of the *Escherichia coli* K12 chromosome: the presence of spacer sequences in the *ori* region carrying information for autonomous replication, *Nucleic Acids Res.,* 10, 3745, 1982.

52. **Oka, A., Sugimoto, K., Sasaki, H., and Takanami, M.,** An in vitro method generating base substitutions in preselected regions of plasmid DNA: application to structural analysis of the replication origin of the *Escherichia coli* K12 chromosome, *Gene,* 19, 59, 1982.

53. **Oka, A., Sasaki, H., Sugimoto, K., and Takanami, M.,** Sequence organization of replication origin of the *Escherichia coli* K12 chromosome, *J. Mol. Biol.,* 176, 443, 1984.

54. **Smith, D. W., Garland, A. M., Herman, G., Enns, R. E., Baker, T. A., and Zyskind, J. W.,** Importance of state of methylation of *oriC* GATC sites in initiation of DNA replication in *Escherichia coli, EMBO J.,* 4, 1319, 1985.

55. **Plasterk, R. H. A., Vrieling, H., and van de Putte, P.,** Transcription initiation of Mu *mom* depends on methylation of the promoter region and a phage-coded transactivator, *Nature (London),* 301, 344, 1983.

56. **Zahn, K. and Blattner, F. R.,** Binding and bending of the lambda replication origin by the phage O protein, *EMBO J.,* 4, 3605, 1985.

57. **Stenzel, T. T., Patel, P., and Bastia, D.,** The integration host factor of *Escherichia coli* binds to bent DNA at the origin of replication of the plasmid pSC101, *Cell,* 49, 709, 1987.

58. **Filutowicz, M. and Appelt, K.,** The integration host factor of *Escherichia coli* binds to multiple sites at plasmid R6K gamma origin and is essential for replication, *Nucleic Acids Res.,* 16, 3829, 1988.

59. **Mukherjee, S., Erickson, H., and Bastia, D.,** Enhancer origin interaction in plasmid R6K involves a DNA loop mediated by initiator protein, *Cell,* 52, 375, 1988.

60. **Chattoraj, D. K., Mason, R. J., and Wickner, S. H.,** Mini-P1 plasmid replication: the autoregulation-sequestration paradox, *Cell,* 52, 551, 1988.

61. **Funnell, B. E., Baker, T. A., and Kornberg, A.,** *In vitro* assembly of a prepriming complex at the origin of the *Escherichia coli* chromosome, *J. Biol. Chem.,* 262, 10327, 1987.

62. **Kogoma, T. and von Meyenberg, K.,** The origin of replication, *oriC,* and the *dnaA* protein are dispensable in stable DNA replication *(sdrA)* mutants of *Escherichia coli* K-12, *EMBO J.,* 2, 463, 1983.

63. **Kogoma, T., Subia, N. L., and von Meyenburg, K.,** Function of ribonuclease H in initiation of DNA replication of *Escherichia coli* K12, *Mol. Gen. Genet.,* 200, 103, 1985.

64. **de Massy, B., Fayet, O., and Kogoma, T.,** Multiple origin usage for DNA replication in *sdrA (rnh)* mutants of *Escherichia coli* K-12: initiation in the absence of *oriC, J. Mol. Biol.,* 178, 227, 1984.

65. **Carl, P. L., Bloom, L., and Crouch, R. J.,** Isolation and mapping of a mutant in *Escherichia coli* with altered levels of ribonuclease H, *J. Bacteriol.,* 144, 28, 1980.

66. **Hansen, E. B., Atlung, T., Hansen, F. G., Skovgaard, O., and von Meyenburg, K.,** Fine structure genetic map and complementation analysis of mutations in the *dnaA* gene of *Escherichia coli, Mol. Gen. Genet.,* 196, 387, 1984.

67. **Atlung, T.,** Allele-specific suppression of *dnaA*(Ts) mutations by *rpoB* mutations in *Escherichia coli, Mol. Gen. Genet.,* 197, 125, 1984.

68. **Schaus, N. A., O'Day, K., and Wright, A.,** Suppression of amber mutations in the *dnaA* gene of *Escherichia coli* K-12 by secondary mutations in *rpoB, ICN-UCLA Symp. Mol. Cell. Biol.,* 22, 315, 1981.

69. **Zyskind, J. W., Deen, L. T., and Smith, D. W.,** Temporal sequence of events during the initiation process in *Escherichia coli* deoxyribonucleic acid replication: roles of the *dnaA* and *dnaC* gene products and ribonucleic acid polymerase, *J. Bacteriol.,* 129, 1466, 1977.

70. **Ream, L. W., Margossian, L., Clark, A. J., Hansen, F. G., and von Meyenburg, K.,** Genetic and physical mapping of *recF* gene of *E. coli* K-12, *Mol. Gen. Genet.,* 180, 115, 1980.

71. **Sako, T. and Sakakibara, Y.,** Coordinate expression of *Escherichia coli dnaA* and *dnaN* genes, *Mol. Gen. Genet.,* 179, 521, 1980.

72. **Burgers, P. J. M., Kornberg, A., and Sakakibara, Y.,** The *dnaN* gene codes for the beta subunit of DNA polymerase III holoenzyme of *E. coli, Proc. Natl. Acad. Sci. U.S.A.,* 78, 5391, 1981.

73. **Armengod, M.-E. and Lambíes, E.,** Overlapping arrangement of the *recF* and *dnaN* operons of *Escherichia coli;* positive and negative control sequences, *Gene,* 43, 183, 1986.

74. **Hansen, F. G., Hansen, E. B., and Atlung, T.,** The nucleotide sequence of the *dnaA* gene promoter and of the adjacent *rpmH* gene, coding for the ribosomal protein L34, of *Escherichia coli, EMBO J.,* 9, 1043, 1982.

75. **Braun, R. E. and Wright, A.,** DNA methylation differentially enhances the expression of one of the two *E. coli dnaA* promoters in vivo and in vitro, *Mol. Gen. Genet.,* 202, 246, 1986.

76. **Donachie, W. D.,** Relationship between cell size and time of initiation of DNA replication, *Nature (London),* 219, 1077, 1968.

77. **Sompayrac, L. and Maaløe, O.,** Autorepressor model for control of DNA replication, *Nature (London),* 241, 133, 1973.

78. **Atlung, T., Løbner-Olesen, A., and Hansen, F. G.,** Overproduction of DnaA protein stimulates initiation of chromosome and minichromosome replication in *Escherichia coli, Mol. Gen. Genet.,* 206, 51, 1987.

79. **Churchward, G., Holmans, P., and Bremer, H.,** Increased expression of the *dnaA* gene has no effect on DNA replication in a *dnaA*⁺ strain of *Escherichia coli, Mol. Gen. Genet.,* 192, 506, 1983.

80. **Bremer, H. and Churchward, G.,** Initiation of chromosome replication in *Escherichia coli* after induction of *dnaA* gene expression from a lac promoter, *J. Bacteriol.,* 164, 922, 1985.

81. **Xu, Y.-C. and Bremer, H.,** Chromosome replication in *Escherichia coli* induced by oversupply of DnaA, *Mol. Gen. Genet.,* 211, 138, 1988.

82. **Rokeach, L. A., Chiaramello, A., Crain, K., Nourani, A., Jannitapour, M., and Zyskind, J.**, Transcription events at the *Escherichia coli* origin of replication, *oriC*, in *DNA Replication and Recombination*, Kelly, T. J. and McMacken, R., Eds., Alan R. Liss, New York, 1987, 415.

83. **Sekimizu, K., Yung, B. Y., and Kornberg, A.**, The dnaA protein of *Escherichia coli* abundance, improved purification, and membrane binding, *J. Biol. Chem.*, 263, 7136, 1988.

84. **Chiaramello, A. and Zyskind, J.**, Growth rate regulation and initiation of DNA replication in *Escherichia coli*, *J. Bacteriol.*, 171, 4272, 1989.

85. **Maaløe, O. and Hanawalt, P. C.**, Thymine deficiency and the normal DNA replication cycle, *J. Mol. Biol.*, 1, 144, 1961.

86. **Lark, D. G., Repko, T., and Hoffman, E. J.**, The effect of amino acid deprivation on subsequent deoxyribonucleic acid replication, *Biochim. Biophys. Acta*, 76, 9, 1963.

87. **Kimura, M., Yura, T., and Nagata, T.**, Isolation and characterization of *Escherichia coli dnaA* amber mutants, *J. Bacteriol.*, 144, 649, 1980.

88. **Schaus, N., O'Day, K., Peters, W., and Wright, A.**, Suppression of amber mutations in the *dnaA* gene of *Escherichia coli* K-12, *J. Bacteriol.*, 145, 904, 1981.

89. **Skarstad, K., Steen, H. B., and Boye, E.**, *Escherichia coli* DNA distributions measured by flow cytometry and compared with theoretical computer simulations, *J. Bacteriol.*, 163, 661, 1985.

90. **Skarstad, K., von Meyenburg, K., and Hansen, F. G.**, Coordination of initiation of chromosome replication in *Escherichia coli:* effects of different *dnaA* alleles, *J. Bacteriol.*, 170, 852, 1988.

91. **Skarstad, K. and Boye, E.**, Perturbed chromosomal replication in *recA* mutants of *Escherichia coli*, *J. Bacteriol.*, 170, 2549, 1988.

92. **Skarstad, K., Steen, H. B., and Boye, E.**, Cell cycle parameters of slowly growing *Escherichia coli* B/r studied by flow cytometry, *J. Bacteriol.*, 154, 656, 1983.

93. **Skarstad, K., Boye, E., and Steen, H. B.**, Timing of initiation of chromosome replication in individual *Escherichia coli* cells, *EMBO J.*, 5, 1711, 1986.

94. **Kogoma, T., Skarstad, K., Boye, E., von Meyenburg, K., and Steen, H. B.**, RecA protein acts at the initiation of stable DNA replication in *rnh* mutants of *Escherichia coli* K-12, *J. Bacteriol.*, 163, 439, 1985.

95. **Lother, H., Lurz, R., and Orr, E.**, DNA binding and antigenic specifications of DNA gyrase, *Nucleic Acids Res.*, 12, 901, 1984.

96. **Orr, E., Fairweather, N. F., Holland, I. B., and Pritchard, R. H.**, Isolation and characterisation of a strain carrying a conditional lethal mutation in the *cou* gene of *Escherichia coli*, *Mol. Gen. Genet.*, 177, 103, 1979.

97. **Filutowicz, M. and Jonczyk, P.**, Essential role of the *gyrB* gene product in the transcriptional event coupled to dnaA-dependent initiation of *Escherichia coli* chromosome replication, *Mol. Gen. Genet.*, 183, 134, 1981.

98. **Filutowicz, M. and Jonczyk, P.**, The *gyrB* gene product functions in both initiation and chain polymerization of *Escherichia coli* chromosome replication: suppression of the initiation deficiency in *gyrB*-ts mutants by a class of *rpoB* mutations, *Mol. Gen. Genet.*, 191, 282, 1983.

99. **Louarn, J., Bouché, J.-P., Pattee, J., and Louarn, J.-M.**, Genetic inactivation of topoisomerase I suppresses a defect in initiation of chromosome replication in *Escherichia coli*, *Mol. Gen. Genet.*, 19, 170, 1984.

100. **Shlomai, J., Arai, K., Arai, N., Kobori, J., Polder, L., Low, R., Hübscher, U., Bertsch, L., and Kornberg, A.**, Enzyme studies of φX174 DNA replication, *ICN-UCLA Symp. Mol. Cell. Biol.*, 14, 137, 1980.

101. **Kobori, J. A. and Kornberg, A.**, The *Escherichia coli dnaC* gene product. III. Properties of the DnaB-DnaC protein complex, *J. Biol. Chem.*, 257, 13770, 1982.

102. **Lanka, E. and Schuster, H.**, The DnaC protein of *Escherichia coli*. Purification, physical properties and interaction with DnaB protein, *Nucleic Acids Res.*, 11, 987, 1983.

103. **Wickner, S. and Hurwitz, J.**, Interaction of *Escherichia coli dnaB* and *dnaC(D)* gene products *in vitro*, *Proc. Natl. Acad. Sci. U.S.A.*, 72, 921, 1975.

104. **Lebowitz, J. H. and McMacken, R.**, The *Escherichia coli* DnaB replication protein is a DNA helicase, *J. Biol. Chem.*, 261, 4738, 1986.

105. **Baker, T. A., Sekimizu, K., Funnell, B., and Kornberg, A.**, Extensive unwinding of the plasmid template during staged enzymatic initiation of DNA replication from the origin of the *Escherichia coli* chromosome, *Cell*, 45, 53, 1986.

106. **Lark, K. G.**, Evidence for direct involvement of RNA in the initiation of DNA replication in *E. coli* 15T⁻, *J. Mol. Biol.*, 64, 47, 1972.

107. **Messer, W.**, Initiation of deoxyribonucleic acid replication in *Escherichia coli* B/r: chronology of events and transcriptional control of initiation, *J. Bacteriol.*, 112, 7, 1972.

108. **Rasmussen, K. V., Atlung, T., Kerszman, G., Hansen, E. B., and Hansen, F. G.**, Conditional change of DNA replication control in an RNA polymerase mutant of *Escherichia coli*, *J. Bacteriol.*, 154, 443, 1983.

109. **Tanaka, M., Ohmori, H., and Hiraga, S.,** A novel type of *E. coli* mutants with increased chromosomal copy number, *Mol. Gen. Genet.,* 192, 51, 1983.

110. **Bagdasarian, M. M., Izakowska, M., and Bagdasarian, M.,** Suppression of the DnaA phenotype by mutations in the *rpoB* cistron of ribonucleic acid polymerase in *Salmonella typhimurium* and *Escherichia coli, J. Bacteriol.,* 130, 577, 1977.

111. **Atlung, T. and Hansen, F. G.,** Effect of *dnaA* and *rpoB* mutations on attenuation in the *trp* operon of *Escherichia coli, J. Bacteriol.,* 156, 985, 1983.

112. **Kohara, Y., Tohodoh, N., Jiang, X.-W., and Okazaki, T.,** The distribution and properties of RNA primed initiation sites of DNA synthesis at the replication origin of *Escherichia coli* chromosome, *Nucleic Acids Res.,* 13, 6847, 1985.

113. **Rokeach, L. A. and Zyskind, J.,** RNA terminating within the *E. coli* origin of replication: stringent regulation and control by DnaA protein, *Cell,* 46, 763, 1986.

114. **Schauzu, M.-A., Kücherer, C., Kölling, R., Messer, W., and Lother, H.,** Transcripts within the replication origin, *oriC,* of *Escherichia coli, Nucleic Acids Res.,* 15, 2479, 1987.

115. **Furth, M. E., Dove, W. F., and Meyer, B. J.,** Specificity determinants for bacteriophage λ DNA replication. III. Activation of replication in λ*ri^c* mutants by transcription of *ori, J. Mol. Biol.,* 154, 65, 1982.

116. **Kitani, T., Yoda, K., Ogawa, T., Okazaki, T.,** Evidence that discontinuous DNA replication in *Escherichia coli* is primed by approximately 10 to 12 residues of RNA starting with a purine, *J. Mol. Biol.,* 184, 45, 1985.

117. **Yoshimoto, M., Kambe-Honjoh, H., Nagai, K., and Tamura, G.,** Early replicative intermediates of *Escherichia coli* chromosome isolated from a membrane complex, *EMBO J.,* 5, 787, 1986.

118. **Yoshimoto, M., Nagai, K., and Tamura, G.,** Asymmetric replication of an *oriC* plasmid in *Escherichia coli, Mol. Gen. Genet.,* 204, 214, 1986.

119. **Lycett, G. W., Orr, E., and Pritchard, R. H.,** Chloramphenicol releases a block in initiation of chromosome replication in a *dnaA* strain of *Escherichia coli* K12, *Mol. Gen. Genet.,* 178, 329, 1980.

120. **Shen, V. and Bremer, H.,** Chloramphenicol-induced changes in the synthesis of ribosomal, transfer, and messenger ribonucleic acids in *Escherichia coli* B/r, *J. Bacteriol.,* 130, 1098, 1977.

121. **Rokeach, L. A., Kassavetis, G. A., and Zyskind, J. W.,** RNA polymerase pauses *in vitro* within the *Escherichia coli* origin of replication at the same sites where termination occurs *in vivo, J. Biol. Chem.,* 262, 7264, 1987.

122. **Meijer, M. and Messer, W.,** Functional analysis of minichromosome replication: bidirectional and unidirectional replication from the *Escherichia coli* replication origin, *oriC, J. Bacteriol.,* 143, 1049, 1980.

123. **Tanaka, M. and Hiraga, S.,** Negative control of *oriC* plasmid replication by transcription of the *oriC* region, *Mol. Gen. Genet.,* 200, 21, 1985.

124. **Stuitje, A. R. and Meijer, M.,** Maintenance and incompatibility of plasmids carrying the replication origin of the *Escherichia coli* chromosome: evidence for a control region of replication between *oriC* and *asnA, Nucleic Acids Res.,* 11, 5775, 1983.

125. **Løbner-Olesen, A., Atlung, T., and Rasmussen, K. V.,** Stability and replication control of *Escherichia coli* minichromosomes, *J. Bacteriol.,* 169, 2835, 1987.

126. **Stuitje, A. R., de Wind, N., van der Spek, J. C., Pors, T. H., and Meijer, M.,** Dissection of promoter sequences involved in transcriptional activation of the *Escherichia coli* replication origin, *Nucleic Acids Res.,* 14, 2333, 1986.

127. **de Wind, N., Parren, P., Stuitje, A. R., and Meijer, M.,** Evidence for the involvement of the 16 kD gene promoter in initiation of chromosomal replication of *Escherichia coli* strains carrying a B/r-derived replication origin, *Nucleic Acids Res.,* 15, 4901, 1987.

128. **Levine, A., Henckes, G., Vannier, F., and Séror, S. J.,** Chromosomal initiation in *Bacillus subtilis* may involve two closely linked origins, *Mol. Gen. Genet.,* 208, 37, 1987.

129. **Séror-Laurent, S. J. and Henckes, G.,** An RNA-DNA copolymer whose synthesis is correlated with the transcriptional requirement for chromosomal initiation in *Bacillus subtilis* contains ribosomal RNA sequences, *Proc. Natl. Acad. Sci. U.S.A.,* 82, 3586, 1985.

130. **Henckes, G., Vannier, F., Buu, A., and Séror-Laurent, S. J.,** Possible involvement of DNA-linked RNA in the initiation of *Bacillus subtilis* chromosomal replication, *J. Bacteriol.,* 149, 79, 1982.

131. **Séror, S. J., Vannier, F., Levine, A., and Henckes, G.,** Stringent control of initiation of chromosomal replication in *Bacillus subtilis, Nature (London),* 321, 709, 1986.

132. **Sarmientos, P. and Cashel, M.,** Carbon starvation and growth rate-dependent regulation of the *Escherichia coli* ribosomal RNA promoter: differential control of dual promoters, *Proc. Natl. Acad. Sci. U.S.A.,* 80, 7010, 1983.

133. **Sarmientos, P., Sylvester, J. E., Contente, S., and Cashel, M.,** Differential stringent control of the tandem *E. coli* ribosomal RNA promoters from the *rrnA* operon expressed in vivo in multicopy plasmids, *Cell,* 32, 1337, 1983.

134. **Gourse, R. L., de Boer, H. A., and Nomura, M.,** DNA determinants of rRNA synthesis in *E. coli:* growth rate dependent regulation, feedback inhibition, upstream activation, antitermination, *Cell,* 44, 197, 1986.

135. **Cashel, M. and Rudd, K. E.,** The stringent response, in *Escherichia coli and Salmonella typhimurium Cellular and Molecular Biology,* Vol. 2, Neidhardt, F. ℃., Ed., American Society for Microbiology, Washington, D.C., 1987, 1410.

136. **Baracchini, E. and Bremer, H.,** Stringent and growth control of rRNA synthesis in *Escherichia coli* are both mediated by ppGpp, *J. Biol. Chem.,* 263, 2597, 1988.

137. **Sarubbi, E., Rudd, K. E., and Cashel, M.,** Basal ppGpp level adjustment shown by new *spoT* mutants affect steady state growth rates and *rrnA* ribosomal promoter regulation in *Escherichia coli, Mol. Gen. Genet.,* 213, 214, 1988.

138. **Geier, G. E. and Modrich, P.,** Recognition sequence of the *dam* methylase of *Escherichia coli* K-12 and mode of cleavage of DpnI endonuclease, *J. Biol. Chem.,* 254, 1408, 1979.

139. **Marinus, M. and Morris, N.,** Isolation of deoxyribonucleic acid methylase mutants of *Escherichia coli* K-12, *J. Bacteriol.,* 114, 1143, 1973.

140. **Marinus, M. G. and Konrad, E. B.,** Hyper-recombination in *dam* mutants of *Escherichia coli* K-12, *Mol. Gen. Genet.,* 149, 273, 1976.

141. **Bale, A., d'Alarcao, M., and Marinus, M. G.,** Characterization of DNA adenine methylation mutants of *Escherichia coli* K-12, *Mutat. Res.,* 59, 157, 1979.

142. **Marinus, M. G., Caraway, M., Frey, A. Z., Brown, L., and Arraj, J. A.,** Insertion mutations in the *dam* gene of *Escherichia coli* K-12, *Mol. Gen. Genet.,* 192, 288, 1983.

143. **Parker, B. and Marinus, M. G.,** A simple and rapid method to obtain substitution mutation in *Escherichia coli:* isolation of a *dam* deletion/insertion mutation, *Gene,* 73, 531, 1988.

145. **Arraj, J. A. and Marinus, M. G.,** Phenotypic reversal in *dam* mutants of *Escherichia coli* K-12 by a recombinant plasmid containing the *dam$^+$* gene, *J. Bacteriol.,* 153, 562, 1983.

146. **Wagner, R. and Meselson, M.,** Repair tracts in mismatched DNA heteroduplexes, *Proc. Natl. Acad. Sci. U.S.A.,* 73, 4135, 1976.

147. **Pukkila, P., Peterson, J., Herman, G., Modrich, P., and Meselson, M.,** Effects of high levels of DNA adenine methylation on methyl directed mismatch repair in *E. coli, Genetics,* 104, 571, 1983.

148. **Lu, A.-L., Clark, S., and Modrich, P.,** Methyl-directed repair of DNA base pair mismatches in vitro, *Proc. Natl. Acad. Sci. U.S.A.,* 80, 4639, 1983.

149. **Sternberg, N.,** Evidence that adenine methylation influences DNA-protein interactions in *Escherichia coli, J. Bacteriol.,* 164, 490, 1985.

150. **Marinus, M.,** Methylation of DNA, in *Escherichia coli and Salmonella typhimurium: Cellular and Molecular Biology,* Neidhardt, F. C., Ingraham, J. L., Low, K. B., Magasanik, B., Schaechter, M., and Umbarger, H. E., Eds., American Society for Microbiology, Washington, D.C., 1987, chap. 46.

151. **Zyskind, J. W. and Smith, D. W.,** Nucleotide sequence of the *Salmonella typhimurium* origin of DNA replication, *Proc. Natl. Acad. Sci. U.S.A.,* 77, 2460, 1980.

152. **Messer, W., Bellekes, U., and Lother, H.,** Effect of *dam* methylation on the activity of the *E. coli* replication origin, *oriC, EMBO J.,* 4, 1327, 1985.

153. **Russell, D. W. and Zinder, N. D.,** Hemimethylation prevents DNA replication in *E. coli, Cell,* 50, 1071, 1987.

154. **Hughes, P., Squali-Houssaini, F.-Z., Forterre, P., and Kohiyama, M.,** In vitro replication of a *dam* methylated and non-methylated *ori-C* plasmid, *J. Mol. Biol.,* 176, 155, 1984.

155. **Eberle, H., Forrest, N., Hrynyszyn, J., and von Knapp, J.,** Regulation of DNA synthesis and capacity for initiation in DNA temperature-sensitive mutants of *Escherichia coli.* I. Reinitiation and elongation, *Mol. Gen. Genet.,* 186, 57, 1982.

156. **Helmstetter, C. and Krajewski, C. A.,** Initiation of chromosome replication in *dnaA* and *dnaC* mutants of *Escherichia coli* B/r, *J. Bacteriol.,* 149, 685, 1982.

157. **Bakker, A. and Smith, D. W.,** Methylation of GATC sites is required for precise timing between rounds of DNA replication in *Escherichia coli, J. Bacteriol.,* in press.

158. **Jacob, F., Brenner, S., and Cuzin, F.,** On the regulation of DNA replication in bacteria, *Cold Spring Harbor Symp. Quant. Biol.,* 191, 460, 1963.

159. **Ogden, G. and Schaechter, M.,** The association of the *Escherichia coli* chromosome with the cell membrane, in *Bacterial Chromatin,* Gualerzi, C., Ed., Springer-Verlag, Berlin, 1986, 45.

160. **Jacq, A., Kohiyama, M., Lother, H., and Messer, M.,** Recognition sites for a membrane-derived DNA binding protein preparation in the *E. coli* replication origin, *Mol. Gen. Genet.,* 191, 460, 1983.

161. **Nagai, K., Hendrickson, W., Balakrishnan, R., Yamaki, H., Boyd, D., and Schaechter, M.,** Isolation of a replication origin complex from *Escherichia coli, Proc. Natl. Acad. Sci. U.S.A.,* 77, 262, 1980.

162. **Hendrickson, W. G., Kusano, T., Yamaki, H., Balakrishnan, R., King, M., Murchie, J., and Schaechter, M.,** Binding of the origin of replication of *Escherichia coli* to the outer membrane, *Cell,* 30, 915, 1982.

163. **Kusano, T., Steinmetz, D., Hendrickson, W. G., Murchie, J., King, M., Benson, A., and Schaechter, M.,** Direct evidence for specific binding of the replicative origin of the *Escherichia coli* chromosome to the membrane, *J. Bacteriol.,* 158, 313, 1984.

164. **Ogden, G. B., Pratt, M. J., and Schaechter, M.,** The replicative origin of the *Escherichia coli* chromosome binds to cell membranes only when hemimethylated, *Cell,* 54, 127, 1988.

165. **MacAlister, T. J., MacDonald, B., and Rothfield, L. I.,** The periseptal annulus: an organelle associated with cell division in gram-negative bacteria, *Proc. Natl. Acad. Sci. U.S.A.,* 80, 1372, 1983.

166. **Sekimizu, K. and Kornberg, A.,** Cardiolipin activation of dnaA protein, the initiation protein of replication in *Escherichia coli, J. Biol. Chem.,* 263, 7131, 1988.

167. **Kogoma, T.,** A novel *Escherichia coli* mutant capable of DNA replication in the absence of protein synthesis, *J. Mol. Biol.,* 121, 55, 1978.

168. **Kogoma, T.,** Mini-review: RNaseH-defective mutants of *Escherichia coli, J. Bacteriol.,* 166, 361, 1986.

169. **Koppes, L. and Nordström, K.,** Insertion of an R1 plasmid into the origin of replication of the *E. coli* chromosome: random timing of replication of the hybrid chromosome, *Cell,* 44, 117, 1986.

170. **Nordström, K., Molin, S., and Light, J.,** Control of replication of bacterial plasmids: genetics, molecular biology, and physiology of the plasmid R1 system, *Plasmid,* 12, 71, 1984.

171. **Hansen, F. G., Hansen, E. B., and von Meyenburg, K.,** The nucleotide sequence of the *dnaA* gene and the first part of the *dnaN* gene of *Escherichia coli* K-12, *Nucleic Acids Res.,* 10, 7373, 1982.

172. **Ohmori, H., Kimura, M., Nagata, T., and Sakakibara, Y.,** Structural analysis of the *dnaA* and *dnaN* genes of *Escherichia coli, Gene,* 29, 159, 1984.

173. **Skovgaard, O. and Hansen, F. G.,** Comparison of *dnaA* nucleotide sequences of *Escherichia coli, Salmonella typhimurium,* and *Serratia marcescens, J. Bacteriol.,* 169, 3976, 1987.

174. **Braun, R. E., O'Day, K., and Wright, A.,** Cloning and characterization of *dnaA*(Cs), a mutation which leads to overinitiation of DNA replication in *Escherichia coli* K-12, *J. Bacteriol.,* 169, 3898, 1987.

175. **Fujita, M. Q., Yoshikawa, H., and Ogasawara, N.,** Structure of *"dnaA"* region of *Pseudomonas putida:* conservation among three bacteria, *B. subtilis, E. coli,* and *P. putida, Mol. Gen. Genet.,* 215, 381, 1989.

176. **Chou, P. Y. and Fasman, G. D.,** Prediction of the secondary structure of proteins from their amino acid sequence, *Adv. Enzymol.,* 47, 45, 1978.

177. **Garnier, J., Osguthorpe, D. J., and Robson, B.,** Analysis of the accuracy and implications of simple methods for predicting the secondary structure of globular proteins, *J. Mol. Biol.,* 120, 97, 1978.

178. **Pabo, C. O. and Sauer, R. T.,** Protein-DNA recognition, *Annu. Rev. Biochem.,* 53, 293, 1984.

179. **Sekimizu, K., Bramhill, D., and Kornberg, A.,** ATP activates dnaA protein in initiating replication of plasmids bearing the origin of the *E. coli* chromosome, *Cell,* 50, 259, 1987.

180. **Finch, P. W. and Emmerson, P. T.,** The nucleotide sequence of the *uvrD* gene of *E. coli, Nucleic Acids Res.,* 12, 5789, 1984.

181. **Ogawa, T., Baker, T. A., van der Ende, A., and Kornberg, A.,** Initiation of enzymatic replication at the origin of the *Escherichia coli* chromosome: contributions of RNA polymerase and primase, *Proc. Natl. Acad. Sci. U.S.A.,* 82, 3562, 1985.

182. **van der Ende, A., Baker, T. A., Ogawa, T., and Kornberg, A.,** Initiation of enzymatic replication at the origin of the *Escherichia coli* chromosome: primase as the sole priming enzyme, *Proc. Natl. Acad. Sci. U.S.A.,* 82, 3954, 1985.

183. **Kaguni, J. M. and Kornberg, A.,** Replication initiated at the origin (*oriC*) of the *E. coli* chromosome reconstituted with purified enzymes, *Cell,* 38, 183, 1984.

184. **Kornberg, A., Baker, T. A., Bertsch, L. L., Bramhill, D., Funnell, B. E., Lasken, R. S., Maki, H., Maki, S., Sekimizu, K., and Wahle, E.,** Enzymatic studies of replication of *oriC* plasmids, in *DNA Replication and Recombination,* Kelly, T. J. and McMacken, R., Eds., Alan R. Liss, New York, 1987, 137.

185. **Sekimizu, K., Bramhill, D., and Kornberg, A.,** Sequential early stages in the *in vitro* initiation of replication at the origin of the *Escherichia coli* chromosome, *J. Biol. Chem.,* 263, 7124, 1988.

186. **Baker, T. A., Funnell, B. E., and Kornberg, A.,** Helicase action of dnaB protein during replication from the *Escherichia coli* chromosomal origin *in vitro, J. Biol. Chem.,* 262, 6877, 1987.

187. **Baker, T. A., Sekimizu, K., Funnell, B. E., and Kornberg, A.,** Extensive unwinding of the plasmid template during staged enzymatic initiation of DNA replication from the origin of the *Escherichia coli* chromosome, *Cell,* 45, 53, 1986.

188. **Seufert, W. and Messer, W.,** DnaA protein binding to the plasmid origin region can substitute for primosome assembly during replication of pBR322 in vitro, *Cell,* 48, 73, 1987.

189. **Seufert, W. and Messer, W.,** Start sites for bidirectional *in vitro* DNA replication inside the replication origin, *oriC,* of *Escherichia coli, EMBO J.,* 6, 2469, 1987.

190. **Enns, R., Eix, S., and Smith, D. W.,** Use of M13-*oriC* chimeric phage: high copy lethal genetic elements and cloning of the *Hemophilus parahemolyticus* origin of replication, *ICN-UCLA Symp. Mol. Cell. Biol.,* 47, 543, 1987.

191. **Firshein, W.,** Two membrane sites for DNA synthesis in pneumococcus, *Mol. Gen. Genet.,* 148, 323, 1976.

192. **Barbeyron, T., Kean, K., and Forterre, P.,** DNA adenine methylation of GATC sequences appeared recently in the *Escherichia coli* lineage, *J. Bacteriol.,* 160, 586, 1984.

193. **Baur, C.-P. and Knippers, R.,** Protein-induced bending of the simian virus 40 origin of replication, *J. Mol. Biol.,* 203, 1009, 1988.

194. **Gingeras, T. R. and Brooks, J. E.,** Cloned restriction/modification system from *Pseudomonas aeruginosa, Proc. Natl. Acad. Sci. U.S.A.,* 80, 402, 1983.

195. **Theriault, G., Roy, P. H., Howard, K. A., Benner, J. S., Brooks, J. E., Waters, A. F., and Gingeras, T. R.,** Nucleotide sequence of the *Pae*R7 restriction/modification system and partial characterization of its protein products, *Nucleic Acids Res.,* 13, 8441, 1985.

196. **Gyurasits, E. B. and Wake, R. G.,** Bidirectional chromosome replication in *Bacillus subtilis, J. Mol. Biol.,* 73, 55, 1973.

197. **Yoshikawa, H. and Sueoka, N.,** Sequential replication of *Bacillus subtilis* chromosome. I. Comparison of marker frequencies in exponential and stationary growth phases, *Proc. Natl. Acad. Sci. U.S.A.,* 49, 559, 1963.

198. **Seiki, M., Ogasawara, N., and Yoshikawa, H.,** Structure of the region of the replication origin of the *Bacillus subtilis* chromosome, *Nature (London),* 281, 699, 1979.

199. **Seiki, M., Ogasawara, N., and Yoshikawa, H.,** Structure and function of the region of the replication origin of the *Bacillus subtilis* chromosome. I. Isolation and characterization of plasmids containing the origin region, *Mol. Gen. Genet.,* 183, 220, 1981.

200. **Ogasawara, N., Seiki, M., and Yoshikawa, H.,** Replication origin region of *Bacillus subtilis* chromosome contains two rRNA operons, *J. Bacteriol.,* 154, 50, 1983.

201. **Ogasawara, N., Mizumoto, S., and Yoshikawa, H.,** Replication origin of the *Bacillus subtilis* chromosome determined by hybridization of the first-replicating DNA with cloned fragments from the replication origin of the chromosome, *Gene,* 30, 173, 1984.

202. **Sueoka, N. and Quinn, W. G.,** Membrane attachment of the chromosome replication origin in *Bacillus subtilis, Cold Spring Harbor Symp. Quant. Biol.,* 33, 695, 1968.

203. **Winston, S. and Sueoka, N.,** DNA-membrane association is necessary for initiation of chromosomal and plasmid replication in *Bacillus subtilis, Proc. Natl. Acad. Sci. U.S.A.,* 77, 2834, 1980.

204. **Yamaguchi, K. and Yoshikawa, H.,** Chromosome-membrane association in *Bacillus subtilis.* III. Isolation and characterization of a DNA-protein complex carrying replication origin markers, *J. Mol. Biol.,* 110, 219, 1977.

205. **Benjamin, P. and Firshein, W.,** Initiation of DNA replication *in vitro* by a DNA/membrane complex extracted from *Bacillus subtilis, Proc. Natl. Acad. Sci. U.S.A.,* 80, 6214, 1983.

206. **Laffan, J. and Firshein, W.,** DNA replication by a DNA-membrane complex extracted from *Bacillus subtilis* site of initiation: *in vitro* and initiation potential of subcomplexes, *J. Bacteriol.,* 169, 2819, 1987.

207. **Ogasawara, N., Moriya, S., and Yoshikawa, H.,** Structure and function of the region of the replication origin of the *Bacillus subtilis* chromosome. IV. Transcription of the *oriC* region and expression of DNA gyrase genes and other open reading frames, *Nucleic Acids Res.,* 13, 2267, 1985.

208. **Lampe, M. F. and Bott, K. F.,** Cloning the *gyrA* gene of *Bacillus subtilis, Nucleic Acids Res.,* 12, 6307, 1984.

209. **Ogasawara, N., Moriya, S., von Meyenburg, K., Hansen, F. G., and Yoshikawa, H.,** Conservation of genes and their organization in the chromosomal replication origin region of *Bacillus subtilis* and *Escherichia coli, EMBO J.,* 4, 3345, 1985.

210. **Ogasawara, N., Moriya, S., Mazza, G., and Yoshikawa, H.,** A *Bacillus subtilis dnaG* mutant harbours a mutation in a gene homologous to the *dnaN* gene of *Escherichia coli, Gene,* 45, 227, 1986.

211. **Tanaka, T. and Sueoka, N.,** Site-specific in vitro binding of plasmid pUB110 to *Bacillus subtilis* membrane fraction, *J. Bacteriol.,* 154, 1184, 1983.

212. **Hoshino, T., McKenzie, T., Schmidt, S., Tanaka, T., and Sueoka, N.,** Nucleotide sequence of *Bacillus subtilis dnaB*: a gene essential for DNA replication initiation and membrane attachment, *Proc. Natl. Acad. Sci. U.S.A.,* 84, 653, 1987.

213. **Moriya, S., Fukuoka, T., Ogasawara, N., and Yoshikawa, H.,** Regulation of initiation of the chromosomal replication by DnaA-boxes in the origin region of the *Bacillus subtilis* chromosome, *EMBO J.,* 7, 2911, 1988.

214. **Hansen, F. G., Koefoed, S., Sørensen, S., and Atlung, T.,** Titration of DnaA protein by *oriC* DnaA-boxes increases *dnaA* gene expression in *Escherichia coli, EMBO J.,* 6, 255, 1987.

215. **Svitil, A. and Zyskind, J.,** unpublished data.

216. **Koppes, L. J. H. and von Meyenburg, K.,** Nonrandom minichromosome replication in *Escherichia coli* K-12, *J. Bacteriol.,* 169, 430, 1987.

217. **Leonard, A. C. and Helmstetter, C. E.,** Cell cycle-specific replication of *Escherichia coli* minichromosomes, *Proc. Natl. Acad. Sci. U.S.A.,* 83, 5101, 1986.

218. **Leonard, A. C. and Helmstetter, C. E.,** Replication patterns of multiple plasmids coexisting in *Escherichia coli, J. Bacteriol.,* 170, 1380, 1088.

219. **Pritchard, R. H., Barth, P. T., and Collins, J.,** Control of DNA synthesis in bacteria, *Symp. Soc. Gen. Microbiol.,* 19, 263, 1969.
220. **Pritchard, R. H.,** Control of DNA replication in bacteria, in *DNA Synthesis: Present and Future,* Molineux, I. and Kohiyama, M., Eds., Plenum Press, New York, 1978, 1.
221. **Brewer, B. J.,** When polymerases collide: replication and the transcriptional organization of the *E. coli* chromosome, *Cell,* 53, 679, 1988.
222. **Wu, H.-Y., Shyy, S., Wang, J. C., and Liu, L. F.,** Transcription generates positively and negatively supercoiled domains in the template, *Cell,* 53, 433, 1988.
223. **Finer-Moore, J. and Stroud, R. M.,** Amphipathic analysis and possible formation of the ion channel in an acetylcholine receptor, *Proc. Natl. Acad. Sci. U.S.A.,* 81, 155, 1984.
224. **Smith, D. W.,** personal observation.
225. **Smith, D. W. and Boerner, P.,** Stimulation of ATP-dependent in vitro DNA replication by factors from the periplasmic space of *Escherichia coli, J. Bacteriol.,* 122, 159, 1975.
226. **Walker, J. E., Gay, N. J., Saraste, M., and Eberle, A. N.,** DNA sequence around the *Escherichia coli unc* operon, *Biochem. J.,* 224, 799, 1984.
227. **von Meyenburg, K. and Hansen, F. G.,** The origin of replication, *oriC,* of the *Escherichia coli* chromosome: genes near *oriC* and construction of *oriC* deletion mutations, *ICN-UCLA Symp. Mol. Cell. Biol.,* 14, 137, 1980.
228. **Lother, H. and Messer, W.,** Promoters in the *E. coli* replication origin, *Nature (London),* 294, 376, 1981.
229. **Cleary, F. M., Smith, D. W., Harding, N. E. and Zyskind, J. W.,** Primary structure of the chromosomal origins *(oriC)* of *Enterobacter aerogenes* and *Klebsiella pneumoniae:* comparisons and evolutionary relationships, *J. Bacteriol.,* 150, 1467, 1982.
230. **Selzer, G., Som, T., Itoh, T., and Tomizawa, J.-I.,** The origin of replication of plasmid p15A and comparative studies on the nucleotide sequences around the origin of related plasmids, *Cell,* 32, 119, 1983.
231. **Moser, D. R., Moser, C. D., Sinn, E., and Campbell, J. L.,** Suppressors of a temperature-sensitive copy-number mutation in plasmid NTP1, *Mol. Gen. Genet.,* 192, 95, 1983.
232. **Hansen, E. B. and Yarmolinsky, M.,** Host participation in plasmid maintenance: dependence upon *dnaA* of replicons derived from P1 and F, *Proc. Natl. Acad. Sci. U.S.A.,* 83, 4423, 1986.
233. **Murakami, Y., Ohmori, H., Yura, T., and Nagata, T.,** Requirement of the *Escherichia coli dnaA* gene function for *ori-2*-dependent mini-F plasmid replication, *J. Bacteriol.,* 169, 1724, 1987.
234. **Hasunuma, K. and Sekiguchi, M.,** Replication of plasmid pSC101 in *Escherichia coli* K-12: requirement for *dnaA* function, *Mol. Gen. Genet.,* 154, 225, 1977.
235. **Gaylo, P. J., Turjman, N., and Bastia, D.,** DnaA protein is required for replication of the minimal replicon of the broad-host-range plasmid RK2 in *Escherichia coli, J. Bacteriol.,* 169, 4703, 1987.
236. **Smith, C. A. and Thomas, C. M.,** Comparison of the nucleotide sequences of the vegetative replication origins of broad host range IncP plasmids R751 and RK2 reveals conserved features of probable functional importance, *Nucleic Acids Res.,* 13, 557, 1985.
237. **Yen, J.-C. and Reznikoff, W. S.,** DnaA, an essential host gene, and Tn5 transposition, *J. Bacteriol.,* 169, 4637, 1987.
238. **Wickner, S. H. and Chattoraj, D. K.,** Replication of mini-P1 plasmid DNA in vitro requires two initiation proteins: the products of P1 *repA* and *E. coli dnaA, Proc. Natl. Acad. Sci. U.S.A.,* 84, 3668, 1987.
239. **Mukherjee, S., Patel, I., and Bastia, D.,** Conformational changes in a replication origin induced by an initiator protein, *Cell,* 43, 189, 1985.
240. **Smith, D. W.,** unpublished observations.
241. **Tuggle, C. K. and Fuchs, J. A.,** Regulation of the operon encoding ribonucleotide reductase in *Escherichia coli:* evidence for both positive and negative control, *EMBO J.,* 5, 1077, 1986.
242. **Ooi, T. and Takanami, M.,** A computer method for construction of secondary structure from polynucleotide sequence: possible structure of the bacterial replication origin, *Biochim. Biophys. Acta,* 655, 221, 1981.
243. **Devereux, J., Haeberli, P., and Marquess, P.,** Sequence analysis software package of the Genetics Computer Group, Version 5, University of Wisconsin, Biotechnology Center, Madison, 1987.
244. **Kohara, Y.,** personal communication, 1987.
245. **Erickson, H. and Zyskind, J.,** unpublished data, 1988.
246. **Boye, E., Løbner-Olesen, A., and Skarstad, K.,** Timing of chromosomal replication in *Escherichia coli, Biochim. Biophys. Acta,* 951, 359, 1988.
247. **Boye, E.,** personal communication, 1989.
248. **Chiaramello, A. and Zyskind, J. W.,** manuscript in preparation, 1989.
249. **Wang, Q. and Kaguni, J. M.,** dnaA protein regulates transcription of the *rpoH* gene of *Escherichia coli, J. Biol. Chem.,* 264, 7338, 1989.
250. **Schauzu, M. A., Kücherer, C., Kölling, R., Messer, W., and Lother, H.,** Transcripts within the replication origin, *oriC* of *Escherichia coli, Nucleic Acids Res.,* 15, 2479, 1987.
251. **Løbner-Olesen, A., Skarstad, K., Hansen, F. G., von Meyenburg, K., and Boye, E.,** The DnaA protein determines the initiation mass of *Escherichia coli* K-12, *Cell,* 57, 881, 1989.

Chapter 3

REPLICATION AND SEGREGATION CONTROL OF *ESCHERICHIA COLI* CHROMOSOMES

Alan C. Leonard and Charles E. Helmstetter

TABLE OF CONTENTS

I. INTRODUCTION

Between the time of birth and subsequent division, a bacterial cell growing in steady state doubles every constituent. The machinery required to synthesize these constituents must be regulated and coordinated within a temporal framework. Two fundamentally important steps toward successful duplication of the bacterial cell are the replication and segregation of chromosomes. We will focus the content of this chapter on these crucial processes.

The mechanisms regulating the replication and segregation of chromosomes in *Escherichia coli* have been difficult to identify due to the large size and complexity of the bacterial chromosome. However, there is now a sense of expectation that this obstacle can be overcome; it is no longer necessary to manipulate the bacterial chromosome itself in order to examine these regulatory systems. Minichromosomes, small plasmids which replicate solely from a resident copy of the chromosomal origin of replication (*oriC*), have become ideal substitutes for chromosomes in physiological and biochemical studies. Using minichromosomes, the successful dissection of ''chromosomal'' replication-segregation control seems a reasonable goal.

The three stages of chromosome replication — initiation, elongation, and termination — are each sufficiently complex to be the topic of a complete review. For this reason, we will not cover in detail the elongation and termination stages of chromosome replication. Readers requiring a comprehensive treatment of these subjects are referred to the reference works available[1-6] (see also, this volume). In addition, our review is limited to *oriC* replication-segregation control and, therefore, the related mechanisms for bacterial plasmids[7-10] will not be included. We will first summarize recent research findings on *E. coli* chromosomes and minichromosomes, and then present our views of the mechanisms controlling initiation of chromosome replication and the segregation process.

II. INITIATION OF REPLICATION FROM *oriC*

A. THE CLONING OF *oriC*

Intact *E. coli* cells are rather uncooperative subjects for the study of replication control. Although not impossible, biochemical analysis of chromosome replication in whole cells is cumbersome and severely limited for study of the initiation phase. This problem was circumvented initially by employing plasmid and phage DNA replication systems to study features of chromosome replication control without using chromosomes themselves (reviewed in References 1 to 3 and 5). In particular, the major biochemical constituents of the elongation phase of replication were identified in cell-free systems using small, single-stranded phage templates.[1,3,5] However, studies of the biochemistry of initiation of chromosome replication were really not possible until the *oriC* region of the chromosome was isolated.

The *E. coli* genome is a circular, double-stranded DNA of 4700 kilobase pairs (kbp).[11,12] Replication initiates from a unique chromosomal site, *oriC*,[13,14] and new replication forks proceed bidirectionally around the chromosome,[15,16] moving at roughly the same rate until they meet within a region designated as the replication terminus *(terC)*[6] (see also, this volume). Replicated chromosomes are equipartitioned into daughter cells.

The natural tertiary state of the *E. coli* chromosome is not simple. Chromosomal DNA is torsionally stressed and compacted more than 1000-fold into a distinct structure called the nucleoid (reviewed in Reference 17 and also this volume), constrained by RNA and histone-like proteins. Up to 50 topologically independent domains of negatively supercoiled chromosomal DNA comprise the nucleoid, and supercoiled DNA is the preferred substrate for the DNA replication proteins. How the nucleoid structure is remodeled during replication is not understood. Nucleoid DNA interacts with proteins of the cell envelope (outer membrane, peptidoglycan layer, and inner membrane) at as many as 20 sites,[17] and some of the interacting DNA is enriched for nucleotide sequences from *oriC*.[18] Although the exact nature of the *oriC*-envelope interaction is unknown, its possible role in replication initiation and segregation will be discussed in later sections of this review.

Small plasmids replicating autonomously from a resident copy of *oriC* provided the material for the direct analysis of *oriC* function *in vivo* without destroying cell viability. These DNA molecules also proved to be ideal templates for development of an *in vitro* system needed to study the biochemistry of initiation of chromosome replication.[19] Extra-chromosomally replicating *oriC* was initially identified on naturally occurring F′ plasmid derivatives maintained in Hfr strains of *E. coli*.[20] Permissive replication in Hfr strains was attributed to the presence of *oriC*-containing chromosomal DNA. However, the large size and complexity of these plasmids prohibited their practical use for studies of replication control.

Recombinant DNA technology was applied to the isolation of the *oriC* region of the chromosome in the late 1970s.[21-23] Restriction endonuclease-cleaved DNA fragments from the *E. coli* chromosome were joined to a nonreplicating fragment derived from an R plasmid which carried an antibiotic resistance determinant. Cells transformed to antibiotic resistance with these DNAs were found to harbor new plasmids whose replication function was supplied by the chromosome-derived DNA. As an alternative approach, specialized transducing phage known to harbor chromosomal DNA from the region encoding asparagine synthetase A (*asnA*), adjacent to *oriC*, were isolated after becoming established as autonomously replicating plasmids in lambda lysogens.[24] Restriction endonuclease maps and nucleotide sequence determinations of the chromosomal DNA found on these *oriC* plasmids, independently isolated in a number of different laboratories,[25-28] revealed the presence and identity of the unique origin of replication of the chromosome. Subsequent deletion analysis proved that only one origin of replication was harbored on the *E. coli* chromosome.[29]

B. THE PHYSIOLOGY OF MINICHROMOSOME REPLICATION

To serve as authentic models for the control of initiation of chromosome replication *in vivo,* minichromosomes should mimic the replication properties of the chromosome in every regard. Therefore, minichromosomes should respond to the same inhibitors, require the same gene products for replication, be timed precisely during the cell division cycle, and be stably maintained at low copy number. Minichromosomes have been shown to fulfill all of these requirements except the last; they are not stably maintained at the low "chromosomal" copy number. However, we will show in later sections of this chapter that elevated minichro-mosome copy numbers and instability are consequences of defective equipartition at division rather than a replication control system which differs fundamentally from that of the chromosome.

1. Replication Requirements and Growth Effects

The synthetic events required to accumulate initiation capacity for chromosome repli-cation are also required for replication to ensue from extrachromosomal copies of *oriC* (see Reference 4 for review). Minichromosomes require the same gene products for replication as the chromosome, determined by inhibition of their replication in temperature-sensitive, initiation-defective mutant strains.[30,31] In spite of this, minichromosome maintenance does not affect cell growth,[23] except when additional chromosomal DNA counterclockwise from *oriC* is included on the plasmids.[32] This DNA encodes the subunits of ATP synthetase whose extrachromosomal expression perturbs the composition of cytoplasmic membrane. Host chromosome replication is also not affected by *oriC* on minichromosomes, even in the presence of many extra copies.[33] This was most clearly shown with chimeric plasmids comprised of *oriC* and pBR322, harbored in *E. coli* hosts at very high copy numbers (between

30 to 80 per genome), because both origins of replication were functional.[34,35] Strains maintaining these chimeric molecules were unaltered in the timing of chromosome replication and cell division. The presence of minichromosomes also did not increase the variance in interinitiation time of the chromosome as measured by density transfer experiments,[36] suggesting minichromosomes and chromosomes do not compete for a single rate-limiting component of the replication machinery. These observations show that *E. coli* cells can initiate replication from many more copies of *oriC* than are normally found in their chromosomal constituents.

2. Replication Timing

Is replication ensuing from extrachromosomal copies of *oriC* timed properly with respect to the cell division cycle and do all copies of *oriC*, chromosomal and minichromosomal, replicate coincidentally? This is a key question with regard to the suitability of minichromosomes as model systems for chromosome replication. Although initial studies suggested minichromosomes replicated throughout the division cycle,[31] refined measurements performed with age-fractionated cells showed unequivocally that minichromosomes replicate with cell cycle specificity.[37] The mean cell age at minichromosome replication coincided with the mean age at initiation of chromosome replication for all growth rates examined and the age distributions of the two events were found to be indistinguishable.[33] Minichromosomes harboring the minimal *oriC* sequence (see Figure 2) within only 327 bp replicated with proper timing, suggesting that the timing mechanism interacts with the minimal *oriC* region of minichromosomes during the division cycle.[33] Thus, the mechanism controlling replication from *oriC*, on chromosomes or minichromosomes, has the capacity to communicate with all copies of *oriC* as if they were part of an interconnected network.

OriC harbored on pBR322 plasmid derivatives also replicates with cell cycle specificity, coincident with the chromosome, indicating that the proper function of *oriC* is not affected by the presence of *ori*pBR322.[35] The *oriC*-pBR322 chimeric plasmids exhibited a composite synthesis pattern: specifically timed replication from *oriC* on a background of pBR322 replication throughout the cycle. Cell-cycle-specific replication was not observed for any other plasmids, including pBR322, pSC101, and F plasmid derivatives replicating from *oriS* or *oriV*.[38]

3. Copy Number Control

The copy number of minichromosomes in *E. coli* hosts was found to be much higher than expected.[30,39] Rather than being harbored at the same number of copies as the chromosomal *oriC*, minichromosomes were found at up to tenfold higher copy numbers (10 to 20 copies per genome). Equally surprising was the observation that despite the elevated copy number, minichromosomes were not stably maintained.[23,30] How is it possible for the copy number to attain a level higher than one per chromosomal *oriC*? This could be accomplished if (1) more than one round of replication took place during the cell division cycle, (2) more than one minichromosome molecule were initially transformed into an *E. coli* host, or (3) the segregation of minichromosomes at division was defective. Since only one round of replication takes place on each copy of *oriC* during each cell division cycle,[40] the first possibility is ruled out. It is equally unlikely that the cells receive 10 to 20 plasmids upon transformation. The high copy number is most easily explained by the third possibility: unequal partition combined with the cellular capacity to replicate large numbers of copies of *oriC*. The details of this concept are presented in Section IV.

C. THE *oriC* REGION OF THE CHROMOSOME
1. The Genes Adjacent to *oriC*

Gene mapping and identification of transcriptional units revealed that nearly all of the DNA adjacent to *oriC* on the chromosome carries genes which are not involved in the

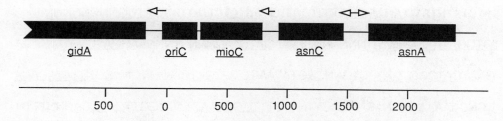

FIGURE 1.   Location of genes and their transcriptional promoters in the *oriC* region. The positions of the coding sequences for proteins expressed from the *oriC* region of the *E. coli* chromosome are shown as filled boxes. The position and direction of transcriptional promoters for these genes are shown by the open arrows. The scale at the bottom is in base pairs. Position 1 is the first nucleotide of the first *Bam*H1 cleavage site left of the minimal *oriC* region.[25-28] The entire leftward *gidA* coding region is not shown.

regulation of DNA replication.[29,30,39,41] The origin of replication appears to reside within the asparagine synthetase (*asn*) operon,[42] an operon responsible for the regulation of nitrogen metabolism. As shown in Figure 1, two genes of this operon are located clockwise of *oriC*: the gene encoding asparagine synthetase A, *asnA*, which converts aspartate and ammonia to asparagine, and *asnC*, whose product is a small multifunctional protein shown to regulate the expression of asparagine synthetase A at the transcriptional level.[43] Interest in *asnC* initially centered on the fact that it was one of two open reading frames whose counterclockwise transcripts proceeded toward *oriC*.[42,43] Although transcripts initiated from the *asnC* promoter enter *oriC*,[44] there is nothing to suggest these transcripts play a regulatory role in *oriC* function. Counterclockwise of *oriC* is another open reading frame designated *gidA*. The product of this gene has been implicated in cell division control due to a division defect in glucose minimal media observed in *gidA* mutant strains.[29] Although the exact function of this gene remains unknown, it was recently shown that synthesis of the GidA protein is regulated by the *asnC* gene product,[42] suggesting *gidA* may also be involved in nitrogen metabolism rather than cell division.

A second open reading frame designated *mioC* (modulation of initiation from *oriC*) or 16 kDa (16 kilodaltons; for the mass of its protein product) is sandwiched between the *asnC* gene and *oriC*.[3,4,34,39] Transcripts from the *mioC* gene proceed into *oriC* and are implicated in the control of minichromosome replication, based on the observation that minichromosomes harboring *mioC* had higher copy numbers than minichromosomes lacking this region.[34,39] Increased minichromosome copy number was dependent on an intact *mioC* transcriptional promoter and not on the presence of an intact protein-encoding region.[39] No function has been assigned to the *mioC* polypeptide.

The *mioC* gene has a number of interesting features. This gene is negatively regulated by *dnaA* gene product,[45,46] a protein absolutely required for initiation of replication from *oriC*. A DnaA protein recognition site lies within the transcriptional promoter region of *mioC*, and it was demonstrated *in vitro* and *in vivo* that transcription from the *mioC* promoter is inhibited by overproduction of the *dnaA* gene product.[45-49] *MioC* gene transcription may be regulated by methylation as well.[46] The strong transcriptional promoter of the *mioC* gene has many nucleotide sequence features associated with stable RNA promoters,[50] suggesting that *mioC* transcription is under stringent regulation. Recently, transcription from the *mioC* promoter was shown to be inhibited by amino acid deprivation in RelA$^+$, but not RelA$^-$, cells and to be stimulated by nutritional upshifts in a manner identical to stringently regulated stable RNA promoters.[47] The stringent regulation of replication from *oriC* has also been reported.[51]

The *mioC* transcripts may be terminated within, or proceed through, *oriC*.[44,46,47,52] Some transcript termination sites coincide with RNA-DNA transition sites,[52,53] considered to be the starting points for DNA polymerization.[54] It was suggested that *mioC* transcripts could

GATCTATNTATTTANAGATCTGTTCTATTGTGATCTCTTATTAGGATCGNNNNNNNNN

TGTGGATAANNNNNNATNNNNNNTTNAAGATCAANNNNNNTNNNAAGGATCNNTANCTGT

GAATGATCGGTGATCCTGNNCNGTATAAGCTGGGATCANAATGNNGGNTTATACACA

NCNCAAAAANNNNACNNCAGTTNTTCTTTGGATAACTACCGGTTGATCCANNNTTTN

NNNCANNNTTATCCAC

FIGURE 2.   The consensus sequence of the chromosomal origin of replication of the Gram-negative bacteria (based on Reference 57). The sequence displayed encompasses the minimal requirement for functional *oriC* (position 23 to 267). The nucleotides specified are found in four of five functional bacterial origins, including *E. coli, S. typhimurium, E. carotovora, E. aerogenes,* and *K. pneumoniae.* N represents nucleotides found in less than three of five origins. The boldface nucleotide sequences represent the 13-mer repeats.[69] Both the GATC sequences and the boxed 9-bp sequences which represent the DnaA protein binding sites are underlined. The last adenosine residue (position 268) of the most clockwise (rightward) DnaA protein binding site is not required for *oriC* function.

serve as primers for initiation of replication at *oriC*.[55] However, this idea was brought into question by the observation that the *mioC* transcript is not required for replication from *oriC* on minichromosomes (or chromosomes),[39] and minichromosomes lacking *mioC* replicate at the proper time during the cell division cycle, coincidentally with the chromosome.[33] Transcripts entering *oriC* either do not normally play a role in the timing of replication from *oriC* or an alternative primer transcript system is activated upon removal of the *mioC* region.

2. The Nucleotide Sequence of *oriC*

The minimal nucleotide sequence required for replication encompasses 245 base pairs (reviewed in References 3 and 4). Comparisons of the *oriC* region of the *E. coli* chromosome with the replication origins of four other members of the Enterobacteriaceae family revealed regions of highly conserved nucleotides regularly spaced between more variable regions.[56] A consensus sequence of the replication origin of the bacterial chromosome was defined;[57] further details are presented in this volume. *OriC* contains sites at which replication proteins interact (conserved sequences)[58,59] and spacer regions (variable sequences)[60] to permit formation of the DNA secondary structure necessary for proper protein recognition. These regions of *oriC* are shown in Figure 2.

The nine-base-pair (bp) sequence 5'-TTAT(C/A)CA(C/A)A-3' (see Figure 2) is found at four positions within the minimal origin of replication in *E. coli*, as well as four other Gram-negative bacteria.[57] Referred to as DnaA boxes, they have been shown to be sites where *dnaA* gene product binds to *oriC* DNA.[61,62] The 9-bp DnaA box sequence is found at many additional chromosomal locations, often in regulatory regions.[63] These locations include the deoxyadenosine methylase *(dam)* gene, *mioC, polA, uvrB, argF, pyrBI,* a number of plasmid DNA origins of replication, the transposon *Tn5*, and the *dnaA* gene itself.[61] In the few cases studied, the DnaA protein negatively regulated gene transcription when bound to its recognition sequence.[63]

In addition to the DnaA box repeats, each of the enteric bacterial origins of known nucleotide sequence contains 9 to 11 GATC sequences clustered within the 245-bp minimal origin of replication.[57] The positions of eight of these GATC sequences are conserved among all five Gram-negative bacterial origins examined (Figure 2).[57] The GATC sequence is the recognition site for the *dam* methylase and the high concentration of sites in *oriC* suggests this enzyme plays a role in the regulation of replication or segregation. DNA methylation may affect the timing of rounds of chromosome replication, perhaps as one of the determinants of the interinitiation interval.[64-67] *OriC* DNA has recently been shown to bind to the outer membrane only when hemimethylated.[68] Thus, methylation of *oriC* sequences could determine the timing and extent of membrane interaction during the cell division cycle (see Section V).

The left end of the minimal sequence of *oriC* is also highly conserved among the members of Enterobacteriaceae. Three tandem repeats of the 13-mer 5'-GATCTNTTNTTTT-3' are found in this AT-rich region of the minimal sequence (Figure 2).[69] These iterated sequences are the sites for initial opening of the duplex DNA during the initiation reaction.[70,71] Disruption of these sequences by deletion can lead to the inactivation of *oriC*.[70] The nature of the proteins interacting at the 13-mer sites during the initiation of replication from *oriC* will be discussed in a later section.

Finally, the minimal origin of replication harbors promoter and termination sequences where transcription by RNA polymerase can begin and end.[41,46,72-74] Short transcripts (100 to 150 bp), initiated from within the minimal *oriC* sequence, have now been characterized *in vitro* and *in vivo*.[46,52] The activity of transcriptional promoters within the minimal *oriC* were found to be 30-fold lower *in vivo* than *in vitro*, suggesting these promoters are tightly regulated in living cells.[46] The role these transcripts may play in the regulation of replication is not yet understood.

D. PROTEIN INTERACTIONS WITHIN *oriC*

What proteins interact at the recognition sites within the minimal *oriC* region, and what roles do they play in the initiation of chromosome replication? Although the picture is not yet complete, genetic and biochemical approaches have identified the major proteins and their temporal sequence of interaction with *oriC* during initiation of chromosome replication.

Prior to the isolation of *oriC*, gene products required for initiation of chromosome replication were identified by selecting and characterizing conditional-lethal (thermosensitive) mutants (reviewed in Reference 3). Initiation-phase mutant strains ceased DNA replication gradually over a period of 40 to 60 min after a shift to nonpermissive temperature, whereas mutants defective in the chain elongation phase of replication ceased replication immediately. Thermosensitive, initiation-defective mutations were mapped in only a few gene regions, designated *dnaA, dnaC, gyrB,* and *dnaB;* the polypeptides encoded by these genes are the major players in the reaction initiating replication forks from *oriC*.[3,4]

The biochemical dissection of replication initiation from *oriC* has proceeded mainly through the effort of the Kornberg laboratory where a cell-free system was developed to replicate the minichromosome DNA template.[19] A number of DNA-protein complexes have been identified in the reconstituted stages preparing minichromosomes for the propagation of bidirectional replication forks.[68]

1. DnaA, DnaB-DnaC, and DNA Gyrase

A key protein found to interact with *oriC* is the *dnaA* gene product,[75,76] a 52-kDa polypeptide present at the rather high levels of 800 to 2100 molecules per cell.[77] This protein is reported to be synthesized continuously during the cell division cycle.[78] The interaction of DnaA protein with *oriC* at the 9-bp DnaA boxes was shown by nuclease protection experiments as well as by filter binding assays.[75] Binding of the DnaA protein at *oriC* is highly cooperative, resulting in the formation of a large nucleoprotein structure comprised of 200 to 250 bp of DNA wrapped around up to 40 molecules of protein.[61,79]

The *dnaA* gene is adjacent to a number of other genes involved in DNA metabolism (*dnaN, recF,* and *gyrB*).[80,81] Nucleotide sequence analysis of the *dnaA* gene has revealed two transcriptional promoters located 150 to 240 bp upstream from the start of the structural gene.[82] A DnaA box is located between the two promoters. Transcription from both promoters is negatively regulated by the binding of DnaA protein to this box.[83-86] Methylation also plays a role in the regulation of transcription from the *dnaA* gene.[87]

Recently, new attributes of the *dnaA* gene product were identified. The DnaA protein binds ATP at a high capacity, and only the ATP form of the protein is functional for the initiation reaction.[88] Slow conversion from the ATP to the inactive ADP form, in the presence of DNA, may be a mechanism to regulate DnaA protein activity during the cell division cycle.[88] The ADP form is reactivated by binding to diphosphatidylglycerol (cardiolipin), a diacidic membrane phospholipid.[89] DnaA protein is recovered directly from membrane fractions of *E. coli* cells.[77] The physiological significance of this interaction is not certain, but factors modulating the membrane association of *dnaA* gene product may play an important role in the regulation of replication.

A number of proteins interact with *dnaA* gene product, as evidenced by the identification of several extragenic suppressors of the *dnaA* temperature-sensitive, initiation-defective phenotype.[3,4] One interesting class of suppressors is found in the *rpo* genes encoding subunits B and B' of RNA polymerase. These suppressor mutations are allele-specific for *dnaA*(TS) mutations, and their existence suggests an interaction between DnaA protein and RNA polymerase.[90-92] Additional proteins interacting with *dnaA* gene product may include the *dnaB* gene product,[93] *dnaZ* gene product,[94] and *groE* gene product.[95] Although the nature of these interactions is unclear, their numbers show that *dnaA* gene product is part of a multicomponent complex involved in DNA replication.

The Kornberg group has described the earliest events of the initiation process; these are the site-specific recognition of *oriC* by DnaA protein and the subsequent unwinding of the DNA template by the concerted activity of *dnaA, dnaB,* and *dnaC* gene products.[69,96] Initially, DnaA protein binds to an adenine nucleotide, ATP or ADP. Mg^+ is essential for this interaction. DnaA protein then binds to each of the four 9-base-pair "boxes" within *oriC*. At 38°C, the ATP form of DnaA protein, in the presence of nucleoside triphosphate and the histone-like protein HU, forms the initial complex with supercoiled minichromosome DNA. The DNA wrapped around the DnaA protein becomes sensitive to P1 nuclease at the counterclockwise edge of the minimal *oriC* sequence.[70] The DnaC and DnaB proteins are then directed, presumably by *dnaA* gene product, to interact with the initial DNA complex at the 13-mer repeats within the counterclockwise edge of the minimal *oriC* sequence. At temperatures above 30°C, the interaction with DnaB and DnaC proteins convert the initial complex into the pre-prepriming complex.

The products of the *dnaB* and *dnaC* genes play a role in both the initiation and elongation phases of replication (reviewed in Reference 3). DnaB gene product (10 to 20 molecules per cell) is a replicative helicase of *E. coli*[97,98] and functions in unwinding duplex DNA in advance of the enzymatic machinery synthesizing new DNA strands. This unwinding activity and movement of the protein along single-stranded DNA is dependent on the hydrolysis of nucleoside triphosphate. DnaB protein has a nucleoside triphosphatase activity in the presence of ribonucleoside triphosphates (rNTPs) and interacts with the *dnaC* gene product to form a stable (dnaB6)-(dnaC6) complex.[99,100] The DnaC protein (100 molecules per cell) is innately unstable and has no known enzymatic activity.[101]

On addition of DNA gyrase and single-stranded binding protein (SSB), the duplex DNA of the pre-prepriming complex is unwound to become the prepriming complex. The swivelase activity of DNA gyrase relieves the topological strain caused by the processive unwinding of the duplex DNA by the helicase activity of DnaB protein. DNA gyrase activity is involved in the initiation, elongation, and termination phases of chromosome replication.[3,102,103] Temperature-sensitive mutations in the *gyrB* subunit of DNA gyrase preferentially inhibit the initiation phase of replication,[102,103] and certain mutations in *gyrB* cannot coexist with some *dnaA*(TS) mutations, even at permissive temperatures.[3] Minichromosome replication is sensitive to alterations in DNA gyrase activity *in vivo,* although the degree of sensitivity is dependent on the genetic composition of the minichromosome.[104] A site for the binding of DNA gyrase within the minimal nucleotide sequence of *oriC* has been identified.[105]

Initiation of minichromosome replication *in vitro* and *in vivo* requires negatively supercoiled DNA template, but the level of superhelicity necessary for efficient initiation is unknown. Transcription from promoters placed adjacent to *oriC* allows minichromosomes with reduced superhelical density to replicate in the Kornberg *in vitro* replication system.[106] Transcriptional activation of *oriC* may play a role in the normal regulation of replication from *oriC* as well. The attainment of a critical topological structure at *oriC,* and the prerequisite unwinding of the DNA duplex for replication to proceed, place great importance on the activities of the topoisomerases and helicases in the cell.

Following helicase-directed DNA unwinding counterclockwise of *oriC* on the prepriming complex, *dnaG* gene product (DNA primase) synthesizes short ribonucleotide primers on the single DNA strands. These primers are required for DNA polymerase III holoenzyme to begin the synthesis of new complementary DNA chains. The details of the priming and polymerization reactions have been extensively reviewed.[1-3]

2. RNA Polymerase

The idea that initiation of replication requires RNA synthesis is based on the observation that rifampicin prevents initiation of chromosome and minichromosome replication in living cells.[4] This belief was reinforced by the isolation of short transcripts from within *oriC*.[46,47] Furthermore, mutations causing increased DNA content *in vivo* have been mapped to *rpoB* or *rpoC* genes which encode subunits of RNA polymerase, and genetic evidence suggests the interaction of RNA polymerase with the *dnaA* gene product.[107,108]

The requirement for RNA polymerase *in vitro* is dependent on the composition of the reaction mix, particularly on the levels of HU protein.[109] At low HU concentrations, RNA polymerase is not required and DNA primase serves as the sole priming function. When DNA is coated with HU, RNA polymerase must be included in the initiation reaction, presumably to activate *oriC*. In the RNA polymerase-dependent system, more *dnaA* gene product is also required for replication, and this system is less tolerant of mutational alterations within the *oriC* sequence.[109] RNA polymerase activity in the presence of high concentrations of HU protein leads to *dnaA*-independent, *oriC*-independent priming of the replication reaction *in vitro* unless specificity factors are also included in the reaction (reviewed in Reference 3). These factors include RNase H and topoisomerase I, which restrict the synthesis of primer transcripts to *oriC* DNA sequences. The role of RNA polymerase as a possible regulator of initiation of replication from *oriC in vivo* will be discussed in Section III.

3. Cellular Envelope Proteins

Analysis of *oriC* DNA attachment to the cell envelope (inner membrane, outer membrane, and peptidoglycan) is one of the most difficult research efforts yet undertaken to dissect the replication-segregation process. Nonetheless, considerable experimental evidence exists to substantiate envelope interaction with the replication origin of the *E. coli* chromosome.[110] Membrane fractions from lysed cells have been found to be enriched in *oriC* DNA.[11,112] A complex containing proteins and a portion of the *E. coli* chromosome spanning *oriC* was isolated by disrupting bacteria with a French pressure cell and fractionating the lysate on sucrose gradients.[111] This DNA-protein complex bound to outer membrane preparations in the presence of magnesium. Purified *oriC* DNA was also found to bind to membrane protein preparations isolated by the same methodology.[113] This binding required divalent cations, took place within the 460-bp *oriC* region, did not depend on a high content of single-stranded DNA, and was enhanced by the presence of one or both 55- and 75-kDa polypeptides. These proteins were unusual; they remained attached to DNA following treatment with 5.5 *M* CsCl. The specific site of attachment to *oriC* or the location of these proteins in the membrane have not been determined. Recently, it has also been reported that *oriC* DNA binds outer membrane preparations only when it is hemimethylated.[68]

Minichromosome DNA has been used to isolate membrane proteins with affinity for *oriC*. One protein designated B′ (60,000 mol wt) was reported to attach specifically to single-stranded DNA, at sites within the minimal *oriC* (Figure 2) and within the *mioC* gene.[114,115] Strains harboring extrachromosomal copies of *oriC* were also found to have increased amounts of a smaller outer membrane protein (31 kDa) as well as increased amounts of *oriC* DNA in their outer membranes.[116] Although the 31-kDa protein was proposed to mediate the attachment of *oriC* to the outer membrane, the direct interaction of the protein with DNA has not been demonstrated.

A recent report on the *in vitro* interaction of *oriC* DNA with particulate membrane fractions has raised a question concerning the exact membrane component interacting with *oriC*.[117] When *oriC*-binding membrane components were fractionated on metrizamide gradients, the site of interaction was found to lie within a minor membrane fraction representing less than 5% of the total envelope, derived from the inner rather than outer membrane. Conceivably, the envelope attachment might take place at junctions between inner membrane, outer membrane, and peptidoglycan, accounting for the involvement of components from both membranes. The binding of *oriC* to this fraction is inhibited by anti-DnaA-protein antibody. Although indirect effects of the antibody are possible, this result implicates the *dnaA* gene product in the interaction of *oriC* with the membrane. In a later section, we will present a model for the segregation of chromosomes in *E. coli* that relies heavily on the existence of *oriC*-envelope interactions.

III. REGULATORY SYSTEMS FOR INITIATION OF CHROMOSOME REPLICATION

Which component(s) of the replication apparatus serve as the timing regulator, and what is the nature of the regulatory loop controlling the timing of initiation of replication from *oriC*? These are difficult questions to answer because the interactions within the replication machinery are very complex. Long before the components of the mechanism which initiates replication from *oriC* were identified, Jacob et al.[118] proposed a simple control circuit. Basically, the synthesis of a replication initiator substance encoded by a chromosomal gene was required for replication to ensue from a distinct chromosomal site, termed the replicator. Although the concept of triggering initiation by accumulating a positive effector remains the foundation for recent models of replication control, it has become evident that a negative control step must also be included.[119] The negative regulatory loop, required to modulate the production of initiation potential during the cell cycle and during transitions in growth states, could take the form of autoregulated synthesis of the positive effector.[120,121] In this section, we will present an overview of the studies which suggest a particular component of the replication machinery serves as the initiation activator. The primary candidates for this role are (1) DnaA protein, (2) RNA polymerase, or (3) a membrane component interacting with *oriC*.

A. REGULATION BY DNAA PROTEIN

The *dnaA* gene product of *E. coli* fits the role of a positive effector whose synthesis is autoregulated. For initiation to be triggered, it is assumed that DnaA protein must fill binding sites on the chromosome, some of which are within *oriC*. Some DnaA protein molecules must also fill the binding sites within the *dnaA* gene itself to affect autoregulation. How, then, does the DnaA protein interact with the proper sites in the proper sequence? Different sites may have different affinities for DnaA protein, as reported in *oriC* and *mioC*, thereby determining the order of binding.[122] Other possibilities come from studies of F and P1 plasmid replication control systems. Two different forms of the effector protein may provide initiator and autorepressor activity, respectively,[123] or DNA looping may permit the effector to bind simultaneously to activator sites in the replication origin and to the autorepressor site.[124]

Since overproduction of a rate-limiting regulator would be expected to induce a burst of initiation from *oriC,* it was of considerable interest to measure the replication of chromosomes and minichromosomes in cells which overproduced DnaA protein. Overproduction was achieved extrachromosomally by placing the *dnaA* coding region under the control of the thermoinducible lambda p_L promoter or the inducible *lacUV5* promoter.[125-127] Overproduction of DnaA protein was shown to cause a transient, threefold stimulation in the replication of both minichromosomes and the chromosomal genes proximal to *oriC.*[125] Although the minichromosomes were replicated completely, all of the induced rounds of chromosome replication were not completed.[125-127] It is not clear why overproduction of DnaA protein leads to stalled chromosomal replication forks, but these results clearly show that increasing the intracellular amount of DnaA protein leads to extra initiation events from *oriC.*

The stimulation of initiation by DnaA protein is consistent with a rate-limiting role, but it must be considered that the method used to overproduce DnaA protein might indirectly affect the rate of replication from *oriC.* The heat shocking of cells is a consequence of one method used, and the effect of heat shock on *oriC* replication has not been examined thoroughly. Heat stress was reported to transiently increase the number of initiation events at *oriC,* and the combination of heat stress and reduced RNA polymerase activity increased chromosome copy number twofold.[128] A shift from 30 to 42°C also resulted in one additional round of chromosome replication during overexpression of the *dnaA* gene product under the control of the *lac* promoter.[129] Thus, heat shock may allow propagation of replication forks that are initiated, but stalled, in the presence of excessive DnaA protein.

It must also be considered that overproduction of DnaA protein may induce initiation indirectly through stimulation of the activity of an alternate molecule which is limiting. An indirect effect of overproducing *dnaA* gene product is to modulate intracellular RNA polymerase activity. The level of free RNA polymerase molecules in the cell is very low, nearly all the RNA polymerase being bound to DNA at nonspecific sites or at the promoters of active chromosomal genes.[130] Therefore, the number of RNA polymerase molecules available to interact with replication-related genes is partially a function of the global gene activity of the cell. Temperature shifts or induction of strong promoters used to overexpress DnaA protein might alter the number and location of active genes in the cell, ultimately affecting replication control. In addition, the transcriptional repressing activity of *dnaA* gene product[63] might increase the availability of RNA polymerase for replication-related reactions during the cell division cycle. Therefore, the molecular basis for the stimulation of initiation by DnaA protein will be difficult to interpret until the role of RNA polymerase in the replication reaction is clear.

B. REGULATION BY RNA POLYMERASE

Although RNA polymerase activity is essential for the initiation of replication from *oriC,* it is unclear whether RNA polymerase activity serves as a determinant of the replication rate. In favor of this possibility is the finding that stimulation of RNA polymerase synthesis, or redistribution of the enzyme to make it available for the replication reaction, also stimulates replication from *oriC* (reviewed in Reference 4). Overinitiation of chromosome replication in *rpoB* or *rpoC* mutant strains also suggests that replication timing may be determined by the activity of RNA polymerase, although intracellular enzyme levels have not been measured in these strains. Some *rpoB* mutants exhibiting overinitiation of replication were reported to have increased levels of DNA supercoiling which might modulate the concentration of DnaA protein required for initiation of replication.[3] This is an important finding if transcription within *oriC* alters the DNA topology of this region prior to initiation. Attainment of a crucial topological state at *oriC* for replication initiation may require the concerted effort of *dnaA* gene product and RNA polymerase activity. If so, fluctuations in the activity of either component would lead to changes in the function of *oriC*.

MioC-less minichromosomes replicate at the proper time during the cell division cycle, but at reduced copy number.[39] The removal of the *mioC* gene from minichromosomes could reduce the average copy number due to either altered segregation (to be discussed in Section IV) and/or replication. Regarding replication defects, loss of the *mioC* gene could result in the inability of every minichromosome molecule to initiate replication during the required time in every cell division cycle. This would obtain if loss of *mioC* caused a limitation for a replication factor not normally limiting for *mioC*-carrying minichromosomes. Based on our knowledge of the *mioC* gene and its regulation, RNA polymerase is a good candidate for the limiting factor. The role of the *mioC* promoter may be to shuttle RNA polymerase to the weak promoters within the minimal *oriC* region or to transcriptionally activate *oriC* to a topologically favorable state for initiation of replication. Loss of the *mioC* gene would lead to less effective competition for available RNA polymerase by the weak promoters within *oriC* or, alternatively, might increase the requirement for *dnaA* gene product to unwind the duplex DNA at *oriC*. If this were the case, some minichromosome copies might not succeed in replicating during every cell division cycle, but those copies that did replicate would be properly timed.

Hypothetically, the rDNA-like transcriptional promoter of the *mioC* gene could provide for efficient coupling of replication to cellular growth conditions, especially during transition states such as nutritional shifts or temperature shifts which redistribute RNA polymerase. The *mioC* transcriptional promoter also allows *oriC* to compete for RNA polymerase with the strong promoters of the ribosomal RNA genes and places the replication reaction under stringent regulation, providing for cell survival under conditions which limit growth. The close association between *dnaA* gene product and RNA polymerase suggests that the concentration of either component may ultimately determine the timing of the replication reaction in a manner dependent on the physiological state of the cell.

C. REGULATION BY A MEMBRANE COMPONENT

Finally, the ultimate signal controlling replication from *oriC* could come from the growing surface of the cell, such as an envelope site whose synthesis is regulated during the cell division cycle to interact with *oriC*. The capacity of the cell to initiate replication simultaneously from many extrachromosomal copies of *oriC* suggests each copy of *oriC* is signaled to replicate as an interconnected network. As discussed previously, there is evidence that *oriC* specifically interacts with the envelope, and membrane phospholipid may play a role in the activation of DnaA protein to a form active for replication. The rate of phospholipid synthesis was reported to increase coincidentally with initiation of chromosome replication.[131] In Sections IV and V, we will discuss the coupling of replication to envelope synthesis and how the location of *oriC*-envelope interaction sites determines the pattern of chromosome and minichromosome segregation at division. In Section VI, we will combine the preceding information on replication regulation with the following discussion on segregation to produce a unified view of replication-segregation control systems.

IV. THE PROCESS OF CHROMOSOMAL DNA SEGREGATION

During steady-state growth of *E. coli,* chromosomal DNA is distributed equally between daughter cells at division. The partitioning of the chromosomes into daughter cells must, therefore, be governed by a very precise allocation system. To understand the system, three key questions need to be answered. (1) Does the allocation system specify which chromosome enters a particular daughter cell at division? (2) If it does, how are the chromosomes discriminated from one another? (3) How is chromosomal equipartition achieved? In this section, we will examine the allocation process by presenting and analyzing information on the segregation processes of both chromosomes and minichromosomes.

A. EXPERIMENTAL FINDINGS
1. Chromosome Segregation

Considerable information has been generated on the allocation of identifiable chromosomal DNA strands into individual *E. coli* cells during many generations of growth and division. DNA strands were generally identified by pulse-labeling cells with radioactive thymidine and observing the distribution of radioactivity among progeny cells during growth in chains.[132-136] Some experiments were also performed using the membrane-elution procedure[137,138] and by differential staining of nuclei in cells grown for various times in 5-bromouracil.[139]

Early studies indicated that *E. coli* chromosomes were chosen at random for partitioning into daughter cells at division, i.e., a given DNA strand had equal probability of being partitioned into either daughter.[132-134] Subsequently, evidence for a complex, nonrandom partitioning mode was reported.[135-139] This apparent contradiction was resolved by the finding that the degree of nonrandomness was a function of growth rate.[136] At rapid growth rates, e.g., in cultures growing with doubling times of about 25 min at 37°C, the chromosomal segregation pattern appeared random, but as the growth rate was reduced, the pattern gradually became more nonrandom. Mechanistically, partitioning must be accomplished in such a way that the nonrandomness of chromosome segregation is observable only at lower growth rates.

A breakthrough concept evolved from experimental measurement of growth rate-dependent nonrandom chromosome segregation, namely, segregation is a probabilistic process.[136,138] To understand this concept, consider the means by which the direction of chromosome segregation can be specified. Every cell has one new polar cap formed at the previous division and one old polar cap formed at some earlier division. Every replicating chromosome has one young template DNA strand formed during the previous round of replication, and one older template DNA strand formed during some earlier replication. Does one of the newly formed sister chromosomes, let us say the one possessing the older template, preferentially segregate toward the older pole? If segregation were totally nonrandom, the older DNA strand would always segregate in the same direction, i.e., the probability of segregating toward the older cell pole would be 1.0. If segregation were random, the probability would be 0.5. The experimental data are now very clear on this issue. At higher growth rates, this probability is close to 0.5 (random), but at lower rates it increases above 0.5. The actual probability value depends on the cell strain, e.g., it reaches a maximum of 0.61 in *E. coli* B/r F and 0.65 in *E. coli* B/r A, growing with doubling times of about 60 min at 37°C.[136-138]

Although the directionality of chromosome segregation is probabilistic in character, the mechanism driving the process is deterministic because each daughter cell is assured one chromosome. There is little information on the chromosomal equipartition mechanism. It may have components similar to those found on various stably maintained bacterial plasmids; namely, genes encoding *cis*-acting centromere-like regions and *trans*-acting partition proteins apparently involved in attachment of the plasmid centromere to the cell envelope.[10,140,141] Chromosomally encoded proteins uniquely involved in partitioning have not been identified, but as described in Section III, *oriC* interacts with either a portion of the cell envelope or with envelope-derived proteins. These DNA-envelope associations could be involved in partitioning, but there is no direct evidence for this being the case. One approach to identifying the components of the partition system would be to isolate mutants defective in some aspect of chromosome segregation. In such mutants, the chromosomes would not separate properly into daughter cells and chromosome-less cells and/or cell division defects would be expected. Mutant strains with defects resembling those anticipated for a malfunctioning segregation mechanism have appeared occasionally (e.g., References 142 and 143), but they have not been well characterized except for DNA gyrase mutants, presumably defective in resolution of double-stranded DNA circles.[143-146] Chromosome-less cells appear in temperature-sensitive DNA replication mutants incubated at nonpermissive temperature and in *recA* mutants (reviewed in Reference 147) and cells deleted for the chromosomal replication terminus.[148] The formation of chromosome-less cells in temperature-sensitive replication mutants is a consequence of continued division in the absence of replication rather than defective segregation. A similar defect in replication-division coordination may also account for the behavior of *recA* mutants. The appearance of chromosome-less cells in mutants with terminus deletions could be related to segregation if this region, along with *oriC*, were involved in partitioning, as has been proposed in a number of the models to be described later.

The attachment of a chromosomal DNA strand to a structural component of the cell during some portion of either replication or segregation cannot result in a permanent association within a fixed region of the cell. A permanent association between a chromosome and a cellular domain, such as a cell pole, would cause totally nonrandom segregation such that the attached strand would always segregate toward the same cell pole. This is inconsistent with the observed probabilistic mode of segregation. The attachment must be temporary or the attachment site must not be fixed in position in the cell. This lack of permanence could be accounted for by the recent findings of Ogden et al.[68] that only hemimethylated *oriC* binds membrane and that methylation requires several minutes after *oriC* replication.

2. Minichromosome Segregation

Since minichromosomes replicate solely from *oriC*, they should prove useful model systems for analysis of segregation. Minichromosomes display a nonrandom pattern of segregation similar to the chromosome. However, despite copy numbers in the range of 10 to 20 per cell, minichromosome-less cells are always present in cultures grown under nonselective conditions. This is partly a function of the makeup of the plasmid, e.g., those containing *mioC* are more stable, but the minichromosome loss rate is still about 5% per generation.[4] The chromosomal equipartition system must be absent or defective in minichromosomes. Minichromosomes also appear to segregate in groups or "clumps" because their measured loss rate is much higher than expected for molecules present at high copy numbers and randomly distributed as individual units.[4]

Attempts have been made to use minichromosomes to isolate and characterize components of plasmid and chromosomal partition systems.[149-153] When genes known to be involved in the partitioning of stably inherited plasmids (*par*) were introduced into minichromosomes, the effects on segregation varied, depending on the source of the *par* genes.[149,150,153] *Par* genes from plasmid pSC101 did not improve stability.[150] On the other hand, F plasmid-derived *par* genes stabilized minichromosomes significantly in rich growth medium, but only slightly in medium supporting growth rates of 0.6 or 0.25 doublings per hour at 37°C.[148,152] Studies on such minichromosome constructs may shed new light on the molecular basis of the chromosomal partition system, but the ability of a chromosomal DNA region to stabilize a minichromosome has yet to be established.

B. THEORETICAL ANALYSIS OF SEGREGATION

From the experimental information summarized in the preceding section, it is evident that chromosome segregation has two important features: (1) nonrandom allocation that varies with growth rate and (2) equipartition. To consider how a segregation mechanism might function to yield these characteristics, we will examine the theoretical expectations for a few hypothetical segregation systems and compare the predictions to the experimental findings for minichromosomes and chromosomes. We will assume that the cell has the capacity to replicate all *oriC*-driven plasmids present, in coordination with the initiation of chromosome replication, up to some plasmid copy-number limit. This assumption is consistent with the finding that minchromosomes with an intact *mioC* gene replicate once per cycle,[40] although it may not be the case for all minichromosomes.

1. Without an Equipartition System

We will begin by considering the simplest case: an *oriC*-driven plasmid that is distributed randomly between daughters at division and is not equipartitioned. Suppose every cell in a population had received one plasmid DNA molecule by transformation and that it replicated in coordination with the initiation of chromosome replication in the cell cycle. At the first division, the apportionment of two plasmids between the daughter cells in the population would be described by the binomial distribution.[154] The fraction of cells which do not receive a plasmid is expressed as $F(0) = (0.5)^n$, where n is the number of plasmids at cell division and is equal to 2 in this case. Therefore, after the first division, 0.25 of the cells would have no plasmid DNA. Since the average copy number per plasmid-containing newborn cell (N) can be expressed as $N = 1/[1 - F(0)]$,[155] N would increase after this first division from 1.0 to 1.33. After the next round of replication division, application of the same formula yields a value of 0.39 for the fraction of cells lacking a plasmid. Thus, after two divisions, the copy number per plasmid-containing newborn cell, N, would have risen to $1/(1 - 0.39) = 1.64$. N would continue to rise in this manner at each division until cells containing the maximum allowable number of plasmids appeared in the population. At this point, saturation of the replication capacity might interfere in cell division (or chromosome replication) and result in either a slower increase in N or a stabilization of N, depending on the precise effects of the increased copy number on cell growth. When plasmid-containing cultures are maintained in media such that cells lacking the plasmids are nonviable, e.g., in the presence of an antibiotic to which the plasmid confers resistance, N actually represents the average plasmid copy number in the population. Thus, a high average copy number is predicted for plasmids that replicate once per cycle and are assorted randomly at division. The predictions for this simplest of hypothetical examples are consistent with the experimental findings for minichromosomes as regards the high copy number, but not with the observed nonrandom distribution or segregation in clumps.

We will next consider the predictions if these same plasmids were not simply distributed at random in the cells, but rather possessed a specific cellular location. This localization could be defined by a variety of compartmentalization schemes, but we will assume it is achieved by attachment to the cell envelope. Two possible arrangements for the attachment sites will be considered: (1) sites located in restricted regions in the cell (e.g., at poles and developing septa) and (2) sites distributed over a large portion of the cell.

(1) If a *small* number of attachment sites were restricted in location to poles and septa, a fraction of the plasmids, equal to the site number, might be assured of allocation into sister cell pairs. If occupation of a site were required for replication, the expected copy number might be low and equal to the site number. Plasmid-less cells would not be expected to appear and, in the absence of any additional constraints, the plasmid distribution in the population as a whole would be random. If a *large* number of sites were located in restricted regions, newly replicated plasmids might reattach in the vicinity of their previous attachment instead of at a distance. This would yield segregation in clumps, plasmid-less cells, and high copy numbers for the reasons described above. This behavior is very similar to the observed segregation pattern for minichromosomes, except that again the distribution in the population would be random which is not consistent with the experimental findings.

(2) If the cells contained attachment sites distributed randomly throughout the cell, the consequences would be similar to those described above for a large number of sites in

restricted locations. On the other hand, if the sites were distributed asymmetrically in the cell, the attached plasmids would necessarily be distributed nonrandomly between daughter cells at division. Thus, the experimental findings for minichromosomes are most consistent with the predictions for a large number of attachment sites distributed asymmetrically in the cell: high copy number, plasmid-less cells, segregation in clumps, and nonrandom distribution into daughter cells.

2. With an Equipartition System

The predictions for an *oriC*-driven replicon possessing an equipartition system would be entirely different from those described above. If one equipartitionable *oriC* plasmid were transformed into a cell and it replicated once coincident with initiation of chromosome replication, the copy number per newborn cell would remain equal to one and there would be no plasmid-less cells in the culture. If the generation time were equal to or longer than C + D min (see Figure 3), i.e., the cell replicated one chromosome during its division cycle, the plasmid copy number would always equal the chromosomal *oriC* copy number. If the generation time were shorter than C + D min, i.e., the cell initiated replication on more than one *oriC* in the cell cycle, the plasmid still might replicate once during each replication window, resulting in copy numbers lower than the chromosomal *oriC* copy number. There are other possible scenarios in fast-growing cells, but the important point is that stable maintenance at low copy number must eventually arise in every case for equipartitioned *oriC* plasmids.

Although this example clearly does not apply to minichromosome segregation, it does reflect chromosome segregation, albeit without the prediction of nonrandom partitioning. To account for the nonrandomness, let us again consider the concept of attachment sites for the chromosome. If the sites were located only at poles and septa, and the newly replicated chromosomes *had* to be distributed between these sites, e.g., attachment at a single site per pole was required for replication, then equipartition must obtain. However, it is not evident how such a model could account for a given DNA strand remaining at a particular site with a probability greater than 0.5 and less than 1.0. If, on the other hand, the sites were distributed asymmetrically in the cell and the chromosome had equal probability of attaching to any one of the sites, then, as with minichromosomes, a nonrandom, probabilistic allocation system would be predicted.

From the preceding analysis, we conclude that both chromosomal and minichromosomal *oriC* are localized asymmetrically in the cell. In the next section, we will first summarize some current models for segregation and then present a physiological explanation for this asymmetry.

V. REGULATORY SYSTEMS FOR SEGREGATION OF *oriC* - DRIVEN REPLICONS

A. SEGREGATION MODELS

Most models of the chromosomal segregation process require DNA to be attached to the cell envelope in order to separate the chromosomes at cell division. The replicon model of Jacob et al.[118] proposed that separation of attachment sites by envelope growth would also separate the DNA attached at the sites. This simple idea for segregation required modification when it became apparent that both membrane and wall follow a pattern of diffuse synthesis and segregation. Koch et al.[156] circumvented this problem for Gram-positive organisms by noting that polar caps are essentially conserved and that chromosomal attachment sites located at the junctions between poles and the lateral envelope could serve as the anchor points for chromosome separation, independent of the mode of lateral wall synthesis. They suggested that the replication origin is attached to a site at the older pole and the terminus to a site at the newer pole.[156,157] After initiation of a round of replication, one of the new origins is displaced and transported to the junction site at the newer pole (or the developing septum), possibly by lateral diffusion of *oriC*-attached membrane proteins through the membrane. The terminus, following displacement by the replication origin, becomes attached to the replication apparatus. Growth of the lateral wall, taking place only between the junction sites, separates the origins. As the origins separate, symmetrical bidirectional replication centers the terminus-replisome complex between them. A new attachment site is formed at this central location, which splits into two junction sites after the end of the round of replication. Formation of this site sets the location of the next septation and division. In a recent modification, Cavalier-Smith[158] proposed that the released origin could be transported to the terminus by remaining associated with a DNA helicase in the replisome complex, rather than by diffusion through the membrane. It was also suggested that the site of septation is defined by the location of the terminus at the time of initiation of septation, rather than an event coupled to completion of replication.

The putative origin and terminus attachment sites could be located at the periseptal annuli which have been detected in Gram-negative bacilli.[159] The annuli consist of two concentric rings of membrane/wall adhesion located at sites of division. Cook and Rothfield[160] have proposed that nascent division sites are formed in proximity to the central division sites (annuli) and then move to the centers of the domains destined to be the complementary sister cells. This could yield the following coupled sequence of events: initiation of replication, partition of the displaced *oriC* to the terminus at the new pole or division site, initiation of formation of a new annulus to which the terminus remains attached, and coordinate centering of the terminus and annulus in the new sister cells.

Alternative modes for the chromosomal separation process have also been proposed. Mendelson[161] presented a scheme based on his theories of *B. subtilis* helical growth. In this model, cell surface growth is accomplished by the insertion of strings of wall material around the circumference of the cell cylinder. As more and more strings are inserted, the entire array is reoriented until a limit is reached such that the strings are ordered lengthwise in the cell surface. If the origins of two new daughter chromosomes were attached to the ends of the first string to be inserted, then at the end of the reorientation process the chromosomes would become separated toward the cell poles. It was also suggested that string reorientation could serve as the clock mechanism to define the timing of the initiation of cell cycle events. The growth processes of the cell could be coupled through coordinate initiation of chromosome replication and the insertion of the first, origin-attached string. At the opposite extreme, the separation process might not rely on the presence of envelope attachment sites for each *oriC*, but could instead be determined entirely within the cytoplasm. Chromosomal separation would be determined by the "space offered by the shape of the growing cell".[162] One newly formed chromosome might remain in place while the other is moved away during replication, based on the availability of cytoplasmic space. A separation of this type could be affected by the transcription/translation of genes along the chromosome.[162] If the genes for ribosomal RNAs and proteins located near the origin of replication were transcribed and translated on both new chromosomes in a "ribosomal assembly microcompartment", expansion of the microcompartment would separate the origin regions without need for envelope attachments.

B. AN EXPLANATION FOR PROBABILISTIC, NONRANDOM SEGREGATION

The models summarized above are reasonable possibilities for the means chromosomes could use to separate from each other, but they offer no obvious explanation for the experimentally observed probabilistic distribution of specific chromosomes into daughter cells. As detailed in Section IV, the oldest chromosomal DNA strand segregates with a probability of between 0.60 and 0.65 toward the oldest cell pole in *E. coli* B/r with a doubling time of C + D min. If a small number of attachment sites were located at fixed positions in the cell, as most models suggest, what would cause the old strand to stay associated with the same pole with a probability other than 1.0 (permanently) or 0.5 (randomly)? In Section IV, we suggested that the nonrandomness of segregation is most likely due to asymmetric distribution of *oriC* attachment sites in the cell. We believe there is a simple physiological explanation for this asymmetric distribution.

Our proposal[163] states that the experimentally measured segregation probabilities are determined by the relative size of the conserved portion of the cell envelope: the polar caps. The simplest way to present this concept is to consider how multicopy objects would be distributed between daughter cells if their locations were restricted by cell geometry. The analysis is shown in Figure 3. We will assume that when these objects replicate, they can be located in any portion of the cell except for the existing cell poles (i.e., in the clear areas and not in the black areas in Figure 3a). If the objects were envelope-attached, then the attachment cannot take place in the older pole. If the molecules reside in the cytoplasm, then there must be a curtain blocking their entrance into the existing poles. After the cell grows and divides (Figure 3 a to c), the objects would be distributed asymmetrically between the left and right halves of each new daughter cell in Figure 3c. The ratio of the number of objects in the cell half with the new pole to the cell half with the old pole is equal to $0.5A_L + A_P/0.5A_L$ if availability of attachment sites were based on envelope area, or $0.5V_L + V_P/0.5V_L$ if based on volume. Thus, objects partitioned at random within the available regions in the cells would show this degree of asymmetric distribution. Since the area (or volume) of a polar cap is roughly one sixth the total area (or volume) of a newborn cell of *E. coli*,[162,164,165] the objects would be distributed in the ratio of 0.6/0.4 between the two halves of the cells shown in Figure 3c.

At subsequent divisions (Figure 3d to f), the objects would always be allocated in this asymmetric 0.6/0.4 ratio between daughter cells. When the objects are replicated before each division, they remain distributed in this fashion relative to the center of the cell and can no longer reside in the existing poles (Figure 3d). When these daughter cells divide at the center line (Figure 3 e to f), one daughter will receive 60% of the objects and the other 40%. As stated above, this is the measured degree of nonrandom segregation for minichromosomes. Thus, this simple idea explains the segregation pattern of minichromosomes.

The nonrandomness of chromosome segregation can also be explained by this same idea, but two additional aspects must be included: (1) the chromosome is present in only one or a few copies and (2) it is equipartitioned. In this case, the restriction on location must relate to the chromosomal *oriC*. When the cell in Figure 3 finishes replicating its chromosome (Figure 3b) and then divides (Figure 3c), both of the daughter cells will contain one chromosome due to equipartition and *oriC* can be located anywhere in the cell, except for the old poles. Thus, on average in a population, 0.6 of the daughter cells would have *oriC* located in the cell half with the new pole and 0.4 would have *oriC* located in the cell half with the old pole (Figure 3c). Now a new round of replication would be initiated on *oriC*s distributed as indicated in Figure 3c on either side of the center line (Figure 3d). The equipartition system must operate such that *oriC* with the younger template stays on its side of the center line and *oriC* with the older template is partitioned across the center line into the domain destined to become the complementary sister cell. Unlike other models in which nonrandom, probabilistic segregation would have to be accounted for by variable strand release and movement into the complementary sister, our model states that it is *always* the *oriC* with the older template that is released and segregated. The nonrandom distribution has already been established by the site distribution before the round of replication initiates. Therefore, the younger *oriC* would have a 0.6 probability of being located in the cell with the younger pole and the older *oriC* would have a 0.6 probability of locating in the daughter with the older pole. This prediction coincides with the experimental findings for the nonrandomness of chromosome segregation.

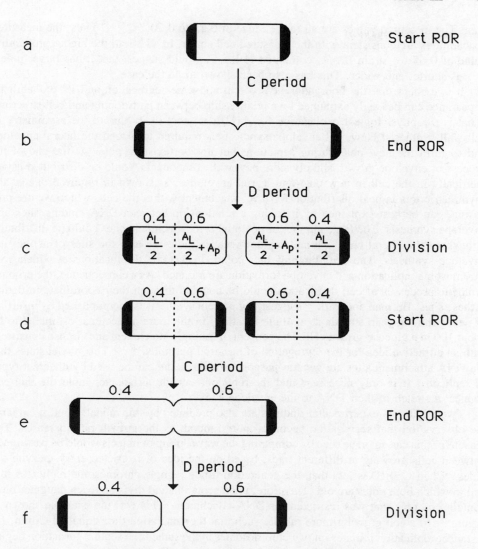

FIGURE 3.   Segregation pattern for replicons with restricted cellular locations. A cell from a culture growing with a generation time of C + D min is shown during two cycles of chromosome replication and cell division. The thinner cell outlines indicate cell envelope that contains attachment sites for the replicons, and the thicker lines show the old poles, which do not contain sites. A round of chromosome replication (ROR) initiates in (a). The round requires C min to complete, and then the cell divides D min later. The relative distribution of the replicons within each daughter cell is given by the formulas inside the cells (c), where A_L is the area of the lateral envelope of a newborn cell and A_P is the area of a polar cap. If it were assumed that A_P equalled one sixth of the total envelope area of a newborn cell, the replicons would be distributed in the fractions indicated above the cells. The next round of chromosome replication and cell division is shown in (d) to (f).

Our model states that the probability a DNA strand will segregate in the same direction two divisions in a row is a function of the relative size of a cell pole. The specific value of the probability depends on pole size and, therefore, would be strain dependent. Strains with larger values for this probability should have larger poles (i.e., be shorter and wider) than cells with smaller probabilities. The average measured value of the probability for *E. coli* strain B/r F growing with a generation time of about 80 min is 0.61.[163] Using the relationships in Figure 3c, this means that the poles must represent 0.18 of the total newborn cell surface area. The measured values are all in the range of 0.15 to 0.20.[162,164,165] Thus, the measured probabilities are consistent with the measured cell shape. In addition, the higher probability value of 0.65 for strain B/r A, compared to B/r F, would suggest that it has larger poles, i.e., is shorter and wider. This has also been shown to be the case.[164]

It is evident that the nonrandomness of chromosome segregation and its probabilistic appearance can be easily explained by a relationship between partitioning and cell structure. What is the physiological explanation for this? It is most likely due to the asymmetry of cell wall synthesis. New, net envelope synthesis is required to extend the lateral envelope and to form the new pole during septation, but not the existing pole. At the end of the process of envelope growth and division, new wall components would be distributed asymmetrically in the cell in new pole and lateral envelopes, as shown in Figure 3. Thus, the asymmetric topological distribution for *oriC* attachment within the cell, which we are proposing, can be best explained by invoking a relationship between DNA-binding sites and envelope synthesis. The large number of potential *oriC* attachment sites could be distributed throughout the lateral envelope and septum because these are also the sites of active, net envelope synthesis. The *oriC* binding sites could, in fact, be the locations at which both chromosome replication and envelope formation are initiated. As a consequence, the primary synthetic processes of cell duplication could be coupled through their coordinate initiation at these sites. Burman and Park[166] suggest that murein synthesis is accomplished by insertion of new peptidoglycan strands at multiple locations in the murein sacculus, estimated to be about 100 in a glucose-grown cell. The concept of multiple attachment sites is also consistent with an attractive idea for the segregation of bacterial plasmids.[140,167] This model states that the DNA attachment sites are general-purpose structures and can be used by different types of replicons. It is only necessary that the replicons encode an adapter molecule that can connect a stretch of their DNA to the envelope site.

Two additional experimental findings are also predicted by this model. First, it explains the observation that segregation becomes more random as the growth rate increases. The reason for this can again be seen by comparing the way a group of objects would be partitioned between cells growing at different rates, based on the idea of restricted entry into the old poles. When C + D is less than the generation time, a single chromosome replicates in a cell in which both poles are old. Therefore, upon septation and division, each daughter must contain one pole that was restricted for DNA attachment. This was the example shown in Figure 3. In a cell growing more rapidly such that the generation time equalled C min, for instance, replication initiates on two chromosomes at the same time septum formation begins due to completion of the previous round of replication. Thus, unlike the slowly growing cell, each of the replicating chromosomes "sees" an old pole on one side and a new pole being formed on the other side. When division is finished, each cell contains a replicating chromosome, one old pole, and one newly formed pole into which *oriC* could be distributed. When this ongoing round of replication ends and each of these cells divides, the two daughter cells are entirely different. In one, both poles were formed during the round of replication and *oriC* could enter *either one* and, therefore, has equal probability of being located anywhere in the cell. In the other daughter, one of the two poles was old when the round started and the distribution of *oriC* would be the same as in the slow-growing cells described in Figure 3. Thus, in this fast-growing population, half of the cells would distribute the old DNA strand 50/50 between daughters and half would distribute the old strand 60/40. Consequently, in the population as a whole, the overall segregation is more random and the average probability an old strand would segregate to an old pole in the entire population would be 0.55 instead of 0.6. As the growth rate is increased further, segregation continues to become more random due to the increased overlap of replication and septation.

Secondly, the unstable maintenance of multicopy minichromosomes, due apparently to segregation in clumps, is also accounted for by the proposed model. Once replication of a minichromosome has been completed, both sister plasmid molecules would be located in the same daughter cell due to the absence of the equipartition system to separate them into the complementary daughters. As a result, attachment sites for the next round of replication of the sister minichromosomes would be more likely to be located within proximity to one another than at a distance. This would result in the development of minichromosome "clones" within the cell and consequent segregation in clumps.

VI. CONCLUSIONS

Figure 4 summarizes our view of regulatory systems involved in *oriC*-directed replication and segregation in *E. coli*. The cell envelope contains a large number of *oriC*-attachment sites, perhaps as many as 100 per chromosomal *oriC*. The attachment sites are distributed over the growing portion of the cell envelope, i.e., everywhere but in the old poles, and may correspond to the sites of envelope synthesis. A replicon that produces an appropriate adapter protein can attach, by means of the adapter, to a site chosen at random from the many available sites. In Figure 4a, attachment is indicated for a chromosomal *oriC*, several minichromosomal *oriC*s, and an additional generic plasmid. In the case of *oriC,* and perhaps other plasmids as well, this adapter could be DnaA protein. Thus, one of the early steps in the initiation process is the binding of sufficient numbers of DnaA protein molecules within *oriC* to affect envelope attachment. We do not rule out the possibility that DnaA protein binds to the envelope sites prior to interacting with the replication origin. The binding of *oriC* to the membrane with DnaA protein provides the scaffold for later replication-related reactions and may also ensure that the active form of DnaA protein is available for the initiation reaction.

The DnaA protein must eventually fill the DnaA boxes in *oriC*. When these boxes are filled, DNA structure at *oriC* becomes permissive for the initiation reaction. However, the filling of the boxes cannot spontaneously trigger initiation of replication. RNA polymerase is also required, and it is the availability of this enzyme for a transcriptional event at *oriC* that ultimately permits initiation of replication from the envelope-bound, DnaA-activated *oriC*. In our view, RNA polymerase becomes more available for replication from *oriC* as the amount of intracellular DnaA protein increases during the division cycle. This regulatory loop is shown in Figure 4b. The binding of DnaA protein within DnaA-regulated chromosomal genes releases RNA polymerase for transcriptional events at different promoters, some of which are in *oriC*. Therefore, as the amount of DnaA protein rises in the cell, *oriC* is prepared to initiate replication (via binding of DnaA protein) coincidentally with the increasing availability of RNA polymerase for replication-related transcription. Initiation will not proceed under conditions where RNA polymerase is unavailable or cannot gain access to *oriC*.

The transcriptional promoters within *oriC* are weak and would not be expected to compete effectively with other nearby promoters for RNA polymerase. To remedy this situation, the strong, stringently regulated promoter with the *mioC* gene may provide *oriC* competitive access to RNA polymerase. RNA polymerase bound to the strong *mioC* promoter would be "shuttled", during counterclockwise transcription of the *mioC* gene, to interact with the promoters within *oriC*. Since DnaA protein inhibits transcription from the *mioC* promoter, thereby switching off the RNA polymerase shuttle, accumulation of DnaA protein in the cell would increase the probability for switch-off of the *mioC* gene. This repression of *mioC* may be accomplished in two ways: (1) increased binding of cytoplasmic DnaA protein at the DnaA box in *mioC* or (2) DNA looping to allow contact of the DnaA box in the *mioC* promoter with DnaA protein bound with *oriC*. Likewise, the stringent regulation of the *mioC* promoter serves to place replication under stringent control as well. Loss of *mioC* function, as with some minichromosomes, would not inhibit replication from *oriC*, but dependence

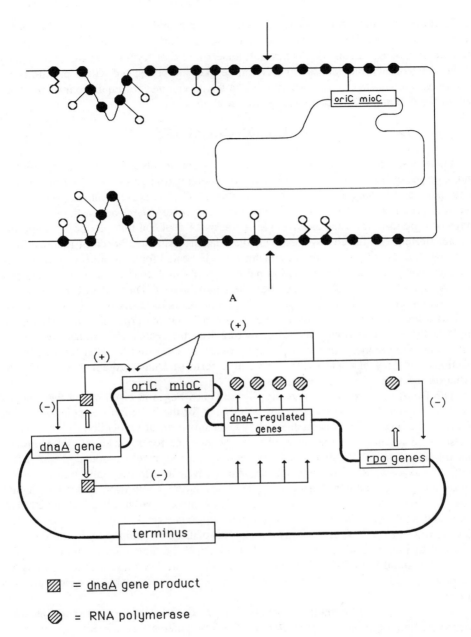

FIGURE 4. Schematic illustration of DNA replication and segregation properties of *E. coli*. (A) Localization of envelope-bound replicons in an *E. coli* cell. One half of a cell growing with a generation time of C + D min, and in the process of division, is shown. The large looped structure represents the chromosome, and the small open circles represent plasmids. The filled circles on the envelope of the cell are the DNA attachment sites located over the entire lateral envelope and division site, but not in the existing polar cap. The plasmids are attached to the general-purpose envelope sites by adapter molecules, with straight lines for *oriC* adapters and crooked lines for adapters of an alternative unspecified plasmid. The arrows identify the location of the next cell division. (B) The DnaA protein-RNA polymerase regulatory loop. The circular *E. coli* chromosome is represented by the thick black line, with important genetic regions identified within boxes. These boxes are not drawn to scale. The boxes labeled *dnaA*-regulated genes and *rpo* genes represent a family of genes whose location on the chromosome is not necessarily contiguous and is not specified in the figure. Thick white arrows represent synthetic pathways and thin dark arrows represent positive (+) or negative (−) regulatory pathways.

of the replication reaction on weak *oriC* promoters would lead to an increased sensitivity to RNA polymerase availability. As a result, rounds of replication might be skipped if *oriC* on *mioC*-less minichromosomes cannot compete effectively for the available RNA polymerase.

There are, then, three critical, potentially limiting stages in the initiation process — *oriC*-envelope attachment, DnaA protein level, and RNA polymerase activity — which form an interconnected regulatory loop. Inactivation of any one would obviously prevent initiation. More importantly, enhancement of the levels of either DnaA protein or RNA polymerase would, according to the interactions shown in Figure 4b, cause a transient stimulation of the initiation reaction, as has been seen experimentally.

At initiation of a round of replication from the *oriC*s in Figure 4a, the chromosomal *oriC* with the new template strand remains within the same half cell (right of the center line) and binds to a site. The *oriC* with the old template is partitioned into the domain destined to be the complementary sister (left of the center line). The probabilistic, nonrandom aspect of chromosome segregation is due to the asymmetric distribution of attachment sites, and not due to variation in the identity of the partitioned *oriC*. The "older" *oriC* may be partitioned by a passive mechanism based on available space in the cell, diffusion of the free *oriC*-protein complex laterally in the membrane, or riding on the replisome. If the latter mechanism exists, this would mean that once the location of the bound *oriC* has been established at the completion of division, the terminus must move to the opposite end of the cell and become fixed.

The minichromosomes are shown clustered in small groups of two or four because they lack a partition system and thus reattach with greater probability close to the previous attachment. Replication would increase the size of the groups. Note that when the next division occurs at the center line (Figure 4a), the daughter cell on the right could be minichromosome-less because of the grouping phenomenon. If the generic plasmid lacked an active partitioning mechanism, it would also be envelope-attached in groups and, therefore, unstably maintained.

We have presented a simplistic conceptual framework for the complex interactions in the replication and segregation of chromosomes in *E. coli* cells. The timing and rate of synthesis of individual components of the replication-segregation apparatus and the temporal interaction of these components during the cell division cycle have yet to be determined. Focus must be placed on the membrane sites at which the processes of replication, segregation, and envelope growth may be united. In addition, the elusive chromosomal equipartition mechanism, whether an active or passive one, must be identified and characterized. We believe that with minichromosomes and existing technology, it should be possible to answer these important questions in the near future.

ACKNOWLEDGMENTS

We wish to express our gratitude to colleagues who were kind enough to send preprints and reprints on their most recent work. Special thanks to Dr. Julia Grimwade Leonard, who drew the figures for this review and provided advice and encouragement when necessary. The work in our laboratories is supported by Public Health Service Grant GM26429 from the National Institutes of Health.

REFERENCES

1. **Kornberg, A.,** *DNA Replication,* W. H. Freeman, San Francisco, 1980.
2. **Kornberg, A.,** *1982 Supplement to DNA Replication,* W. H. Freeman, San Francisco, 1982.
3. **McMacken, R., Silver, L., and Georgopoulos, C.,** DNA replication, in *Escherichia coli and Salmonella typhimurium: Cellular and Molecular Biology,* Ingraham, J. L., Low, K. B., Magasanik, B., Schaechter, M., and Umbarger, H. E., Eds., American Society for Microbiology, Washington, D.C., 1987, 564.
4. **von Meyenburg, K. and Hansen, F.,** Regulation of chromosome replication, in *Escherichia coli and Salmonella typhimurium: Cellular and Molecular Biology,* Ingraham, J. L., Low, K. B., Magasanik, B., Schaechter, M., and Umbarger, H. E., Eds., American Society for Microbiology, Washington, D.C., 1987, 1555.
5. **Marians, K.,** Enzymology of DNA replication in prokaryotes, *CRC Crit. Rev. Biochem.,* 17, 153, 1984.
6. **Kuempel, P. and Henson, J.,** The terminus region of the chromosome, in *The Molecular Biology of Bacterial Growth,* Schaechter, M., Neidhardt, F. C., Ingraham, J. L., and Kjeldgaard, N. O., Eds., Jones and Bartlett, Boston, 1985, 325.
7. **Scott, J. R.,** Regulation of plasmid replication, *Microbiol. Rev.,* 48, 1, 1984.
8. **Filutowicz, M., McEachern, M. J., Mukopadhyay, P., Greener, A., Yang, S., and Helinski, D. R.,** DNA and protein interactions in the regulation of plasmid replication, in *The Seventh John Innes Symposium: Virus Replication and Genome Interactions, J. Cell. Sci.* (Suppl.), 7, 15, 1987.
9. **Nordstrom, K., Molin, S., and Light, J.,** Control of replication of bacterial plasmids: genetics, molecular biology, and physiology of the plasmid R1 system, *Plasmid,* 12, 71, 1984.
10. **Friedman, S., Martin, K., and Austin, S.,** The partition system of the P1 plasmid, *Branbury Rep.,* 24, 285, 1986.
11. **Cairns, J.,** The bacterial chromosome and its manner of replication as seen by autoradiography, *J. Mol. Biol.,* 6, 208, 1963.
12. **Kohara, Y., Akiyama, K., and Isono, K.,** The physical map of the whole *E. coli* chromosome: application of a new strategy for rapid analysis and sorting of a large genomic library, *Cell,* 50, 495, 1987.
13. **Bird, R., Louarn, J., Martuscelli, J., and Caro, L.,** Origin and sequence of chromosome replication in *Escherichia coli, J. Mol. Biol.,* 70, 549, 1972.
14. **Marsh, R. C. and Worcel, A.,** A DNA fragment containing the origin of replication of the *Escherichia coli* chromosome, *Proc. Natl. Acad. Sci. U.S.A.,* 74, 2720, 1977.
15. **Prescott, D. M. and Kuempel, P. L.,** Bidirectional replication of the chromosome in *Escherichia coli, Proc. Natl. Acad. Sci. U.S.A.,* 69, 2842, 1972.
16. **Masters, M. and Broda, P.,** Evidence for the bidirectional replication of the *Escherichia coli* chromosome, *Nature (London) New Biol.,* 232, 137, 1971.
17. **Drlica, K.,** The nucleoid, in *Escherichia coli and Salmonella typhimurium: Cellular and Molecular Biology,* Ingraham, J. L., Low, K. B., Magasanik, B., Schaechter, M., and Umbarger, H. E., Eds., American Society for Microbiology, Washington, D.C., 1987, 91.
18. **Hendrickson, W. G., Kusano, T., Yamaki, H., Balakrishnan, R., King, M., Murchie, J., and Schaechter, M.,** Binding of the origin of replication of *Escherichia coli* to the outer membrane, *Cell,* 30, 915, 1982.
19. **Fuller, R. S., Kaguni, J., and Kornberg, A.,** Enzymatic replication of the origin of the *Escherichia coli* chromosome, *Proc. Natl. Acad. Sci. U.S.A.,* 78, 7370, 1981.
20. **Hiraga, S.,** Novel F prime factors able to replicate in *Escherichia coli* Hfr strains, *Proc. Natl. Acad. Sci. U.S.A.,* 73, 198, 1976.
21. **Yasuda, S. and Hirota, Y.,** Cloning and mapping of the replication origin of *Escherichia coli, Proc. Natl. Acad. Sci. U.S.A.,* 74, 5458, 1977.
22. **Messer, W., Bergmans, H. E. N., Meijer, M., Womack, J. E., Hansen, F. G., and von Meyenburg, K.,** Minichromosomes: plasmids which carry the *E. coli* replication origin, *Mol. Gen. Genet.,* 162, 269, 1978.
23. **Leonard, A. C., Weinberger, M., Munson, B. R., and Helmstetter, C. E.,** The effects of *oriC* containing plasmids on host cell growth, *ICN-UCLA Symp. Mol. Cell. Biol.,* 19, 171, 1980.
24. **von Meyenburg, K., Hansen, F. G., Nielsen, L. D., and Riise, E.,** Origin of replication, *oriC,* of the *Escherichia coli* chromosome on specialized transducing phages lambda *asn, Mol. Gen. Genet.,* 160, 287, 1978.
25. **Sugimoto, K., Oka, A., Sugisaki, H., Takanami, M., Nishimura, A., Yasuda, S., and Hirota, Y.,** Nucleotide sequence of the replication origin of *Escherichia coli* K12, *Proc. Natl. Acad. Sci. U.S.A.,* 76, 575, 1979.
26. **Messer, W., Meijer, M., Bergmans, H. E. N., Hansen, F. G., von Meyenburg, K., Beck, E., and Schaller, H.,** Origin of replication, *oriC,* of the *Escherichia coli* K12 chromosome: nucleotide sequence, *Cold Spring Harbor Symp. Quant. Biol.,* 43, 139, 1979.

27. **Zyskind, J. W., Harding, N. E., Takeda, Y., Cleary, J. M., and Smith, D. W.,** The DNA replication origin of the *Enterobacteriaceae, ICN-UCLA Symp. Mol. Cell. Biol.,* 22, 13, 1981.

28. **Meijer, M., Beck, E., Hansen, F. G., Bergmens, H. E. N., Messer, W., von Meyenburg, K., and Schaller, H.,** Nucleotide sequence of the origin of replication of the *Escherichia coli* K-12 chromosome, *Proc. Natl. Acad. Sci. U.S.A.,* 76, 580, 1979.

29. **von Meyenburg, K. and Hansen, F. G.,** The origin of replication, *oriC,* of the *Escherichia coli* chromosome: genes near *oriC* and construction of *oriC* deletion mutations, *ICN-UCLA Symp. Mol. Cell. Biol.,* 19, 137, 1980.

30. **von Meyenburg, K., Hansen, F. G., Riise, E., Bergmans, H. E. N., Meijer, M., and Messer, W.,** Origin of replication, *oriC,* of the *Escherichia coli* K12 chromosome: genetic mapping and minichromosome replication, *Cold Spring Harbor Symp. Quant. Biol.,* 43, 121, 1979.

31. **Leonard, A. C., Hucul, J. A., and Helmstetter, C. E.,** Kinetics of minichromosome replication in *Escherichia coli* B/r, *J. Bacteriol.,* 149, 499, 1982.

32. **von Meyenburg, K., Jorgensen, B. B., and van Deurs, B.,** Physiological and morphological effects of overproduction of membrane-bound ATP synthase in *Escherichia coli, EMBO J.,* 3, 1791, 1984.

33. **Helmstetter, C. E. and Leonard, A. C.,** Coordinate initiation of chromosome and minichromosome replication in *Escherichia coli, J. Bacteriol.,* 169, 3489, 1987.

34. **Stuitje, A. R. and Meijer, M.,** Maintenance and incompatibility of plasmids carrying the replication origin of the *Escherichia coli* chromosome: evidence for a control region of replication between *oriC* and *asnA, Nucleic Acids Res.,* 11, 5775, 1983.

35. **Leonard, A. C., Wasielewski, A., and Helmstetter, C. E.,** Cell cycle-specific replication of minichromosomes and *oriC*-chimeric plasmids of *Escherichia coli,* paper presented at Molecular Basis of Bacterial Growth and Division, EMBO workshop, Segovia, September 27 to October 1, 1987, 15.

36. **Koppes, L. J. H.,** *OriC* plasmids do not affect timing of chromosome replication in *Escherichia coli* K12, *Mol. Gen. Genet.,* 209, 188, 1987.

37. **Leonard, A. C. and Helmstetter, C. E.,** Cell cycle-specific replication of *Escherichia coli* minichromosomes, *Proc. Natl. Acad. Sci. U.S.A.,* 83, 5101, 1986.

38. **Leonard, A. C. and Helmstetter, C. E.,** Cell cycle replication of multiple plasmids coexisting in *Escherichia coli, J. Bacteriol.,* 170, 1380, 1988.

39. **von Meyenburg, K., Hansen, F. G., Atlung, T. A., Boe, L., Clausen, I. G., van Deurs, B., Hansen, E. B., Jorgensen, B. B., Jorgensen, F., Koppes, L., Michelsen, O., Nielsen, J., Pedersen, P. E., Rasmussen, K. V., Riise, E., and Skovgaard, O.,** Facets of the chromosomal origin of replication, *oriC,* of *E. coli,* in *The Molecular Biology of Bacterial Growth,* Schaechter, M., Neidhardt, F. C., Ingraham, J. L., and Kjeldgaard, N. O., Eds., Jones and Bartlett, Boston, 1985, 260.

40. **Koppes, L. J. H. and von Meyenburg, K.,** Nonrandom minichromosome replication in *Escherichia coli* K-12, *J. Bacteriol.,* 169, 430, 1987.

41. **Hansen, F. G., Koefoed, S., von Meyenburg, K., and Atlung, T.,** Transcription and translation events in the *oriC* region of the *E. coli* chromosome, *ICN-UCLA Symp. Mol. Cell. Biol.,* 22, 37, 1981.

42. **Kolling, R., Gielow, A., Seufert, W., Kucherer, C., and Messer, W.,** *AsnC,* a multifunctional regulator of genes located around the replication origin of *Escherichia coli, oriC, Mol. Gen. Genet.,* 212, 99, 1988.

43. **Kolling, R. and Lother, H.,** *AsnC:* an autogenously regulated activator of asparagine synthetase A transcription in *Escherichia coli, J. Bacteriol.,* 164, 310, 1985.

44. **Chiaramello, A. and Zyskind, J. W.,** Regulation of RNA transcription entering *Escherichia coli oriC,* paper presented at Molecular Basis of Bacterial Growth and Division, EMBO workshop, Segovia, September 27 to October 1, 1987, 5.

45. **Lother, H., Koelling, R., Kuecherer, C., and Schauzu, M.,** DnaA protein-regulated transcription: effects on the *in vitro* replication of *Escherichia coli* minichromosomes, *EMBO J.,* 4, 555, 1985.

46. **Schauzu, M. A., Kucherer, C., Kolling, R., Messer, W., and Lother, H.,** Transcripts within the replication origin, *oriC,* of *Escherichia coli, Nucleic Acids Res.,* 15, 2479, 1987.

47. **Rokeach, L. A. and Zyskind, J. W.,** RNA terminating within the *E. coli* origin of replication: stringent regulation and control by DnaA protein, *Cell,* 46, 763, 1986.

48. **Stuitje, A. R., deWind, N., van der Spek, J. C., Pors, T. H., and Meijer, M.,** Dissection of promoter sequences involved in transcriptional activation of the *Escherichia coli* replication origin, *Nucleic Acids Res.,* 14, 2333, 1986.

49. **Rokeach, L. A., Chiaramello, A., Junker, D. E., Crain, K., Nourani, A., Jannatipour, M., and Zyskind, J. W.,** Effects of DnaA protein on replication and transcription events at the *Escherichia coli* origin of replication, *oriC, UCLA Symp. Mol. Cell Biol. New Ser.,* 47, 415, 1987.

50. **Lamond, A. I. and Travers, A. A.,** Stringent control of bacterial transcription, *Cell,* 41, 6, 1985.

51. **Seror, S., Vannier, F., Levine, A., and Henckes, G.,** Stringent control of initiation of chromosomal replication in *B. subtilis, Nature (London),* 321, 709, 1986.

52. **Junker, D. E., Rokeach, L. A., Ganea, D., Chiaramello, A., and Zyskind, J.,** Transcription termination within the *Escherichia coli* origin of DNA replication, *oriC, Mol. Gen. Genet.,* 203, 101, 1986.

53. **Rokeach, L., Kassavetis, G. A., and Zyskind, J. W.,** RNA polymerase pauses *in vitro* within the *Escherichia coli* origin of replication at the same sites where termination occurs *in vivo*, *J. Biol. Chem.,* 262, 7264, 1987.

54. **Kohara, Y., Todoh, N., Jiang, X., and Okazaki, T.,** The distribution and properties of RNA primed initiation sites of DNA synthesis at the replication origin of *Escherichia coli* chromosome, *Nucleic Acids Res.,* 13, 6847, 1985.

55. **Zyskind, J. W. and Smith, D. W.,** The bacterial origin of replication, *oriC, Cell,* 46, 489, 1986.

56. **Zyskind, J. W., Harding, N. E., Takeda, Y., Cleary, J. M., and Smith, D. W.,** The DNA replication origin region of the *Enterobacteriaceae, ICN-UCLA Symp. Mol. Cell. Biol.,* 22, 13, 1981.

57. **Zyskind, J. W., Smith, D. W., Hirota, Y., and Takanami, M.,** The consensus sequence of the bacterial origin, *ICN-UCLA Symp. Mol. Cell. Biol.,* 22, 26, 1981.

58. **Hirota, Y., Oka, A., Sugimoto, K., Asada, K., Sasaki, K., Sasaki, H., and Takanami, M.,** *Escherichia coli* origin of replication: structural organization of the region essential for autonomous replication and the recognition frame model, *ICN-UCLA Symp. Mol. Cell. Biol.,* 22, 1, 1981.

59. **Oka, A., Sasaki, H., Sugimoto, K., and Takanami, M.,** Sequence organization of the replication origin of the *Escherichia coli* K-12 chromosome, *J. Mol. Biol.,* 176, 443, 1984.

60. **Asada, K., Sugimoto, K., Oka, A., Takanami, M., and Hirota, Y.,** Structure of replication origin of the *Escherichia coli* K12 chromosome: the presence of spacer sequences in the *ori* region carrying information for autonomous replication, *Nucleic Acids Res.,* 10, 3745, 1982.

61. **Fuller, R. S., Funnell, B. E., and Kornberg, A.,** The DnaA protein complex with the *E. coli* chromosomal replication origin *oriC* and other sites, *Cell,* 38, 889, 1984.

62. **Matsui, M., Oka, A., Takanami, M., Yasuda, Y., and Hirota, Y.,** Sites of DnaA protein-binding in the replication origin of the *E. coli* K-12 chromosome, *J. Mol. Biol.,* 184, 529, 1985.

63. **Patterson, T., Popplewell, A., Pringle, J. H., Begg, K., and Masters, M.,** The effect of altering DnaA protein levels on transcription from promoters in the vicinity of DnaA boxes: *in vivo* studies, paper presented at Molecular Basis of Bacterial Growth and Division, EMBO workshop, Segovia, September 27 to October 1, 1987, 103.

64. **Messer, W., Bellekes, U., and Lother, H.,** Effect of dam methylation on the activity of the *E. coli* replication origin, *oriC, EMBO J.,* 4, 1327, 1985.

65. **Smith, D. W., Garland, A. M., Herman, G., Enns, R. E., Baker, T. A., and Zyskind, J. W.,** Importance of the state of methylation of *oriC* GATC sites in initiation of DNA replication in *Escherichia coli, EMBO J.,* 4, 1319, 1985.

66. **Hughes, P., Squali-Houssaini, F., Forterre, P., and Kohiyama, M.,** *In vitro* replication of *dam* methylated and nonmethylated *oriC* plasmid, *J. Mol. Biol.,* 176, 155, 1984.

67. **Russell, D. W. and Zinder, N. D.,** Hemimethylation prevents DNA replication in *E. coli, Cell,* 50, 1071, 1987.

68. **Ogden, G. B., Pratt, M. J., and Schaechter, M.,** The replicative origin of the *Escherichia coli* chromosome binds to cell membranes only when hemimethylated, *Cell,* 54, 127, 1988.

69. **Kornberg, A., Baker, T. A., Bertsch, L. L., Bramhill, D., Funnell, B. E., Lasken, R. S., Maki, H., Sekimizu, K., and Wahle, E.,** Enzymatic studies of replication of *oriC* plasmids, *UCLA Symp. Mol. Cell. Biol. New Ser.,* 47, 137, 1987.

70. **Bramhill, D. and Kornberg, A.,** Duplex opening by DnaA protein at novel sequences in initiation of replication at the origin of the *E. coli* chromosome, *Cell,* 52, 743, 1988.

71. **Umek, R. M., Eddy, M. J., and Kowalski, D.,** DNA sequences required for unwinding prokaryotic and eukaryotic replication origins, in *Cancer Cells, 6/Eukaryotic DNA Replication,* Kelly, T. and Stillman, B., Eds., Cold Spring Harbor Laboratory, Cold Spring Harbor, NY, 1988, 473.

72. **Lother, H. and Messer, W.,** Promoters in the *E. coli* replication origin, *Nature (London),* 294, 376, 1981.

73. **Morelli, G., Buhk, H.-J., Fisseau, C., Lother, H., Yoshinaga, K., and Messer, W.,** Promoters in the region of the *E. coli* replication origin, *Mol. Gen. Genet.,* 184, 255, 1981.

74. **Morita, M., Sugimoto, K., Oka, A., Takanami, M., and Hirota, Y.,** Mapping of promoters in the replication origin region of the *E. coli* chromosome, *ICN-UCLA Symp. Mol. Cell. Biol.,* 22, 29, 1981.

75. **Fuller, R. S. and Kornberg, A.,** Purified DnaA protein in initiation of replication of the *Escherichia coli* chromosomal origin of replication, *Proc. Natl. Acad. Sci. U.S.A.,* 80, 5817, 1983.

76. **Atlung, T., Rasmussen, K. V., Clausen, E., and Hansen, F. G.,** Role of the DnaA protein in control of DNA replication, in *The Molecular Biology of Bacterial Growth,* Schaechter, M., Neidhardt, F. C., Ingraham, J. L., and Kjeldgaard, N. O., Eds., Jones and Bartlett, Boston, 1985, 282.

77. **Sekimizu, K., Yung, B. Y., and Kornberg, A.,** The DnaA protein of *E. coli*: abundance, improved purification and membrane binding, *J. Biol. Chem.,* 263, 7136, 1988.

78. **Sakakibara, Y. and Yuasa, S.,** Continuous synthesis of the *dnaA* gene product of *Escherichia coli* in the cell cycle, *Mol. Gen. Genet.,* 186, 87, 1982.

79. **Funnell, B. E., Baker, T. A., and Kornberg, A.,** *In vitro* assembly of a prepriming complex at the origin of the *Escherichia coli* chromosome, *J. Biol. Chem.,* 262, 10327, 1987.

80. **Miki, T., Kimura, T., Hiraga, S., Nagata, T., and Yura, T.,** Cloning and physical mapping of the *dnaA* region of the *Escherichia coli* chromosome, *J. Bacteriol.,* 140, 817, 1979.

81. **Sakakibara, Y., Tsukano, H., and Sako, T.,** Organization and transcription of the *dnaA* and *dnaN* genes of *Escherichia coli, Gene,* 13, 47, 1981.

82. **Hansen, F. G. and Atlung, T.,** The nucleotide sequence of the *dnaA* gene promoter and the adjacent *rpmH* gene, coding for the ribosomal protein L34, of *Escherichia coli, EMBO J.,* 1, 1043, 1982.

83. **Braun, R. E., O'Day, K., and Wright, A.,** Autoregulation of the DNA replication gene *dnaA* in *E. coli* K-12, *Cell,* 40, 159, 1985.

84. **Kucherer, C., Lother, H., Kolling, R., Schauzu, M. A., and Messer, W.,** Regulation of transcription of the chromosomal *dnaA* gene of *Escherichia coli, Mol. Gen. Genet.,* 205, 115, 1986.

85. **Atlung, T., Clausen, E. S., and Hansen, F. G.,** Autoregulation of the *dnaA* gene of *Escherichia coli* K12, *Mol. Gen. Genet.,*200, 442, 1985.

86. **Wang, Q. and Kaguni, J. M.,** Transcriptional repression of the *dnaA* gene of *Escherichia coli* by DnaA protein, *Mol. Gen. Genet.,* 209, 518, 1987.

87. **Braun, R. E. and Wright, A.,** DNA methylation differentially enhances the expression of one of the two *E. coli dnaA* promoters *in vivo* and *in vitro, Mol. Gen. Genet.,* 202, 246, 1986.

88. **Sekimizu, K., Bramhill, D., and Kornberg, A.,** ATP activates *dnaA* protein in initiating replication of plasmids bearing the origin of the *E. coli* chromosome, *Cell,* 50, 259, 1987.

89. **Sekimizu, K. and Kornberg, A.,** Cardiolipin activation of DnaA protein, the initiation protein of replication in *Escherichia coli, J. Biol. Chem.,* 263, 7131, 1988.

90. **Schaus, N., O'Day, K., and Wright, A.,** Suppression of amber mutations in the *dnaA* gene of *Escherichia coli* K-12 by secondary mutations in *rpoB, ICN-UCLA Symp. Cell. Mol. Biol.,* 22, 315, 1981.

91. **Bagdasarian, M., Izakowska, M., and Bagdasarian, M.,** Suppression of the DnaA phenotype by mutations in the *rpoB* cistron of ribonucleic acid polymerase in *Salmonella typhimurium* and *Escherichia coli, J. Bacteriol.,* 130, 577, 1977.

92. **Atlung, T.,** Analysis of seven *dnaA* suppressor loci in *Escherichia coli, ICN-UCLA Symp. Cell. Mol. Biol.,* 22, 297, 1981.

93. **Frey, J., Chandler, M., and Caro, L.,** Overinitiation of chromosome and plasmid replication in a *dnaAcos* mutant of *Escherichia coli* K-12: evidence for *dnaA-dnaB* interactions, *J. Mol. Biol.,* 179, 171, 1984.

94. **Walker, J. R., Haldenwang, W. G., Ramsey, J. A., and Blinkowa, A.,** Evidence that the *Escherichia coli dnaZ* product, a polymerization protein, interacts *in vivo* with the *dnaA* product, an initiation protein, *ICN-UCLA Symp. Cell. Mol. Biol.,* 22, 325, 1981.

95. **March, J. and Masters, M.,** Suppression of temperature sensitive *E. coli dnaA* mutants by over-expression of the heat shock genes *groEL* and *groES,* paper presented at Molecular Basis of Bacterial Growth and Division, EMBO workshop, Segovia, September 27 to October 1, 1987, 59.

96. **Sekimizu, K., Bramhill, D., and Kornberg, A.,** Sequential early stages in the *in vitro* initiation of replication at the origin of the *Escherichia coli* chromosome, *J. Biol. Chem.,* 263, 7124, 1988.

97. **Baker, T. A., Funnell, B. E., and Kornberg, A.,** Helicase action of *dnaB* protein during replication from the *Escherichia coli* chromosomal origin *in vitro, J. Biol. Chem.,* 262, 6877, 1987.

98. **LeBowitz, J. H. and McMacken, R.,** The *Escherichia coli* DnaB replication protein is a DNA helicase, *J. Biol. Chem.,* 261, 4738, 1986.

99. **Kobori, J. A. and Kornberg, A.,** The *Escherichia coli dnaC* gene product. III. Properties of the DnaB-DnaC protein complex, *J. Biol. Chem.,* 257, 13770, 1982.

100. **Wickner, S. and Hurwitz, J.,** Interaction of *Escherichia coli dnaB* and *dnaC(D)* gene products *in vitro, Proc. Natl. Acad. Sci. U.S.A.,* 72, 921, 1975.

101. **Kobori, J. A. and Kornberg, A.,** The *Escherichia coli dnaC* gene product. II. Purification, physical properties and role in replication, *J. Biol. Chem.,* 257, 13763, 1982.

102. **Filutowicz, M. and Jonczyk, P.,** Essential role of the *gyrB* gene product in the transcriptional event coupled to the *dnaA*-dependent initiation of *Escherichia coli* chromosome replication, *Mol. Gen. Genet.,* 183, 134, 1981.

103. **Filutowicz, M.,** Requirement of DNA gyrase for the initiation of chromosome replication in *Escherichia coli* K-12, *Mol. Gen. Genet.,* 177, 301, 1980.

104. **Leonard, A. C., Whitford, W. G., and Helmstetter, C. E.,** Involvement of DNA superhelicity in minichromosome maintenance in *Escherichia coli, J. Bacteriol.,* 161, 687, 1985.

105. **Lother, H., Lurz, R., and Orr, E.,** DNA binding and antigenic specifications of DNA gyrase, *Nucleic Acids Res.,* 12, 901, 1984.

106. **Kornberg, A.,** DNA replication, *J. Biol. Chem.,* 263, 1, 1988.

107. **Tanaka, M., Ohmori, H., and Hiraga, S.,** A novel type of *E. coli* mutants with increased chromosomal copy number, *Mol. Gen. Genet.,* 192, 51, 1983.

108. **Rasmussen, K. V., Atlung, T., Kerzman, G., Hansen, G. E., and Hansen, F. G.,** Conditional change of DNA replication control in an RNA polymerase mutant of *Escherichia coli, J. Bacteriol.,* 154, 443, 1983.

109. **Ogawa, T., Baker, T. A., van der Ende, A. and Kornberg, A.,** Initiation of enzymatic replication at the *E. coli* chromosome: contributions of RNA polymerase and primase, *Proc. Natl. Acad. Sci. U.S.A.,* 82, 3562, 1985.

110. **Ogden, G. B. and Schaechter, M.,** Chromosomes, plasmids, and the bacterial cell envelopes, in *Microbiology — 1985,* Leive, L., Ed., American Society for Microbiology, Washington, D.C., 1985, 282.

111. **Nagai, K., Hendrickson, W., Balakrishnan, R., Yamaki, H., Boyd, D., and Schaechter, M.,** Isolation of a replication origin complex from *Escherichia coli, Proc. Natl. Acad. Sci. U.S.A.,* 77, 262, 1980.

112. **Yoshimoto, M., Kambe-Honjoh, H., Nagai, K., and Tamura, G.,** Early replicative intermediates of *E. coli* chromosome isolated from a membrane complex, *EMBO J.,* 5, 787, 1986.

113. **Kusano, T., Steinmetz, D., Hendrickson, W. G., Murchie, J., King, M., Benson, A., and Schaechter, M.,** Direct evidence for specific binding of the replication origin of the *Escherichia coli* chromosome to the membrane, *J. Bacteriol.,* 158, 313, 1984.

114. **Jacq, A., Lother, H., Messer, W., and Kohiyama, M.,** Isolation of a membrane protein having an affinity to the replication origin of *E. coli, ICN-UCLA Symp. Mol. Cell. Biol.,* 19, 189, 1980.

115. **Jacq, A., Kohiyama, M., Lother, H., and Messer, W.,** Recognition sites for a membrane-derived DNA binding protein preparation in the *E. coli* replication origin, *Mol. Gen. Genet.,* 191, 460, 1983.

116. **Wolf-Watz, H. W. and Masters, M.,** Deoxyribonucleic acid and outer membrane: strains diploid for the *oriC* region show elevated levels of deoxyribonucleic acid-binding protein and evidence for specific binding of the *oriC* region to the outer membrane, *J. Bacteriol.,* 140, 50, 1979.

117. **Rothfield, L. and Chakraborti, A.,** Isolation and characterization of a membrane subfraction responsible for the *oriC*-binding activity of the *E. coli* cell envelope, paper presented at Molecular Basis of Bacterial Growth and Division, EMBO workshop, Segovia, September 27 to October 1, 1987, 180.

118. **Jacob, F. S., Brenner, S., and Cuzin, F.,** On the regulation of DNA replication in bacteria, *Cold Spring Harbor Symp. Quant. Biol.,* 28, 329, 1963.

119. **Prichard, R. H.,** Control of DNA replication in bacteria, in *The Microbial Cell Cycle,* Nurse, P. and Streiblova, E., Eds., CRC Press, Boca Raton, FL, 1984, 19.

120. **Sompayrac, L. and Maaloe, O.,** Autorepressor model for control of DNA replication, *Nature (London) New Biol.,* 241, 133, 1973.

121. **Margalit, H., Rosenberger, R. F., and Grover, N. B.,** Initiation of DNA replication in bacteria: analysis of an autorepressor control model, *J. Theor. Biol.,* 111, 183, 1984.

122. **Hansen, F. G., Atlung, T., and von Meyenburg, K.,** The initiator titration model for control of bacterial chromosome replication, paper presented at Molecular Basis of Bacterial Growth and Division, EMBO workshop, Segovia, September 27 to October 1, 1987, 13.

123. **Tratwick, J. D. and Kline, B. C.,** A two-stage molecular model for control of mini-F replication, *Plasmid,* 13, 59, 1985.

124. **Chattoraj, D. K., Mason, R. J., and Wickner, S. H.,** Mini-P1 plasmid replication: the autoregulation-sequestration paradox, *Cell,* 52, 551, 1988.

125. **Atlung, T., Lobner-Olesen, A., and Hansen, F. G.,** Overproduction of DnaA protein stimulates initiation of chromosome and minichromosome replication in *Escherichia coli, Mol. Gen. Genet.,* 206, 51, 1987.

126. **Pierucci, O., Helmstetter, C. E., Rickert, M., Weinberger, M., and Leonard, A. C.,** Overexpression of the *dnaA* gene in *Escherichia coli* B/r: chromosome and minichromosome replication in the presence of rifampicin, *J. Bacteriol.,* 169, 1871, 1987.

127. **Churchward, G., Holmans, P., and Bremer, H.,** Increased expression of the *dnaA* gene has no effect on DNA replication in a dnaA⁺ strain of *Escherichia coli, Mol. Gen. Genet.,* 192, 506, 1983.

128. **Guzman, E. C., Jimenez-Sanchez, A., Orr, E., and Pritchard, R. H.,** Heat stress in the presence of low RNA polymerase activity increases chromosome copy number, *Mol. Gen. Genet.,* 212, 203, 1988.

129. **Xu, Y.-C. and Bremer, H.,** Chromosome replication in *Escherichia coli* induced by oversupply of *dnaA, Mol. Gen. Genet.,* 211, 138, 1988.

130. **Bremer, H. and Dalbow, D. G.,** Regulatory state of ribosomal genes and physiological changes in the concentration of free ribonucleic acid polymerase in *Escherichia coli, Biochem. J.,* 150, 9, 1975.

131. **Pierucci, O.,** Phospholipid synthesis during the cell division cycle of *Escherichia coli, J. Bacteriol.,* 138, 453, 1979.

132. **Ryter, A., Hirota, Y., and Jacob, F.,** DNA-membrane complex and nuclear segregation in bacteria, *Cold Spring Harbor Symp. Quant. Biol.,* 33, 669, 1986.

133. **Chai, N. C. and Lark, K. G.,** Cytological studies of deoxyribonucleic acid replication in *Escherichia coli* 15T⁻: replication at slow growth rates and after a shift-up into rich medium, *J. Bacteriol.,* 104, 401, 1970.

134. **Lin, E. C. C., Hirota, Y., and Jacob, F.,** On the process of cellular division in *Escherichia coli.* VI. Use of a methocel autoradiographic method for the study of cellular division in *Escherichia coli, J. Bacteriol.,* 108, 375, 1971.

135. **Pierucci, O. and Zuchowski, C.,** Non-random segregation of DNA strands in *Escherichia coli* B/r, *J. Mol. Biol.,* 80, 477, 1973.

136. **Cooper, S. and Weinberger, M.,** Medium-dependent variation of deoxyribonucleic acid segregation in *Escherichia coli, J. Bacteriol.,* 130, 118, 1977.

137. **Pierucci, O. and Helmstetter, C. E.,** Chromosome segregation in *Escherichia coli* B/r at various growth rates, *J. Bacteriol.,* 128, 708, 1976.

138. **Cooper, S., Schwimmer, M., and Scanlon, S.,** Probabilistic behavior DNA segregation in *Escherichia coli, J. Bacteriol.,* 134, 60, 1978.

139. **Canovas, J. L., Tresguerres, E. F., Yousif, A. M. E., Lopez-Saez, J. F., and Navarrete, M. H.,** DNA segregation in *Escherichia coli* cells with 5-bromodeoxyuridine-substituted nucleoids, *J. Bacteriol.,* 158, 128, 1984.

140. **Austin, S. and Abeles, A.,** Partition of unit-copy miniplasmids to daughter cells. II. The partition region of miniplasmid P1 encodes an essential protein and a centromere-like site at which it acts, *J. Mol. Biol.,* 169, 373, 1983.

141. **Gustafsson, P., Wolf-Watz, H., Lind, L., Johansson, K., and Nordstrom, K.,** Binding between the *par* region of plasmids R1 and pSC101 and the outer membrane fraction of host bacteria, *EMBO J.,* 2, 27, 1983.

142. **Hirota, Y., Ryter, A., and Jacob, F.,** Thermosensitive mutants of *E. coli* affected in the process of DNA synthesis and cellular division, *Cold Spring Harbor Symp. Quant. Biol.,* 33, 677, 1968.

143. **Hussain, K., Begg, K., Salmond, G. P. C., and Donachie, W. D.,** *ParD:* a new gene coding for a protein required for chromosome partitioning and septum localization in *E. coli, Mol. Microbiol.,* 1, 73, 1987.

144. **Orr, E., Fairweather, N. F., Holland, I. B., and Pritchard, R. H.,** Isolation and characterization of a strain carrying a conditional lethal mutation in the *cou* gene of *Escherichia coli* K12, *Mol. Gen. Genet.,* 177, 103, 1979.

145. **Steck, T. R. and Drlica, K.,** Bacterial chromosome segregation: evidence for DNA gyrase involvement in decatenation, *Cell,* 36, 1081, 1984.

146. **Norris, V., Alliotte, T., Jaffe, A., and D'Ari, R.,** DNA replication termination in *Escherichia coli parB* (a *dnaG* allele), *parA,* and *gyrB* mutants affected in DNA distribution, *J. Bacteriol.,* 168, 494, 1986.

147. **Helmstetter, C. E.,** Timing of synthetic activities in the cell cycle, in *Escherichia coli and Salmonella typhimurium: Cellular and Molecular Biology,* Ingraham, J. L., Low, K. B., Magasanik, B., Schaechter, M., and Umbarger, H. E., Eds., American Society for Microbiology, Washington, D.C., 1987, 1594.

148. **Henson, J. M. and Kuempel, P. L.,** Deletion of the terminus region (340 kilobase pairs of DNA) from the chromosome of *Escherichia coli, Proc. Natl. Acad. Sci. U.S.A.,* 82, 3766, 1985.

149. **Ogura, T., Miki, T., and Hiraga, S.,** Copy-number mutants of the plasmid carrying the replication origin of the *Escherichia coli* chromosome: evidence for a control region of replication, *Proc. Natl. Acad. Sci. U.S.A.,* 77, 3993, 1980.

150. **Hinchliffe, E., Kuempel, P. L., and Masters, M.,** *Escherichia coli* minichromosomes containing the pS101 partitioning locus are not stably inherited, *Plasmid,* 9, 286, 1983.

151. **Kunze, H. and Messer, W.,** Regions around the *E. coli* origin of replication which affect the stability of plasmids, in *Plasmids in Bacteria,* Helinski, D. R., Cohen, S. N., Clewell, D. B., Jackson, D. A., and Hollaender, A., Eds., Plenum Press, New York, 1985, 865.

152. **Masters, M., Pringle, J. H., and Oliver, I. R.,** A cloned fragment of *Escherichia coli* chromosomal DNA which promotes accurate plasmid partitioning, *UCLA Symp. Cell. Mol. Biol. New Series,* 1986, 181.

153. **Lobner-Olesen, A., Atlung, T., and Rasmussen, K. V.,** Stability and replication control of *Escherichia coli* minichromosomes, *J. Bacteriol.,* 169, 2835, 1987.

154. **Nordstrom, K., Molin, S., and Aagaard-Hansen, H.,** Partitioning of plasmid R1 in *Escherichia coli.* I. Kinetics of loss of plasmid derivatives deleted of the *par* region, *Plasmid,* 4, 215, 1980.

155. **Grover, N.,** personal communication, 1988.

156. **Koch, A. L., Mobley, H. L. T., Doyle, R. J., and Streips, U. N.,** The coupling of wall growth and chromosome replication in Gram-positive rods, *Microbiol. Lett.,* 12, 201, 1981.

157. **Sonnenfeld, E. M., Koch, A. L., and Doyle, R. J.,** Cellular location of origin and terminus of replication in *Bacillus subtilis, J. Bacteriol.,* 163, 895, 1985.

158. **Cavalier-Smith, T.,** Bacterial DNA segregation: its motors and positional control, *J. Theor. Biol.,* 127, 361, 1987.

159. **MacAlister, T. J., Macdonald, B., and Rothfield, L. I.,** The periseptal annulus: an organelle associated with cell division, *Proc. Natl. Acad. Sci. U.S.A.,* 80, 1372, 1983.

160. **Cook, W. R. and Rothfield, L. I.,** Polarization: a new step in the biogenesis of cell division, paper presented at Molecular Basis of Bacterial Growth and Division, EMBO workshop, Segovia, September 27 to October 1, 1987, 178.

161. **Mendelson, N. H.,** A model of bacterial segregation based upon helical geometry, *J. Theor. Biol.,* 112, 25, 1985.

162. **Woldringh, C. L. and Nanninga, N.,** Structure of the nucleoid and cytoplasm in the intact cell, in *Molecular Cytology of Escherichia coli,* Nanninga, N., Ed., Academic Press, London, 1985, 161.

163. **Helmstetter, C. E. and Leonard, A. C.,** Mechanism for chromosome and minichromosome segregation in *Escherichia coli, J. Mol. Biol.,* 197, 195, 1987.
164. **Pierucci, O.,** Dimensions of *Escherichia coli* at various growth rates: model for envelope growth, *J. Bacteriol.,* 135, 559, 1978.
165. **Trueba, F. J. and Woldringh, C. L.,** Changes in cell diameter during the division cycle of *Escherichia coli, J. Bacteriol.,* 142, 869, 1980.
166. **Burman, L. G. and Park, J. T.,** Molecular model for elongation of the murein sacculus of *Escherichia coli, Proc. Natl. Acad. Sci. U.S.A.,* 81, 1844, 1984.
167. **Austin, S. J.,** The P1 partition system, paper presented at Molecular Basis of Bacterial Growth and Division, EMBO workshop, Segovia, September 27 to October 1, 1987, 133.

Chapter 4

TERMINATION OF REPLICATION IN *BACILLUS SUBTILIS*, *ESCHERICHIA COLI*, AND R6K

Peter L. Kuempel, Thomas M. Hill, and Anthony J. Pelletier

TABLE OF CONTENTS

I. INTRODUCTION

Bidirectional replication of circular chromosomes concludes with the meeting of the two opposing replication forks. For several prokaryotic chromosomes that have been extensively studied, this meeting occurs in a specific region called the terminus. The meeting of forks in this region is not left to chance: the termination of replication is confined to this region by loci that inhibit further movement of replication forks. Although much remains to be learned about completion of the replication cycle and separation of daughter chromosomes to daughter cells, considerable progress has recently been made in characterizing the loci that halt replication. The purpose of this review is to acquaint the reader with the background and current understanding of these inhibitory loci in the chromosomes of *Bacillus subtilis*, *Escherichia coli*, and plasmid R6K.

II. *BACILLUS SUBTILIS*

A. THE BASIC PATTERN OF CHROMOSOME REPLICATION

Bacillus subtilis is a Gram-positive, transformable, spore-forming eubacterium, and the unique properties of this organism have been exploited elegantly in studies of chromosome replication. Determination of the basic pattern of chromosome replication was facilitated by combined use of DNA transformation to determine gene frequencies and spore outgrowth to obtain synchronous replication. Such studies demonstrated that bidirectional replication in *B. subtilis* 168 was initiated near *purA*, and the last genes to be replicated were in the vicinity of the *gltA-citK* loci (Figure 1).[1,2] Termination of replication consequently occurred in that region.

Radioactive labeling of the last DNA replicated during the replication cycle established that termination was actually confined to a small, specific region of the chromosome. An interesting labeling procedure was used, based on several properties of this spore-forming organism. Sporulating cells were labeled briefly with ³H-thymidine and then treated with the potent DNA synthesis inhibitor 6-(hydroxyphenylazo)uracil. Since spores can only form from cells in which replication is completed, the only labeled DNA present in spores was from chromosomes that terminated during the brief labeling period.[3] Studies with this procedure demonstrated that replication was completed in a region of approximately 200 kb,[4,5] which consisted of a discrete set of restriction fragments.[6,7]

An unambiguous restriction map of 150 kb of the terminus region DNA was constructed by Weiss and Wake, based on DNA labeled by the procedure described above.[8] The amount of label in the fragments changed abruptly part way through this region, giving the first indication of the exact location at which replication forks meet. It was proposed that a specific block to replication, called *terC*, was at the transition point. On the lightly labeled side, replication forks approached in the clockwise direction and arrived at the block site first. Therefore, most were already arrested when label was added. The more heavily labeled DNA was replicated by the counterclockwise forks, which arrived at *terC* within 5 min after the clockwise forks. Insertion of extra DNA (prophage SPβ, 126 kb) to the left of this region did not alter the apparent location of *terC*.[9]

A similar restriction map of the terminus has also been constructed by Monteiro et al.[10] Much of the terminus region has been cloned,[10-12] and restriction mapping of these fragments is consistent with the original map of Weiss and Wake. Transformation with these fragments has demonstrated that *terC* is very near the gene for *gltA*, in the interval between *gltA* and *citK* (Figure 1).[12,13]

B. STUDIES WITH ASYMMETRIC CHROMOSOMES

The replication pattern of the chromosome of *B. subtilis* 168 is slightly asymmetric, with the clockwise forks arriving in the terminus region several minutes in advance of the

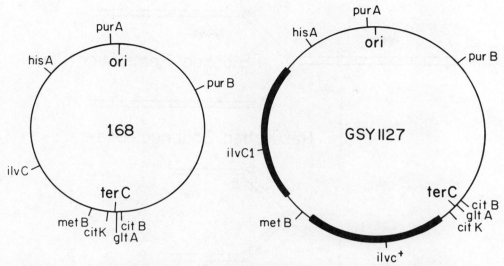

FIGURE 1.   Genetic maps of *B. subtilis* 168[52] and stable merodiploid strain GSY1127[14] showing relevant loci. The highlighted regions on the map for GSY1127 show the nontandem duplication.

counterclockwise forks. This asymmetry simplified the initial location of *terC* and suggested that *terC* actively inhibited replication forks. In studies of termination, however, it has often proved advantageous to introduce large differences in the lengths of DNA that the two replication forks must travel before they meet. If termination still occurs at the same locus, one can exclude the possibility that termination simply occurs wherever replication forks happen to meet. In addition, if one fork reaches the terminus well in advance of the other, the isolation and analysis of stalled replication forks is simplified. Asymmetric chromosome replication has been very useful for analysis of termination in both *B. subtilis* and *E. coli* (see below).

A strain which has been of considerable importance in studies of termination in *B. subtilis* is strain GSY1127, which contains a nontandem duplication of 25% of the chromosome (Figure 1).[14] The duplication occurs in the part of the chromosome containing *ilvC*, which is replicated by the counterclockwise-traveling replication fork. Clockwise replication forks consequently reach the *citK-gltA* region, and therefore the presumed location of *terC*, well in advance of counterclockwise replication forks. *IlvC* is now directly opposite the origin. O'Sullivan and Anagnostopoulos were the first to use this strain to study termination, using transformation to determine the replication pattern during spore outgrowth.[14] Their data indicated that, in spite of the duplication, *citK* and *gltA* were still replicated later than *ilvC*$^+$. This indicated, of course, that *terC* inhibited replication and that the block site was located near *gltA*. Specifically, they proposed that *terC* was near *gltA*, on the *citB* side. As mentioned above, more sensitive analyses using restriction fragments have demonstrated that the block site is actually between *gltA* and *citK*.

C. STUDIES OF BLOCKED REPLICATION FORKS

Weiss and Wake have also used the asymmetry present in GSY1127 to study replication.[15] Their innovative approach, however, was not based on the more traditional marker frequency analysis, and it has led to a very fruitful series of investigations. Their previous experiments had suggested that termination occurred in a 24.8-kb *Bam*HI fragment.[8,9] Weiss and Wake reasoned that if replication was indeed inhibited in that fragment, it should exist transiently as a Y-shaped structure (Figure 2). These would travel slower in agarose electrophoresis and could be identified by Southern blot hybridizations. Such structures were detected in

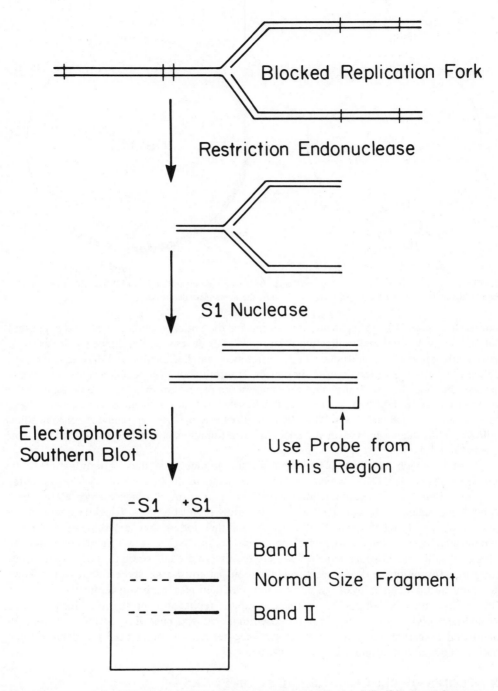

FIGURE 2.   Agarose gel/Southern hybridization assay for blocked replication forks.[15] Y-shaped fragments can be identified by their slower migration during agarose gel electrophoresis (band I). Digestion with S1 nuclease releases the arm from the Y (band II). The size of band II locates the block site. This technique has been useful in identifying the location of replication fork inhibition in both *B. subtilis* and *E. coli*.

cells with unrearranged chromosomes, but the asymmetry present in GSY1127 simplified their detection and analysis: the frequency of chromosomes that had the clockwise-traveling replication forks blocked at this site, awaiting the arrival of the counterclockwise replication forks, was increased considerably.

DNA from the 24.8-kb *Bam*HI fragment containing *terC* was detected at three different positions in Southern hybridizations (Figure 2). Besides the 24.8-kb fragment (fragments that did not contain forks), there were slower (band I) and faster (band II) migrating forms. The kinetics of appearance and disappearance of band I during a synchronous replication cycle, as well as its sensitivity to digestion by S1, indicated that band I was indeed the Y-shaped structure caused by a blocked replication fork. The properties of band II (15.4 kb) indicated it was an arm from the Y structure, and the length was used to locate the block site within the 24.8-kb fragment. Using restriction enzymes that cut closer to the inhibition site, Smith et al.[13] have identified Y forks in a 1.75-kb fragment.

Several papers have subsequently been published by Wake's laboratory, presenting further characterization of bands I and II. Band I from *Bam*HI-digested DNA has been purified from gels and its single-strand composition was that expected of replication forks blocked 15.4 kb from the upstream end of the *Bam*HI fragment.[16] Electron microscopy of the gel-purified material also demonstrated the presence of Y-fork structures of the appropriate size. A more detailed analysis of the sensitivity of purified band I to single-strand-specific nuclease has also been conducted.[17] Band I was very sensitive to digestion with nuclease P1, and digestion gave rise to fragments of 24.8 (normal size fragment) and 15.4 kb (band II). Further digestion gave rise to a 9.7-kb fragment, which is the stem of the Y, produced by cutting the 24.8-kb fragment at the single-strand gap or nick present at the former site of the replication fork.

Recently, the sequence of 1297 bp of the *terC* region was reported.[18] The sequence contains two long, inverted repeats that extend for 47 to 48 nucleotides, show 77% homology, and are separated by 59 nucleotides (Figure 7). The block site for clockwise replication forks, which approach from the right in the figure, is located either immediately before inverted repeat II or possibly in it. It was proposed that the inverted repeat contained the signal that inhibited replication. This entire region is AT-rich, and the inverted repeats are located between two open reading frames designated ORF1 and ORF2. ORF2 is required for termination since inhibition does not occur if ORF2 is inactivated.[54]

The replication block site of *B. subtilis* can be deleted, and this is providing the basis for further genetic manipulation of this region. Using induction of a temperature-sensitive, integrase-deficient mutant of prophage SPβ, Zahler[19] had constructed strains that appeared to lack the terminus region, based on the loss of genetic markers flanking *terC*. Iismaa and Wake have tested these strains and demonstrated that *terC* is indeed deleted.[20] In the strain with the smallest deletion (approximately 230 kb), the fusion fragment was identified. The other deletions ended outside the region that has been mapped and cloned at present. All of these strains exhibited normal morphology and growth rate, although they did not sporulate.

Iismaa and Wake have also constructed a smaller deletion, which removed 11.2 kb of DNA spanning *terC* by crossing a plasmid-constructed deletion into the chromosome.[20] This strain was capable of sporulation. Iismaa et al. have subsequently used this strain to reinsert a 1.75-kb *terC* fragment at a location 25 kb from its original location.[21] The insert was in the normal orientation and *terC* functioned at this location. Further experiments will undoubtedly test other locations and orientations, and this approach can also be used for site-specific mutagenesis of the *terC* sequence.

III. *ESCHERICHIA COLI*

A. BASIC PROPERTIES OF THE TERMINUS REGION

The terminus region of *E. coli* was originally identified by Masters and Broda[22] and

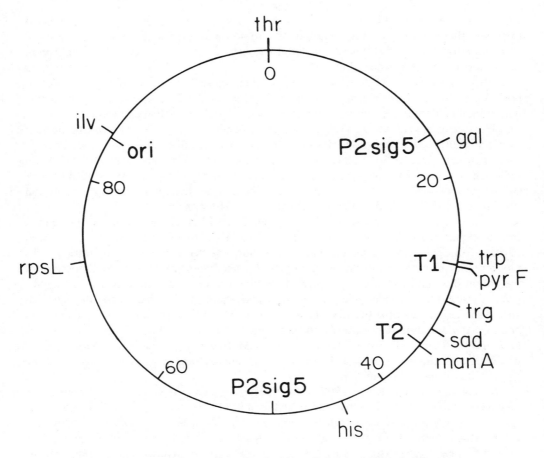

FIGURE 3.   Genetic map of *Escherichia coli* showing relevant loci.[53]

Bird et al.,[23] who used marker frequency assays to demonstrate bidirectional replication of
the chromosome. The origin of replication *(oriC)* was determined to be located near min
84, and the lowest marker frequency was directly opposite on the circular chromosome,
between *trp* (min 27) and *his* (min 44) (Figure 3). This established that replication forks
met somewhere in that interval.

In 1977, it was demonstrated that the terminus region of *E. coli* impeded the passage
of both clockwise- and counterclockwise-traveling replication forks. This established that
termination was constrained to occur in a particular region. To demonstrate inhibition, the
laboratories of both Kuempel[24] and Louarn[25] used a similar approach. Instead of using cells
with altered chromosome structure, as in *B. subtilis*, asymmetric replication was achieved
by positioning origins near the terminus. Prophage P2*sig5*[24] and a derivative of plasmid
R100.1[25] were used as origins. The results of these initial and subsequent[26,27] studies demonstrated that the inhibition of replication forks from both directions occurred somewhere
in the interval between *trp* (min 27) and *manA* (min 36).

A problem that hampered further investigations of the terminus region was ignorance
of its physical structure. In sharp contrast to the rest of the chromosome, there were very
few loci in the interval between *trp* and *man*. Since this region had only been mapped by
conjugation, it was even possible that the 8-min interval was caused by a reduced rate of
conjugal transfer and that *man* and *trp* were physically adjacent to each other. This uncertainty
was resolved when transposon insertions were used to construct cotransduction maps that
extended 8 min across the terminus region.[28-30] The strong selections provided by these

insertions also aided the mapping of other loci that had only been provisionally located in the terminus.

A major advance in the study of the terminus region was the construction by Bouché of a restriction map of over 450 kb of terminus region DNA.[31] This map established the relationship between known genetic markers and the physical map,[32] and it paved the way for identifying and analyzing terminus region functions using the techniques of molecular biology. In subsequent experiments, Bejar and Bouché used a cosmid vector to clone almost all of the DNA between min 30.5 and 34.0,[33] and additional regions were soon available. This increased the number of probes from the terminus region that could be used in marker frequency assays to identify the location of the block site(s). A restriction map and set of cloned fragments from the min 29.7 to 33.2 region have also been constructed by Asada et al.[34]

The terminus region is largely nonessential for viability and, as will be seen in the next section, this has allowed termination functions to be studied by deletion mapping. The largest deletion that has been obtained removed 340 kb of DNA (7% of the chromosome) between minutes 28.5 and 35.9 (deletion 1608, Figure 5).[35] Inhibition of replication forks did not occur in strains harboring this deletion. Although these strains were very filamentous, this was not due to loss of the termination system. More recent experiments have demonstrated that filamentation was caused by induction of the SOS system due to loss of a locus near min 33.5.[55] Inactivation of the termination system itself does not lead to a filamentous phenotype.

B. THE TERMINUS REGION CONTAINS TWO SEPARATE LOCI THAT INHIBIT REPLICATION

The next significant step in understanding replication termination in *E. coli* was the independent reports from de Massy et al.[36] and Hill et al.[37] that in *E. coli*, unlike *B. subtilis*, there are two widely separated sites that inhibit replication. The loci were identified through the use of replication origins located near the terminus and marker frequency assays based on DNA-DNA hybridization. These loci, designated T1 and T2, are located at the two edges of the terminus region (Figure 3) and act in a polar fashion. T1 inhibits counterclockwise forks, but has no effect on clockwise forks, and T2 inhibits clockwise forks, but has no effect on counterclockwise forks. The sites, therefore, act as a replication fork trap: they permit forks to enter, but not leave, the terminus region.

The marker frequency assay employed by both groups showed a broad curve of replication inhibition, extending over a region of 30 to 50 kb at either termination locus. Figure 4A shows data from our laboratory for the T1 region.[38] It was uncertain whether sequences that caused partial inhibition were distributed throughout this interval or whether there was a single inhibitory site, possibly at the top of the curve, and that the observed pattern was a secondary effect. Both groups placed T1 at min 28.5, based on the marker frequency assay and deletions that did not affect T1 function.[36,37]

The placement of the T2 locus was somewhat more ambiguous. DeMassy et al. mapped the T2 locus at min 33.5,[36] whereas the data of Hill et al. located it between min 34.5 and 35.7.[37] The latter group tested deletions that removed virtually the entire region between min 32.2 and 34.5, and these had no effect on the inhibition of replication forks at T2 (see Figure 5 for details). Other deletions that removed inhibition placed T2 in the 60-kb region between min 34.5 and 35.7. More recent experiments have demonstrated that the strains used by the two groups have differences in the T2 region.[56]

A subsequent deletion study refined the mapping of the T1 and T2 loci.[38] T2 was located between kb 438 and 442 (min 35.6). Although T1 was not deleted in this study, available deletions confined its location to the kb 80 to 100 interval (near min 28.1 in Figure 4A). The deletions used in these studies demonstrated that the loci required for inhibition were

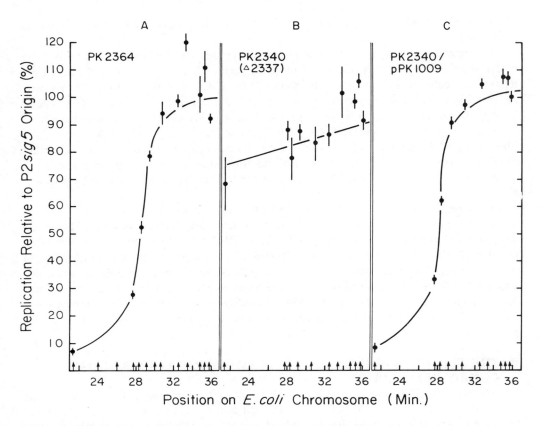

FIGURE 4.   Inhibition of counterclockwise-traveling replication forks at T1 and dependence on *tus*.[38] Replication was initiated at P2*sig*5 integrated at min 46. (A) Strain PK2364, which exhibits normal termination. (B) No inhibition in PK2340, in which a 10-kb deletion (kb 433 to 443; min 35.7) removed *tus*. (C) Restoration of T1 function when *tus* was supplied to PK2340 on plasmid pPK1009. (From Hill, T. M., Kopp, B. J., and Kuempel, P. L., *J. Bacteriol.*, 170, 662, 1988. With permission.)

actually located at the bottom of the broad inhibition curve seen in the marker frequency assay. Furthermore, deletion 2369 (Figure 5), which removed 90 kb just upstream from T2, had no effect on replication inhibition. If upstream sites are present, this demonstrates that they are not necessary in the termination process.

C. SITES OF ARREST OF REPLICATION AT T1 AND T2

The marker frequency assay, when combined with various deletions, could determine the locations of the termination sequences (T1 and T2) required for inhibition. It did not accurately demonstrate, however, where replication forks were actually arrested. The terminator loci need not be coincident with the sites at which replication is actually arrested. This is a subtle distinction, but different techniques must be used to identify these various sites.

Recently, the precise location of the sites of arrest near T1 and T2 were determined by Pelletier et al.[39] An assay similar to that of Weiss and Wake[15] was used, in which the released arms of "Y" forks are detected by Southern hybridization analysis (Figure 2). As in previous experiments, replication was initiated from P2*sig*5 integrated near the terminus. The site of arrest for T1 was determined to be at kb 90 (min 28.1), near the *pyrF* gene, and the site of arrest for T2 is at kb 442 (min 35.6). The use of these assays also permitted several other observations to be made. The most important was that only a single site of arrest was observed in the T1 and T2 regions. The regions 50 kb upstream of T1 and 30 kb upstream

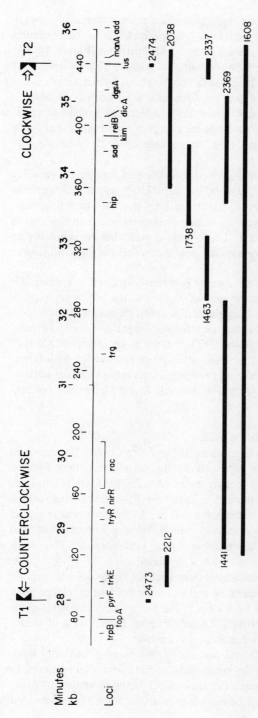

FIGURE 5. Map of the terminus region of *E. coli* showing physical coordinates (kb 70 to 450),[31] relevant deletions and genetic markers,[32,38] and location of T1 and T2.[49,50] Deletions 1441, 1463, and 1738 did not affect T2; deletion 2038 removed T2. This located T2 to the kb 390 to 450 interval.[37] More recent deletions, 2473[49] and 2474,[40] have located T1 and T2 to much smaller regions.

of T2 were tested and no additional sites were observed. In addition, the time at which the arrested forks were observed indicated that the forks were not slowed or significantly inhibited until these sites were reached.

The absence of upstream sites was an unexpected result since marker frequency assays had consistently indicated that replication was inhibited over a broad region that included the indicated sites of arrest, but which also extended 30 to 50 kb upstream (Figure 4). We had expected to find sites of arrest distributed throughout this region. The difference in results is presumably due to the difference in the assays. The Southern hybridization assay is quite sensitive and can detect the arrested forks in the first cells in which this occurs. To detect inhibition with the marker frequency assay, however, a substantial amount of replication must have occurred on the upstream side of the block site. The broad inhibition curve obtained with the marker frequency is presumably due to stacking up of the multiple replication forks that are induced by P2*sig*5.

The use of the Southern blot assay has also permitted studies on exponentially growing cells in which replication was initiated at *oriC* instead of the P2*sig*5 origin.[39] Fragments corresponding to both sites of arrest were observed, which indicates that the block sites were used in many of the replication cycles. However, the actual frequency at which replication is arrested at the sites cannot yet be determined. To do this, it will also be necessary to determine the rate at which daughter chromosomes are resolved once the opposing forks meet at the site of arrest.

Since the sites of arrest are located within the regions previously defined to contain the termination loci T1 and T2, this indicated that the sites of arrest and the inhibitory signals were coincident or nearly so. This allowed the construction of small deletions in the region where replication forks are arrested to locate further the inhibitory signals. T1 and T2 have now been removed with deletions of 1 kb (deletion 2473)[49] and 1.6 kb (deletion 2374),[40] respectively (Figure 5). It should be noted that previous attempts to remove T1 with larger deletions had been unsuccessful. Although almost all of the terminus region can be deleted, it appears that some unidentified, essential functions are probably located between *pyrF* and *trkE*.

D. *TUS*, TERMINUS UTILIZATION SUBSTANCE

A third component of termination that has been identified in *E. coli* is *tus*, the terminus utilization substance.[38] Its presence was first indicated by the observation that T1 was inactivated by deletions in the T2 region, even though these were 350 kb apart (Figure 4A and B). Using a set of overlapping deletions, the site required for T1 function was located between kb 438 and 442. Further experiments established that the T1 function was regained in these deletion mutants when this region was provided on a low-copy-number plasmid (Figure 4C) that had no homology with the deletion-containing chromosome. This demonstrated a requirement for a *trans*-acting factor, which was named *tus*.

Complementation studies with small plasmids have been used to locate *tus* with greater accuracy: 3000 bp of this region have been sequenced and shown to contain two large open reading frames, ORF1 and ORF2, separated by 64 bp.[40] *Tus* is encoded in ORF2 since insertion of a kanamycin resistance gene inactivated the *tus* function. Inactivation of ORF1 had no detectable effect on termination and its function is unknown.

Due to the very tight linkage between T2 and *tus*, up to this point it had only been shown that T1 required *tus*. We have recently demonstrated that the insertional mutation in *tus* inactivates T2 as well as T1.[40] Furthermore, the loss of T2 function is due to a *tus* requirement and not inactivation of T2 itself: T2 function is restored when *tus* is provided on a plasmid.

It is not known at present whether proteins other than *tus* are required for T1 and T2 functions. T2 still functions in strains harboring the deletions that removed T1.[57] If other proteins are involved, they are not located immediately adjacent to T1, in a fashion analogous to the location of *tus* and T2.

IV. TERMINATION IN R6K AND PLASMIDS WITH SUBCLONED TERMINATION SITES

The other well-characterized prokaryotic replication that exhibits inhibition of replication is plasmid R6k. Recent studies have shown that termination in R6K and *E. coli* are similar, and the basic properties of R6K need to be presented before discussion of termination sequences, which will be described in the last section.

R6K contains three origins of replication that are used to different extents *in vivo* and *in vitro*.[41,42] Initiation occurs in a unidirectional fashion and the replication fork proceeds to the asymmetrically located terminus, where it is inhibited.[43] Initiation in the opposite direction then occurs at the same origin, and that replication fork proceeds until it, too, halts at the terminus. Daughter chromosomes then separate, and they exist transiently in an open circular form possessing a single-stranded discontinuity near the terminus.[44] This pattern of replication has been described as sequentially, asymmetrically bidirectional.

The R6K terminus is functional when inserted into ColEI-derived replicons. This has simplified its characterization since ColEI replication is unidirectional. Using ColEI/R6K hybrid plasmids, Kolter and Helinski demonstrated that replication forks were severely inhibited when they reached the inserted R6K terminus DNA.[45] This effect was observed for both orientations of the terminus. Replication forks eventually proceeded through the inserted terminus and the replication cycle was then completed. Bastia et al. have subcloned from a 2-kb fragment containing the R6K terminus and demonstrated that a 216-bp fragment was sufficient to inhibit replication forks.[44] They sequenced this fragment, but observed no obvious sequence that could be identified as the terminator.[46]

The R6K terminus functions *in vitro* and functional extracts can be obtained from cells that had not harbored the plasmid.[47] Germino and Bastia consequently suggested that if a protein was required for termination, it was supplied by the host. No further characterization of this interesting system has been reported until recently. Horiuchi et al. have used agarose gels to visualize inhibited plasmid replication.[48] This approach simplifies termination studies since all earlier studies had relied on electron microscopy to characterize inhibition.

We have recently used agarose gels to test whether T1 and T2 function in ColEI plasmids and have demonstrated that inhibition occurred in the appropriate fragments.[49] As expected for T1 and T2, inhibition was both orientation and *tus* dependent. This assay has been used in further studies of T1 and T2, and it will facilitate a number of future projects such as studies of the mechansim of *tus* function and analyses of various mutated terminator sequences.

V. THE TERMINATOR SEQUENCE

We have recently identified a 23-bp sequence that is present in both the T1 and T2 regions.[50] The sequence contains only one mismatch in 23 bp, and it is located within 100 bp of the sites of arrest. We propose that this is the *tus*-dependent terminator sequence. The sequence, shown in Figure 6A, was identified by a computer comparison of sequences from the T2 and T1 regions. The T2 region has been sequenced in our laboratory[40] and the sequence of the T1 region was available from studies of the *pyrF* gene.[51]

The proposed terminator sequence is not palindromic and has a directed orientation consistent with the observed polarity of inhibition. Figure 6A shows the T2 sequence in the orientation it has in the genetic map shown in Figure 5, and it blocks clockwise-traveling replication forks which approach from the left (5′ side). To simplify comparison of the sequences, T1 is shown with the same orientation as T2. In the chromosome, of course, T1

HOMOLOGY COMPARISONS TO T1 AND T2

Source	Aligned Sequence (5´-3´)

A)

T1   A A T T A G T A T G T T G T A A C T A A A G T

T2   A A T A A G T A T G T T G T A A C T A A A G T

R6K Left   C T C T T G T G T G T T G T A A C T A A A T C

R6K Right   C T A T T G A G T G T T G T A A C T A C T A G

Consensus   G – – T G T T G T A A C T A

B)

B. subtilis I   A C T G A A C A T T T G G T A C A T A G T T T

B. subtilis II   A T T G A A T A T T T A G T A C A T A G T G T

FIGURE 6. Terminator sequences. (A) T2 is shown 5'→3' with the same orientation as in Figure 5. It blocks replication forks that approach from the left. T1 is inverted for sake of comparison. R6K left and R6K right correspond to nucleotides 768 to 791 and 867 to 889, respectively, in the published sequence of the R6K terminus.[46] R6K left is shown inverted for sake of comparison. The boxes surround sequences common to *E. coli* and R6K; these are shown separately as the consensus. (B) The 23-bp regions of the 48-bp inverted repeats I and II of *B. subtilis* that show homology with the *E. coli* terminator. *B. subtilis* I corresponds to nucleotides 689 to 667 and *B. subtilis* II corresponds to 784 to 806 in the sequence of the *B. subtilis* terminus region.[18] The boxes surround base pairs homologous to the *E. coli* terminator sequence.

is present in the inverted orientation and blocks counterclockwise-traveling replication forks, which approach from the right. Arranged in this fashion, the two sequences provide a replication fork trap. Their orientation can be considered to be an inverted repeat, with 352 kb of DNA between the repeats (Figure 7).

As a definitive test of the terminator sequence, we have synthesized an oligomer containing the 23-bp T2 sequence and inserted this in either orientation into a ColEI-derived plasmid. The insert blocks replication forks, but only when present in the correct orientation and only in a *tus*⁺ strain.[50] This demonstrates that the sequence is sufficient to cause termination.

A computer search also identified a core sequence of the terminator that was present as an inverted repeat[50] in the previously published sequence of the R6K terminus region.[46] The more 5' repeat (here called R6K left) is identical with the terminator over 13 bp and R6K right is identical over 11 bp (Figure 6A). The orientation of the sequences is the same as in *E. coli*, except that the inverted repeats are only separated by 80 bp (Figure 7). This arrangement suggests that R6K right blocks forks that approach from the left and R6K left blocks forks that approach from the right. A replication fork trap of this type would therefore function independent of its orientation. This is consistent with the observation that the R6K terminus functions in ColEI plasmids in either orientation.[45,48] The presence of similar sequences in the T1 and T2 regions, as well as in R6K, leads to a prediction: termination in R6K should be *tus* dependent. We have tested ColEI/R6K hybrid plasmids and determined that replication is inhibited at the R6K terminus sequences in a *tus*⁺ strain, but not in a *tus*⁻ strain.[49]

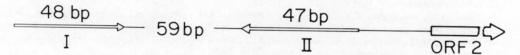

E. coli

23 bp
T1

--- ~ 352,000 bp ---

23 bp
T2

tus

R6K

20 bp
R6K left

73 bp

20 bp
R6K right

B. subtilis

48 bp
I

59 bp

47 bp
II

ORF 2

FIGURE 7.   Terminus regions of *E. coli*, R6K, and *B. subtilis*. The terminator sequences for *E. coli* and R6K are shown by open arrows; replication forks traveling in the direction opposite that of the arrow are inhibited. It is proposed that the inverted repeats of *B. subtilis*[18] have a comparable function and inhibit replication forks in an analogous fashion.

Using the Genbank sequence analysis programs and databank, sequences similar to an 11-bp consensus sequence of T1, T2, R6K left, and R6K right (Figure 6A) have been found at other locations in the *E. coli* chromosome and in several plasmids.[50] It is not yet known if any of these other occurrences are functional.

Unexpectedly, sequences similar to the terminator have also been identified in the terminus of *B.subtilis*. As mentioned previously, this region contains an inverted repeat 48 bp long in which the two sequences exhibit 77% homology.[18] Within the region of highest homology between these repeats, there is a 23-bp region that resembles the terminator sequence. In inverted repeat I, the homology is 52% and in inverted repeat II, it is 61% (Figure 6B). If these sequences are functionally similar to those in *E. coli,* this indicates that the *B. subtilis* terminus region is also arranged as a replication fork trap (Figure 7). The trap is similar to that in R6K in that the sequences are separated by less than 100 bp instead of 352 kb. The *B. subtilis* terminus also appears to encode a protein in ORF2 immediately adjacent to the right repeat.[18] As mentioned previously, this gene product is required for termination[54] and indicates further similarity between *B. subtilis* and *E. coli.*

The particular advantage that a termination system supplies to a cell is still unknown since mutants of *E. coli* and *B. subtilis* that lack the termination system can be readily obtained and appear to grow normally. The presence of similar systems in such different organisms, however, indicates that it confers a selective advantage and that the appropriate conditions simply have not yet been identified.

REFERENCES

1. **Harford, N., Lepesant-Kejzlarova, J., Lepesant, J. A., Hamers, R., and Dedonder, R.,** Genetic circularity and mapping of the replication origin region of the *Bacillus subtilis* chromosome, in *Microbiology — 1976,* Schlesinger, D., Ed., American Society for Microbiology, Washington, D.C., 1976, 28.
2. **Hye, R. J., O'Sullivan, M. A., Howard, K., and Sueoka, N.,** Membrane association of origin, terminus and replication fork in *Bacillus subtilis,* in *Microbiology — 1976,* Schlesinger, D., Ed., American Society for Microbiology, Washington, D.C., 1976, 83.
3. **Sargent, M. G.,** Chromosome replication in sporulating cells of *Bacillus subtilis, J. Bacteriol.,* 142, 491, 1980.
4. **Sargent, M. G.,** Specific labeling of the *Bacillus subtilis* chromosome terminus, *J. Bacteriol.,* 143, 1033, 1980.
5. **Adams, R. T. and Wake, R. G.,** Highly specific labeling of the *Bacillus subtilis* chromosome terminus, *J. Bacteriol.,* 143, 1036, 1980.
6. **Weiss, A. S., Hariharan, I. K., and Wake, R. G.,** Analysis of the terminus region of the *Bacillus subtilis* chromosome, *Nature (London),* 293, 673, 1981.
7. **Sargent, M. G. and Monteiro, M. J.,** Characterization of the chromosomal terminus of *Bacillus subtilis* and its attachment to the cell membrane, in *Molecular Cloning and Gene Regulation in Bacillus,* Ganesan, A. T., Chang, S., and Hoch, J. A., Eds., Academic Press, New York, 1982, 181.
8. **Weiss, A. S. and Wake, R. G.,** Restriction map of DNA spanning the replication terminus of the *Bacillus subtilis* chromosome, *J. Mol. Biol.,* 171, 119, 1983.
9. **Weiss, A. S. and Wake, R. G.,** Impediment to replication fork movement in the terminus region of the *Bacillus subtilis* chromosome, *J. Mol. Biol.,* 179, 745, 1984.
10. **Monteiro, M. J., Sargent, M. G., and Piggot, P. J.,** Characterization of the replication terminus of the *Bacillus subtilis* chromosome, *J. Gen. Microbiol.,* 130, 2403, 1984.
11. **Weiss, A. S., Smith, M. T., Iismaa, T. P., and Wake, R. G.,** Cloning DNA from the replication terminus region of the *Bacillus subtilis* chromosome, *Gene,* 24, 83, 1983.
12. **Iismaa, T. P., Smith, M. T., and Wake, R. G.,** Physical map of the *Bacillus subtilis* replication terminus region: its confirmation, extension and genetic orientation, *Gene,* 32, 171, 1984.
13. **Smith, M. T., Aynsley, C., and Wake, R. G.,** Cloning and localization of the *Bacillus subtilis* chromosome replication terminus, *terC, Gene,* 38, 9, 1985.
14. **O'Sullivan, M. A. and Anagnostopoulos, C.,** Replication terminus of the *Bacillus subtilis* chromosome, *J. Bacteriol.,* 151, 135, 1982.
15. **Weiss, A. S. and Wake, R. G.,** A unique DNA intermediate associated with termination of chromosome replication in *Bacillus subtilis, Cell,* 39, 683, 1984.
16. **Weiss, A. S., Wake, R. G., and Inman, R. B.,** An immobilized fork as a termination of replication intermediate in *Bacillus subtilis, J. Mol. Biol.,* 188, 199, 1986.
17. **Hanley, P. J. B., Carrigan, C. M., Rowe, D. B., and Wake, R. G.,** Breakdown and quantitation of the forked termination of replication intermediate of *Bacillus subtilis, J. Mol. Biol.,* 196, 721, 1987.
18. **Carrigan, C. M., Haarsma, J. A., Smith, M. T., and Wake, R. G.,** Sequence features of the replication terminus of the *Bacillus subtilis* chromosome, *Nucleic Acids Res.,* 15, 8501, 1987.
19. **Zahler, S. A.,** Specialized transduction in *Bacillus subtilis* in *The Molecular Biology of the Bacilli,* Vol. 1 *Bacillus subtilis,* Dubnau, D. A., Ed., Academic Press, New York, 1982, 269.
20. **Iismaa, T. P. and Wake, R. G.,** The normal replication terminus of the *Bacillus subtilis* chromosome, *terC,* is dispensible for vegetative growth and sporulation, *J. Mol. Biol.,* 195, 299, 1987.
21. **Iismaa, T. P., Carrigan, C. M., and Wake, R. G.,** Relocation of the replication terminus, *terC,* of *Bacillus subtilis* to a new chromosomal site, submitted, *Gene,* 67, 183, 1988.
22. **Masters, M. and Broda, P.,** Evidence for the bidirectional replication of the *Escherichia coli* chromosome, *Nature (London),* 232, 137, 1971.
23. **Bird, R. E., Louarn, J., Martuscelli, J., and Caro, L.,** Origin and sequence of chromosome replication in *Escherichia coli, J. Mol. Biol.,* 70, 549, 1972.
24. **Kuempel, P. L., Duerr, S. A., and Seeley, N. R.,** Terminus region of the chromosome in *Escherichia coli* inhibits replication forks, *Proc. Natl. Acad. Sci. U.S.A.,* 74, 3927, 1977.
25. **Louarn, J., Patte, J., and Louarn, J. M.,** Evidence for a fixed termination site of chromosome replication in *Escherichia coli* K12, *J. Mol. Biol.,* 115, 295, 1977.
26. **Kuempel, P. L. and Duerr, S. A.,** Chromosome replication in *Escherichia coli* is inhibited in the terminus region near the *rac* locus, *Cold Spring Harbor Symp. Quant. Biol.,* 43, 563, 1979.
27. **Louarn, J., Patte, J., and Louarn, J. M.,** Map position of the replication terminus on the *Escherichia coli* chromosome, *Mol. Gen. Genet.,* 172, 7, 1979.
28. **Bitner, R. M. and Kuempel, P. L.,** P1 transduction map spanning the replication terminus of *Escherichia coli* K12, *Mol. Gen. Genet.,* 184, 208, 1981.

29. **Bitner, R. M. and Kuempel, P. L.,** P1 transduction mapping of the *trg* locus in *rac⁺* and *rac* strains of *Escherichia coli* K-12, *J. Bacteriol.,* 149, 529, 1982.

30. **Fouts, K. E. and Barbour, S. D.,** Insertion of transposons through the major cotransduction gap in *Escherichia coli* K-12, *J. Bacteriol.,* 149, 106, 1982.

31. **Bouché, J. P.,** Physical map of a 470 × 10³ base-pair region flanking the terminus of DNA replication in the *Escherichia coli* K12 genome, *J. Mol. Biol.,* 154, 1, 1982.

32. **Bouché, J. P., Gélugne, J. P., Louarn, J., and Louarn, J. M.,** Relationships between the physical and genetic maps of a 470 × 10³ base-pair region around the terminus of *Escherichia coli* K12 DNA replication, *J. Mol. Biol.,* 154, 21, 1982.

33. **Bejar, S. and Bouché, J. P.,** Molecular cloning of the region of the terminus of *Escherichia coli* K-12 DNA replication, *J. Bacteriol.,* 153, 604, 1983.

34. **Asada, K., Nakatani, S., and Takanami, M.,** Cloning of the contiguous 165-kilobase-pair region around the terminus of *Escherichia coli* K-12 DNA replication, *J. Bacteriol.,* 163, 398, 1985.

35. **Henson, J. M. and Kuempel, P. L.,** Deletion of the terminus region (340 kilobase pairs of DNA) from the chromosome of *Escherichia coli*, *Proc. Natl. Acad. Sci. U.S.A.,* 82, 3766, 1985.

36. **De Massy, B., Bejar, S., Louarn, J., Louarn, J. M., and Bouché, J. P.,** Inhibition of replication forks exiting the terminus region of the *Escherichia coli* chromosome occurs at two loci separated by 5 min, *Proc. Natl. Acad. Sci. U.S.A.,* 84, 1759, 1987.

37. **Hill, T. M., Henson, J. M., and Kuempel, P. L.,** The terminus region of the *Escherichia coli* chromosome contains two separate loci that exhibit polar inhibition of replication, *Proc. Natl. Acad. Sci. U.S.A.* 84 1754, 1987.

38. **Hill, T. M., Kopp, B. J., and Kuempel, P. L.,** Termination of DNA replication in *Escherichia coli* requires a *trans*-acting factor, *J. Bacteriol.,* 170, 662, 1988.

39. **Pelletier, A. J., Hill, T. M., and Kuempel, P. L.,** Location of the sites that inhibit progression of replication forks in the terminus region of *Escherichia coli*, *J. Bacteriol.,* 170, 4293, 1988.

40. **Hill, T. M., Tecklenburg, M., Pelletier, A. J., and Kuempel, P. L.,** *tus*, the *trans*-acting factor required for termination of DNA replication in *Escherichia coli*, is a DNA-binding protein, *Proc. Natl. Acad. Sci. U.S.A.,* 86, 1989.

41. **Crosa, J. H.,** Three origins of replication are active in vivo in R plasmid RSF1040, *J. Biol. Chem.,* 255, 11075, 1980.

42. **Inuzuka, N., Inuzuka, M., and Helinski, D.,** Activity in vitro of three replication origins of the antibiotic resistance plasmid RSF1040, *J. Biol. Chem.,* 255, 11071, 1980.

43. **Lovett, M. A., Sparks, R. B., and Helinski, D. R.,** Bidirectional replication of plasmid R6K DNA in *Escherichia coli*: correspondence between origin of replication and position of single-strand break in relaxed complex, *Proc. Natl. Acad. Sci. U.S.A.,* 72, 2905, 1975.

44. **Bastia, D., Germino, J., Crosa, J., and Hale, P.,** Molecular cloning of the replication terminus of the plasmid R6K, *Gene,* 14, 81, 1981.

45. **Kolter, R. and Helinski, D. R.,** Activity of the replication terminus of plasmid R6K in hybrid replicons in *Escherichia coli*, *J. Mol. Biol.,* 124, 425, 1978.

46. **Bastia, D., Germino, J., Crosa, J., and Ram, J.,** The nucleotide sequence surrounding the replication terminus of R6K, *Proc. Natl. Acad. Sci. U.S.A.,* 78, 2095, 1981.

47. **Germino, J. and Bastia, D.,** Termination of DNA replication in vitro at a sequence-specific replication terminus, *Cell,* 23, 681, 1981.

48. **Horiuchi, T., Hidaka, M., Akiyama, M., Nishitani, H., and Sekiguchi, M.,** Replication intermediate of a hybrid plasmid carrying the replication terminus *(ter)* site of R6K as revealed by agarose gel electrophoresis, *Mol. Gen. Genet.,* 210, 394, 1987.

49. **Pelletier, A. J., Hill, T. M., and Kuempel, P. L.,** Termination sites T1 and T2 from the *Escherichia coli* chromosome inhibit DNA replication in ColE1-derived plasmids, *J. Bacteriol.,* 171, 1739, 1989.

50. **Hill, T. M., Pelletier, A. J., Tecklenburg, M. L., and Kuempel, P. L.,** Identification of the DNA sequence from the *E. coli* terminus region that halts replication forks, *Cell,* 55, 459, 1988.

51. **Turnbough, C. L., Jr., Kerr, K. H., Funderburg, W. R., Donahue, J. P., and Powell, F. E.,** Nucleotide sequence and characterization of the *pryF* operon of *Escherichia coli* K12, *J. Biol. Chem.,* 262, 10239, 1987.

52. **Piggot, P. J. and Hoch, J. A.,** Revised genetic linkage map of *Bacillus subtilis*, *Microbiol. Rev.,* 49, 158, 1985.

53. **Bachmann, B. J.,** Linkage map of *Escherichia coli* K-12, Edition 7, *Microbiol. Rev.,* 47, 180, 1983.

54. **Smith, M. T. and Wake, R. G.,** DNA sequence requirements for replication fork arrest at *terC* in *Bacillus subtilis*, *J. Bacteriol.,* 170, 4083, 1988.

55. **Lim, D. F., Hill, T. M., and Kuempel, P. L.,** unpublished experiments.

56. **Francois, V., Louarn, J., Rebollo, J.-E. and Louarn, J.-M.,** Replication termination, non-divisible zones and structure of the *Escherichia coli* chromosome, in *The Bacterial Chromosome*, Drlica, K. and Riley, M., Eds., American Society for Microbiology, Washington, D. C., in press.

57. **Pelletier, A. J.,** unpublished experiments.

Section II. Viral Chromosomes

INTRODUCTION

The variety of ways in which a DNA or RNA genome can be packaged in a protective protein coat is strikingly revealed by the viruses of animals, plants, and bacteria. As the chapters in this section demonstrate, the modes of interaction of viral genomes and proteins show as many differences as the organisms that are infected. Their simplicity of structure makes viruses ideal model systems for studies of nucleoprotein assembly and nucleic acid replication and transcription. Viral genomes are chromosomes since they encode the genetic information required to perpetuate the virus particles. Some animal viruses, such as polyoma and SV40, contain true minichromosomes since they consist of double-helical DNA complexed with histones to form nucleosomes. With these viruses, interactions of the DNA with the capsid proteins are minimal, compared to the RNA plant viruses. Bacterial viruses such as λ, T4, T7, P1, and P22 contain large amounts of highly condensed DNA that is largely free of bound protein. Questions of interest for these viruses concern the mechanisms of DNA packaging during virus assembly and DNA ejection during the process of infection.

It is not only the simplicity of virus structure that has attracted researchers to viruses as model systems. Replication and transcription can be profitably investigated since these processes are less complicated than for eukaryotes and prokaryotes. This is due to the limited genome size of viruses, which requires that the genetic material be used efficiently. And it is due to the parasitic nature of viral infection which utilizes the host cell systems of transcription and replication. Because of these factors, only a handful of genes (six for polyoma virus) are needed for the smallest viruses to produce a successful infection with progeny virions. Viruses with chromosomes which code for many more proteins (165 genes for T4) are also widely studied. Understanding their replication and assembly sheds light on how the component elements of subcellular processes interact. During assembly of bacteriophage T4 particles, for example, three different pathways, representing the DNA-containing head, tail, and tail fibers, interact to produce the completed particles.

In addition to these attractions of viruses as model systems, the field is important because of the relevance of virus research to human health. Many diseases are caused by infection of human cells with viruses, including some cancers. Information of health relevance is also derived from studies of viruses such as polyoma and SV40, which produce tumors in animals.

A fundamental insight of virus research is that all viruses, no matter what hosts are infected, use similar processes in their replication and follow similar rules in their assembly. The first evidence for the universality of the principles governing virus assembly was derived from X-ray crystallography and electron microscopy. The protein shells of simple spherical and rod-shaped plant viruses were discovered to be built of highly symmetric arrangements of protein subunits. The protein capsids of "spherical" viruses such as SBNV (southern bean mosaic virus) and CCMV (cowpea chlorotic mottle virus) were found to possess icosahedral symmetry with precisely 180 subunits. Other plant viruses are built of a helical arrangement of protein subunits. The organization of the viral genomes inside the completed virions is largely determined by the capsid structure for those RNA and DNA genomes which closely interact with protein subunits. Investigations of animal and bacterial viruses have also revealed that the same rules of construction apply, a remarkable revelation.

Chapter 5

POLYOMA AND SV40 CHROMOSOMES

Marie-Hélène Kryszke and Moshe Yaniv

TABLE OF CONTENTS

I. INTRODUCTION

Polyoma and SV40 are DNA tumor viruses that grow in mouse cells and simian cells, respectively. They can transform nonpermissive cells *in vitro* and induce tumors *in vivo* in a variety of rodents.

The viruses are easy to grow and to purify, and the viral genomes are easy to manipulate. These small DNA molecules (5297 base pairs [bp] for polyoma, 5243 bp for SV40) were the first eukaryotic genomes to be entirely sequenced. Their limited coding capacity makes their expression completely dependent on cellular functions. Thus, polyoma and SV40 are very attractive tools to investigate the mode of regulation of gene expression in mammalian cells. For example, transcriptional enhancers were found in these viruses, affording great advances in the study of these crucial eukaryotic regulatory sequences.

The viral DNA is a circular, double-stranded molecule associated with cellular histones into a minichromosome containing 24 nucleosomes, with properties comparable to those of the nuclear chromatin. The topological constraints imposed by this structure and the varying distribution of the nucleosomes along the minichromosome play a crucial role in the process of viral expression.

Another interesting particularity of SV40 and polyoma is the way they have adapted their functional organization and their strategies of expression to the small size of their genomes. Although their primary DNA sequences are quite different, both viruses exhibit a similar genetic organization. The early genes, expressed soon after infection, belong to a single transcription unit covering one half of the genome. The late genes correspond to the other half of the DNA molecule, and they are expressed after the onset of viral DNA replication. Between the two divergent transcription units lies the short noncoding region which comprises all the regulatory signals required for transcription and replication control. Nevertheless, in some respects both viruses have chosen distinct strategies to promote their gene expression.[1,2]

II. AN OVERVIEW OF THE FUNCTIONAL ORGANIZATION OF POLYOMA AND SV40 GENOMES

A. THE VIRAL EARLY CODING SEQUENCES

The three overlapping polyoma early genes are transcribed, from 72 to 25.8 map units, into a 22 S nuclear RNA which gives rise, after differential splicing, to three distinct 15 to 16 S mRNAs. The mRNA coding for large T antigen (T Ag) is spliced between nucleotides (nt) 410 and 794, the mRNA coding for middle T antigen (mT Ag) between nt 747 and 808, and the mRNA coding for small t antigen (t Ag) between nt 747 and 794 (Figure 1). Thus, one messenger encodes one distinct protein. Two alternative donor splice sites and two alternative acceptor splice sites are present in the primary transcript, three of the four possible combinations being used by the splicing machinery. The N-terminal moieties of the three early proteins are encoded in the same reading frame, starting at a unique AUG codon. Their unrelated C-terminal domains result from the use of the three reading frames, a fact rarely observed in eukaryotes. Such transcription and processing patterns demonstrate how a short DNA sequence can provide sufficient information for the synthesis of several proteins.[3]

A comparable strategy is employed by SV40 for the synthesis of its two early proteins (Figure 2). One nuclear transcript is alternatively spliced into two distinct mature mRNAs by joining each of the two donor splice sites present (at nt 4917 or 4640) to a unique acceptor splice site (at nt 4572). One of the mRNAs encodes the large T Ag and the other encodes the small t Ag; no protein equivalent to polyoma mT Ag is encoded by SV40. The same reading frame is also used to encode the N termini of T and t Ags. Only large T Ag sequences are encoded in the second reading frame beyond the splice junction.[1]

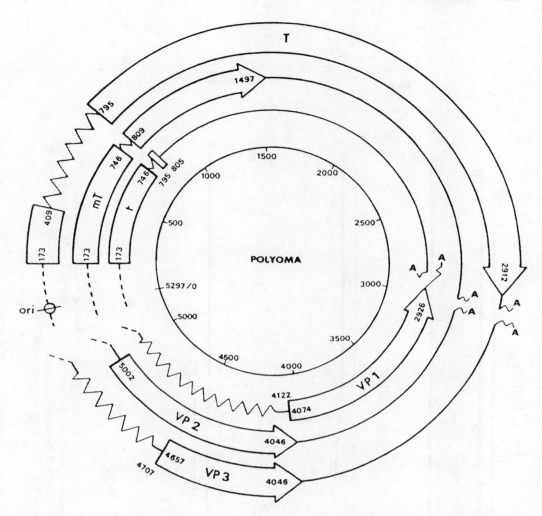

FIGURE 1.   Organization of polyoma virus genome. The viral DNA is schematized by a circle, with nucleotide positions according to the numbering system of Soeda et al.[140] Its transcription pattern is described. 3' untranslated RNA sequences are represented by continuous lines, 5' untranslated leader sequences by dashed lines, spliced sequences by broken lines, and the coding sequences by open arrows. Poly A sequences are schematized at the mRNA 3' ends (A). The origin of DNA replication is marked by a barred circle (ori).

T and t Ags are structurally and functionally similar in polyoma and SV40. Both T Ags are phosphorylated nuclear proteins with ATPase, nucleotide-binding, and DNA-binding activities. SV40 T Ag contains a very basic amino acid sequence responsible for the nuclear targeting of the protein. Polyoma T Ag possesses two such signals, one of which is very homologous to the SV40 signal. Polyoma and SV40 T Ags control viral DNA replication and are capable of immortalizing primary cells. SV40 T Ag interacts with the cellular protein p53 and can transform cells, whereas in the case of polyoma, the transforming function is fulfilled by mT Ag, a membrane-associated protein which forms a complex with the cellular protein pp60[c-src]. Thus, the same roles in lytic infection and in cell transformation are played by the viral early proteins of polyoma and SV40, but with a different distribution, between two polypeptides in the case of SV40 or three polypeptides in the case of polyoma virus.[2]

An interesting feature is that in the SV40 genome, a silent open reading frame (ORF) is present downstream of T Ag coding sequences. A shift from T Ag ORF to this new ORF

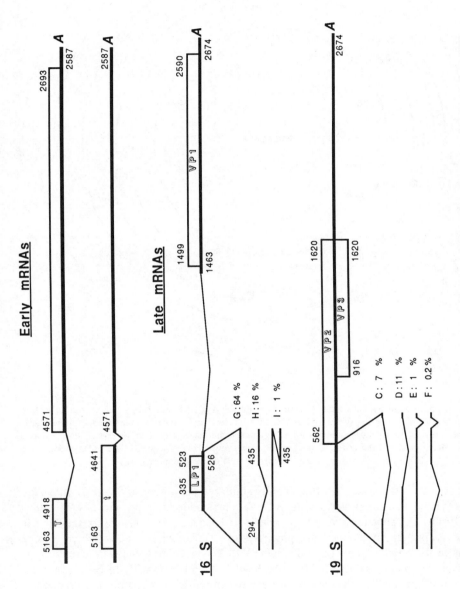

FIGURE 2.   Structure of the SV40 early and late mRNAs. Bold lines correspond to the sequences present in the mature mRNAs, with coding sequences represented by open boxes. Their limits are shown according to Tooze's nucleotide numbering system.[1] Spliced sequences are indicated by broken lines. Poly A sequences are schematized at the mRNA 3′ ends (A). Relative proportions of the different late mRNA species are given with a detailed representation of their 5′ regions.

would give rise to a fusion protein with a C terminus containing 60% hydrophobic amino acids adjacent to six acidic residues present in the T Ag normal sequence. Such a situation is reminiscent of the Py mT Ag structure which comprises a cluster of six glutamic acid residues followed by a hydrophobic sequence responsible for membrane anchorage of the protein. However, no translation product of this additional ORF could be naturally detected in cells infected by SV40. Furthermore, Lewis et al.[4] deleted nt 2899 of SV40 to induce this frameshift, but the mutant T Ag synthesized was deficient in both replication and transformation. It cannot be completely ruled out, however, that this additional ORF could have some biological significance since it is also found in the genome of the related human papovavirus BKV.

Another ORF is present in the SV40 early transcripts synthesized during the late phase of the lytic cycle. These RNAs initiate more upstream than those in the early phase of infection. They consequently include one or two additional AUG codons and an ORF corresponding to a sequence of 23 amino acids. The 2.7-kilodalton (kDa) product of this ORF has been detected in SV40-infected cells. Its counterpart is also produced in human cells infected by the papovaviruses BKV or JVC. It shows remarkable conservation of an arginine- and proline-rich domain. The use of the more upstream AUG codons decreases that of the AUG for T and t Ags, and could account, in part, for the reduced synthesis of these two proteins later in infection.[5] Although polyoma early transcripts also undergo a shift in the choice of the start sites after the onset of DNA replication, no ORF is present in their leader sequences.

B. THE VIRAL LATE CODING SEQUENCES

The polyoma late transcription unit extends from 66 to 25.3 map units. Three mRNAS —16 S, 18 S and 19 S — are produced by differential splicing of the late primary transcripts. They encode the viral late proteins VP1, VP3, and VP2, respectively (Figure 1). The polyoma late mRNAs are characterized by the existence, within their leader region, of a tandemly repeated 57-nt sequence which is represented only once in the viral genome (nt 5020 to 5076).[6] This tandem repeat in the mature RNAs arises from multiple splicing of giant transcripts that cover several genomic lengths and appear late in the lytic cycle due to inefficient transcription termination.[7] No further splicing is required to generate the 19 S mRNA, but the production of the 18 S and 16 S mRNAs requires the excision of an intron (nt 5019 to 4708 or 5019 to 4123, respectively). Three different AUG codons direct the synthesis of VP1, VP2, and VP3 in the three corresponding ORFs (4074 to 2926, 5002 to 4046, and 4657 to 4046, respectively).[8]

Since the 3' ends of early and late mRNAs overlap by 50 nt in polyoma virus (88 nt in the case of SV40), it has been suggested that viral transcription could be regulated at this level, the abundance of late mRNA synthesis after DNA replication hindering early transcription. Sequences important for early and late transcription, including polyadenylation signals, are located in the region corresponding to the mRNA 3' ends.[2]

The SV40 late mRNAs have no repeated leader sequences, but result from a complex series of splicing events. They fall into two classes, 19 S and 16 S (Figure 2). The longer precursor RNAs include unspliced transcripts, as well as RNAs that have been spliced between nt 294 and 435, upstream of the coding sequences. The 19 S mRNAs are spliced from nt 294, 373, or 526 to an acceptor at nt 558 and encode the minor capsid proteins VP2 and VP3. The 16 S mRNAs are spliced bewteen nt 526 and 1463. They encode the major capsid protein VP1 as well as a short polypeptide, called LP1 or agnoprotein, which is specific for SV40. The biological role of LP1 has not yet been assigned.[9] Thus, each SV40 late mRNA is bifunctional. Recent studies have clarified the manner in which these particular messengers are translated, taking into account the fact that translation initiation at internal AUGs is uncommon in eukaryotes.

Equivalent amounts of VP1 and LP1 are normally produced from the 16 S mRNAs. A sequence that makes ribosomes pause longer on the mRNA was inserted upstream of the initiator codon for LP1. By enhancing the efficiency of recognition of this AUG, this sequence increased the synthesis of agnoprotein, while decreasing the synthesis of VP1. In the same construct, mutating the LP1 AUG to UUG restored quantitative synthesis of VP1. These results suggest that regulation of translation of the 16 S mRNAs involves a leaky scanning mechanism: ribosomes scan the RNA and start translation at the first initiator encountered, with an efficiency depending on the relative strength of this signal. When the first AUG is not used, ribosomes scan farther, allowing translation of a downstream ORF.[10] Besides, since LP1 and VP1 ORFs do not overlap, part of VP1 synthesis can be explained by the occurrence of reinitiation events after the end of translation of LP1 on the same RNA molecule.

In the case of the 19 S mRNAs, the efficiency of translation of VP3 is 2.5-fold higher than that of VP2, in agreement with the relative strengths of their respective initiator signals. This correlates with the observation that deletion of the VP2 AUG increases the synthesis of VP3 by 50%.[11] In the 19 S mRNA D species (Figure 2), the AUG for LP1 is also present. Its utilization reduces the VP2/VP3 ratio, compared to that obtained with the other species of 19 S RNA because the LP1 ORF ends 48 nt downstream of the VP2 initiator sequence. VP2 translation is thus inhibited by LP1 translation since no ribosome scanning can take place within sequences in the process of translation. All the 19 S mRNA species also contain the VP1 ORF, but do not allow its translation because very few ribosomes are capable of scanning the RNA all the way to the VP1 initiator.[12]

III. THE NONCODING REGIONS OF POLYOMA AND SV40

A. EARLY PROMOTERS

Transcription start sites in the polyoma genome have been mapped precisely using indirect methods (S1 nuclease mapping, primer extension) and by direct analysis of the 5′ ends of capped mRNAs. They are highly heterogeneous (Figure 3). Major early start sites are localized between nt 145 and 156, 23 to 36 bp downstream of a TATA sequence (nt 120 to 127). Minor start sites are found within the early coding region (nt 298 to 302, TATA box between nt 270 and 276), as well as upstream of the major sites (nt 16 to 20, TATA box between 5281 and 5286). In all cases, a CCAAT-like sequence is present 70 to 96 bp upstream of the transcription start sites.[13,14]

The sequences required for efficient transcription *in vivo* and *in vitro* have been identified by deletion studies. The CCAAT box, the TATA box, and the region corresponding to the major early start sites play a role *in vivo* only in the absence of the sequences located further upstream. However, *in vitro*, they are the only sequences required. Deletion of the TATA box decreases transcription efficiency *in vitro* without altering the choice of the start sites. The sequences that are crucial for early transcription *in vivo* are located between nt 5022 and 5267. In addition, sequences located between nt 4362 and 4841 also seem to influence transcription efficiency *in vivo*.[15-18]

SV40 early transcription start sites fall into three groups: the early-early start sites (EES; nt 5230 to 5233 and nt 5235 to 5237) utilized during the early phase of the viral cycle, the late-early start sites (LES; nt 21 to 22, nt 30 to 31, and nt 34 to 35) utilized only after the onset of DNA replication, and downstream start sites (DSB) located between nt 5220 and 5225 (Figure 3). The early-early promoter includes an AT-rich region with a TATA-like sequence (5′ TATTTAT 3′, nt 15 to 21). This sequence drives the transcription machinery to initiate RNA synthesis at the EES. Mutation of the TATTTAT sequence to TATGCAT greatly reduces transcription from the EES and affects the general early transcription rate. Both the GC-rich 21-bp repeats upstream of the TATA box and the 72-bp repeats are required

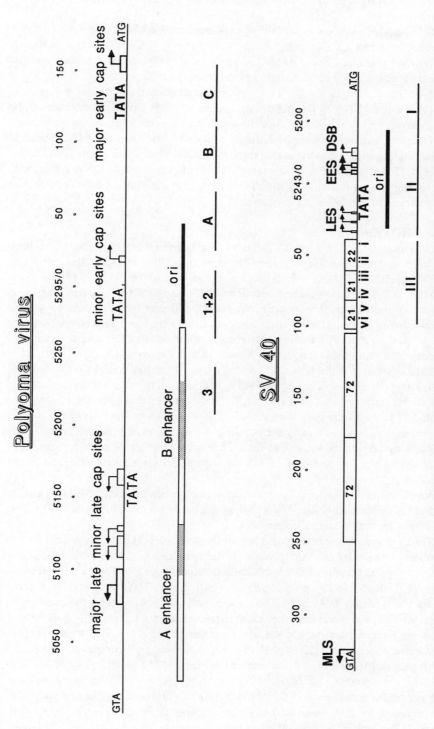

FIGURE 3.   Functional map of polyoma and SV40 noncoding regions. Indicated on the linear viral DNA sequences are the start of VP1 and T Ag coding sequences (ATG), the TATA sequences present in early and late promoters, and the early and late transcription initiation sites (major sites indicated by bold boxes and arrows). The origins of DNA replication are represented by bold lines (ori) and the T Ag binding sites are shown (A, B, C, 1 + 2, and 3 in the case of polyoma; I, II, and III in the case of SV40). Polyoma enhancers are schematized by open boxes, with the core sequences shadowed. Represented on the SV40 sequence are the early-early, late-early, downstream, and major late transcription start sites (EES, LES, DSB, and MLS, respectively); the 21-bp repeats (open boxes marked 21,21,22) with their six GC-motifs (i to vi); the 72-bp repeats (open boxes marked 72).

for efficient early transcription. No equivalent of the GC-rich region is present in the polyoma early promoter, suggesting a partially different mechanism of early transcription regulation for the two viruses.

The SV40 late-early promoter is tenfold weaker than the early-early promoter. It has no genuine TATA box, but AT elements interspersed within the GC-rich 21-bp repeats can serve as TATA-box substitutes. Some of the GC motifs in this region are important and the 72-bp repeats are required for efficient transcription from the LES.

In the late phase of the lytic cycle, SV40 and polyoma early transcription is repressed by T Ag, which binds to DNA in the noncoding regions of both viruses. Activation of the SV40 late-early promoter, or utilization of upstream start sites in the case of polyoma, could be explained either by a change in template structure following the onset of DNA replication or by increased binding of T Ag at the major early start sites. Repression of the SV40 early-early promoter by T Ag, together with the template amplification due to DNA replication, may account for the predominance of late-early transcripts at this time of infection, despite the relative weakness of the late-early promoter.[19]

B. LATE PROMOTERS

In the polyoma genome, at least 15 different late transcription start sites have been localized between nt 5075 and 5168 (Figure 3). More than 90% of these sites are found between nt 5075 and 5099. Minor sites appear at positions 5107 to 5122, 5125 to 5129, and 5157 to 5168. A TATA box is present between nt 5150 and 5158.[20] Transient expression studies have shown that the only sequences required for efficient late transcription *in vivo* are located downstream of the initiation sites, a quite unusual situation for a RNA polymerase II promoter. A basal element is contained between nt 5022 and nt 5130 and a stimulatory element is comprised within the sequence 4900 to 5022.[21] *In vitro*, a unique start site is used (around 5127) since it is the only one correctly spaced relative to the TATA box.[22,23]

The SV40 late promoter is devoid of a TATA box and is characterized by an extreme heterogeneity of transcription start sites (Figure 3). These sites extend over 300 nt, between positions 28 and 325, the major one being nt 325. The utilization of the different start sites is dependent on sequences immediately surrounding each of them and the use of the major site is influenced by sequences located at least 25 bp downstream of nt 325. Sequences located farther than 19 bp upstream of nt 325 have no effect on the determination of the start sites.[24]

Two domains included in each 72-bp repeat influence the late transcription rate. Domain I (nt 232 to 265) is implicated before DNA replication, whereas domain II (nt 179 to 204) is important during the late phase of infection.[25]

The SV40 late promoter is rather inefficient in the absence of T Ag and is activated in a direct or indirect fashion by the viral protein. It has been shown that late transcription is stimulated by T Ag independently of DNA replication. Whether the SV40 origin of replication is active or mutated, the SV40 late promoter is more efficient in COS cells— which express T Ag constitutively— than in CV1 cells. The same 5′ ends are generated for the transcripts, but in the presence of T Ag minor start sites close to the origin are preferentially utilized.[26,27]

Two elements were defined within the late promoter which are active in the presence of T Ag. One element represents 25 to 35% of the total activity and comprises element III (nt 200 to 270) plus the origin of DNA replication (element I). The second element (element A, nt 168 to 200) is responsible for 65 to 75% of the promoter activity and functions in an orientation-dependent manner, even in the absence of origin and of T Ag binding sites. The A element is required for T-Ag-activated late transcription. A third sequence, element II, included in the 21-bp repeats, has an effect in the absence of elements III and A. Element I + III requires T Ag specific interaction with its origin-binding sites (see below) to be functional whereas element A seems to display a more general mechanism of action, perhaps involved in the indirect pathway of activation by T Ag.[28] Besides, recent studies demonstrate that T Ag suppresses the negative control of late transcription exerted, before DNA replication, by the 21-bp repeats.[29]

Some data suggest that the polyoma late promoter is also activated by T Ag during the late phase of the lytic cycle. Activation is more efficient on episomal polyoma genomes than on the integrated genomes found in transformed cells. The polyoma late promoter, when cloned upstream of the bacterial CAT gene, is functional in both situations. Thus, one possible reason for its lack of activity in transformed cells could be this lack of stimulation by T Ag. But another model has been proposed to account for the absence of synthesis of late proteins in transformed cells. It implies post-transcriptional control rather than regulation at the level of transcription initiation. According to this model, late mRNA stability would be greatly reduced in the absence of the repeated leader, which cannot be synthesized from an integrated genome, since readthrough producing giant transcripts is necessary. In the same way, during the early phase of lytic infection, the high level of early transcription would prevent late transcriptional readthrough, and the late mRNA produced would be very unstable. Later in the viral cycle, repression of early mRNA synthesis would allow abundant late transcriptional readthrough and formation of late mRNAs with repeated leader sequences. Together with template amplification caused by DNA replication, this would account for the shift from abundant early transcription to predominant late transcription after the onset of DNA replication.[30] Nevertheless, recent data suggest that the activation of transcription by polyoma T Ag can occur in a rather unspecific fashion, independently of the presence of particular T Ag binding sites. Furthermore, leader repeats have been found in the 5′ region of the mRNAs produced from the polyoma late promoter linked to the CAT gene in transient expression assays, without any correlation between the number of repeats present and the expression rate observed.[21]

C. ORIGINS OF DNA REPLICATION

The viral sequences required in *cis* for DNA replication represent the minimal origin of replication. The polyoma origin is located at 0.72 map units in the noncoding region of the viral genome. It was defined by deletion analysis. The origin core extends from nt 5270 to nt 42. It includes eight consecutive A:T bp, a GC-rich palindrome, and at least part of the following 18-bp palindrome.[31] Point mutations have helped identify essential nucleotides within the 34-bp GC-rich palindrome (Figure 4).[32] Outside the core, auxiliary elements behave as replication activators. They are called α (nt 5097 to 5126) and β (nt 5172 to 5202). They can substitute for each other. The β + core configuration is functional, with the distance between the two elements varying up to 200 bp. The α + core structure is also active, provided the α element is maintained in its native orientation relative to the core.[33,34] According to Veldman et al.,[35] four redundant sequences activate polyoma replication: A (5108 to 5130), B (5179 to 5214), C (5148 to 5179), and D (5022 to 5198). None of them is indispensable, but two elements together are sufficient to activate replication: A + D, A + C, or B + C, as well as two copies of the A element in the absence of any other activator sequence.

The transition point between discontinuous and continuous DNA synthesis on each strand defines the genuine origin of bidirectional replication (OBR). This point was assigned to nt 20 on the early strand and to nt 37 on the late strand. Thus, in the polyoma genome, the origin of replication contains a 16-nt region in which no initiation of DNA synthesis occurs. It seems that the sequences of the core as well as sequences located upstream on the late side determine the position of the OBR on the early strand, which in turn conditions the position of the OBR on the opposite strand.[36]

The SV40 origin is also located in the viral noncoding region (at 0.67 map units). The minimal sequences required in *cis* for DNA replication are contained within a 65-bp segment

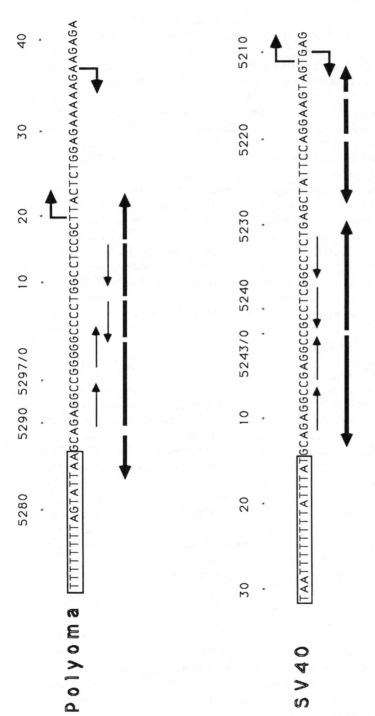

FIGURE 4.   Sequence of polyoma and SV40 origins of replication. On each nucleotide sequence (early strand only) are indicated the AT-rich cluster (boxed) and the origin of bidirectional replication (convergent arrows on both sides of the sequence). Below each sequence are represented the four pentanucleotide motifs recognized by T Ag (thin arrows) and the palindromes (bold arrows).

(nt 5208 to 30), including a perfect 27-bp palindrome flanked by a continuous stretch of 17 A:T bp (encompassing the early-early TATA box) and an imperfect 15-bp inverted repeat. The 17-bp AT block could act on replication by facilitating initiation through bending of the template. The 21-bp repeat region (see Section III.B) has a stimulatory effect on DNA replication *in vivo*. Increasing the distance between the 21-bp repeats and the origin core impairs DNA replication. The positive influence of the 72-bp repeats (see Section III.B) can be demonstrated *in vivo* in the absence of the 21-bp repeats.[37,38]

In SV40, the OBR coincides on both DNA strands (nt 5210/5211) so that initiations on the early strand occur within the origin core and initiations on the late strand take place on the early side of the core.[39]

D. T ANTIGEN BINDING SITES

As already mentioned, T Ag regulates SV40 and polyoma transcription and replication through binding to the viral DNA in its noncoding region. The availability of purified T Ag allowed the precise identification of its binding sites on polyoma and SV40 genomes, using DNase I footprinting *in vitro*.

In the case of polyoma, three high-affinity binding sites are found outside the origin of replication (Figure 3). Site A (nt 25 to 75) overlaps the early boundary of the origin core. Site B (nt 85 to 120) is located just upstream of the early TATA box. Site C (nt 124 to 163) includes the TATA box and the major early transcription start sites. Three lower-affinity binding sites, protected against DNase I digestion at higher protein concentrations or at lower ionic strength, are found within and on the late side of the origin. Sites 1 and 2 constitute a single interaction region in the origin core (nt 5260 to 10); site 3 is on the late side of the core (nt 5207 to 5239). An identical motif (5' GAGGC 3' or 5' GGGGC 3') appears several times within each T Ag binding domain. The affinity of each site for the protein depends on the number and spacing of the pentanucleotide motifs. The optimal configuration consists of three motifs, each separated by one helical turn.[40] Interference experiments using dimethylsulfate (DMS) have shown which G residues in the DNA make contact with bound T Ag. The results obtained were further confirmed by the study of point mutants. In sites A, B, C, or 3, every G residue contained in the GAGGC or GGGGC pentanucleotides on both DNA strands is involved in DNA-protein interaction. Site B includes two pentanucleotides and sites A and C contain three pentanucleotides each, the groups of essential guanines being aligned on the same side of the DNA helix. Such an organization suggests that multiple interactions involving one T Ag promoter each can be stabilized by protein-protein contacts within a binding site. However, no cooperativity is observed between distinct sites. In the palindromic site 1 + 2, two pentanucleotides are found in tandem on each strand, separated by only 2 bp. These tandem repeats overlap at the point of symmetry. In this configuration, the groups of G residues are spread all around the DNA helix; they must induce different protein-protein contacts. It can be suggested that in this particular case, only two T Ag molecules bind to the outer pentanucleotides and that contacts with the central sequences involve a domain of T Ag distinct from the one responsible for pentanucleotide recognition. Site 1 + 2 is not bound by a subpopulation of T Ag different from the former one because competition for T Ag binding can be observed between site 1 + 2 and the other sites.[41]

SV40 noncoding region includes three high-affinity T Ag binding sites (Figure 3). Site 1 (nt 5175 to 5211), located on the early side of the origin of replication, is mainly involved in the repression of early transcription. Site II (nt 5212 to 36) overlaps the 27-bp palindrome of the origin and plays a crucial role in DNA replication. Site III (nt 41 to 99) is outside the origin. The basic consensus DNA sequence recognized by SV40 T Ag is also 5' (G/T)(A/G)GGC 3'. Site I contains three pentanucleotides arranged in tandem on the late strand, plus one pentanucleotide on the opposite strand overlapping two of the three tandem repeats.

Site III has six tandemly repeated pentanucleotides on the early strand. Site II includes four motifs spaced by only one nucleotide and positioned as two pairs in an inverted repeat (Figure 4). In every case, all the G residues included in the pentanucleotides are protected against methylation by the binding of T Ag.[42,43] Site II has been more precisely studied. Directed mutagenesis experiments demonstrated that every nucleotide in the pentanucleotides is essential, while adjacent sequences have no effect. The four pentanucleotides together are required for efficient T Ag binding and DNA replication. Their spacing is crucial: deletion of one base pair between pentanucleotides is more deleterious than its substitution. Inversion of one repeat inhibits DNA replication. Thus, each pentanucleotide cannot be considered as an independent functional unit. T Ag binds in a stepwise fashion to the core DNA and the arrangement of all the T Ag monomers bound must fulfill correct conditions of distance, rotation, and orientation to each other to make the complete site functional.[44] No cooperativity is observed between binding at sites I and II. In the absence of ATP, T Ag binds only to site I. In the presence of ATP— or of a nonhydrolyzable analog of ATP— binding to site II is obtained and the whole binding affinity (I + II) is increased.[45]

Polyoma sites A, B, and C are not required for DNA replication. Comparing their location to that of sites I, II, and III in SV40, we observe a concomitant displacement of the T Ag binding sites and of the early promoter relative to the origin of replication. Thus, these sites are most probably responsible for the control of early transcription. In particular, site C, which overlaps the TATA box and the major early start sites, could repress early transcription. However, it is not the only site involved in the repression of early transcription by T Ag since its deletion or its replacement by a heterologous viral promoter does not affect this regulation.[46] The location of site 1 + 2 suggests a role in replication initiation. At present, transactivation of the polyoma late promoter could not be correlated with the presence of T Ag binding sites in the DNA.[21]

E. ENHANCERS

The polyoma 246-bp *Bcl*I-*Pvu*II restriction fragment and the SV40 72-bp repeats, located in the viral noncoding regions on the late side of the origins of replication, behave as transcriptional enhancers: they are *cis*-acting regulatory sequences, capable of stimulating the transcription of a gene, even heterologous, when placed upstream or downstream of the transcription unit. Their effect (a 200-fold activation of transcription of the rabbit β-globin gene by the SV40 enhancer in HeLa cells) is still observed at a distance of several kilobases and is independent of the orientation of the enhancer relative to the direction of transcription.[47,48] The presence of a functional enhancer increases the amount of RNA polymerase molecules on the DNA template.[49,50] When two promoters are linked to an enhancer, the higher stimulation is exerted on the proximal promoter.[51] Enhancers are present in other viral genomes and can be found upstream of a great number of cellular genes.

A detailed analysis of polyoma and SV40 enhancers has shown that both of them are comprised of several redundant functional domains. The polyoma 246-bp fragment (nt 5022 to 5267) can be subdivided into two independent enhancers (Figure 5): enhancer A (*Bcl*I-*Pvu*II fragment, nt 5022 to 5130) and enhancer B (*Pvu*II-*Pvu*II fragment, nt 5131 to 5267). The A enhancer is the major enhancer in mouse fibroblasts: it is three times more active than the B enhancer. The minimal sequences required for enhancer function have been defined by 5' and 3' deletions: the core of enhancer A corresponds to nt 5096 to 5129 and the core of enhancer B to nt 5175 to 5229.[52] The polyoma enhancer shares sequence homologies with other viral or cellular enhancers. The A enhancer contains a sequence that is homologous to the adenovirus 5-E1A gene enhancer and to the immunoglobulin heavy chain gene (IgH) enhancer. In the A enhancer are also found two sequences homologous to the SV40 and to the c-*fos* cellular proto-oncogene enhancers, respectively. The B enhancer includes the enhancer core sequence of SV40 first described by Weiher et al.[53] (see below).

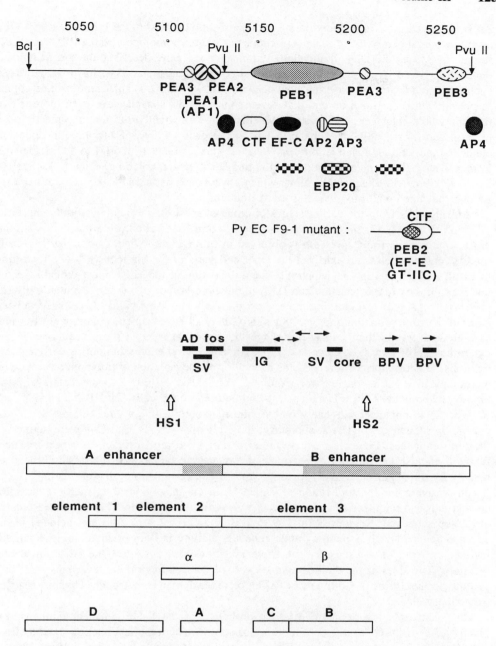

FIGURE 5.   DNA-protein interactions on the polyoma enhancer. The enhancer sequence is schematized, with vertical arrows for the *Bcl*I and *Pvu*II restriction sites. Interactions of factors PEA1, PEA2, PEA3, PEB1, and PEB3 on the wild-type enhancer were described in our laboratory.[90,94,114] Interactions of CTF, AP2, AP3, and AP4 were described by Tijan.[96] EBP20 was purified in McKnight's laboratory: its major binding site is shown, flanked by two minor binding sequences.[121] Binding of PEB2 or CTF is indicated on the PyEC F9-1 sequence.[122-124] Below are represented inverted and direct repeats present in the enhancer (horizontal arrows), homologies with other enhancers, DNase I hypersensitive sites HS1 and HS2 (vertical open arrows),[82] the A and B enhancers with their core sequences shadowed,[52] the three enhancer elements defined by Mueller et al.,[54] the α and β replication activators described by Hassell et al.,[34] and the enhancer domains A, B, C, and D defined by Veldman et al.[35]

An auxiliary element in the polyoma B enhancer is homologous to the IgH enhancer. In the origin-proximal part of the B enhancer, a tandem repeat is homologous to the BPV enhancer. In their study on viral DNA replication, Veldman et al.[35] have described the four redundant domains — A, B, C, and D — that build the polyoma enhancer. They have shown that a synthetic oligonucleotide covering domain A (nt 5109 to 5130) is sufficient in three copies to provide enhancer function and that seven copies of this sequence are even more active than the whole enhancer. According to Mueller et al.,[54] a slightly different organization can be proposed for the polyoma enhancer. This enhancer is subdivided into three independent elements, individually inactive, but functional as pairs: element 1 (nt 5057 to 5073), element 2 (nt 5074 to 5130), and element 3 (nt 5131 to 5229). Natural polyoma variants (Ts48, P16, TOR, NG59R, MV) bear duplications which always include nt 5114 to 5137, further supporting the functional importance of the A domain.[55]

Murine cells from the earlier stages of embryogenesis are refractive to polyoma virus infection. Viral expression follows the appearance of differentiated cell types.[56,57] A similar block to polyoma virus expression is observed in cultured embryonal carcinoma (EC) cells (undifferentiated cells derived from the inner cell mass of the blastocyst).[58] The inhibition takes place after adsorption, penetration, and uncoating of the virus, most probably at the levels of viral early transcription and DNA replication. Polyoma host-range mutants (Py EC mutants), capable of overcoming the expression block and growing in EC cells, have been isolated in different laboratories.[59-67] Strikingly, they all have sequence rearrangements and/ or mutations within their enhancers, suggesting a crucial role for these sequences in the determination of viral host range and in the control of viral expression during early embryogenesis. Other nonpermissive systems have allowed the isolation of polyoma mutants with alterations in their enhancers; for example, trophoblast cells (Py Tr mutants),[67] Friend erythroleukemia cells (Py FL mutants),[68] and neuroblastoma cells (Py NB mutants).[69]

The SV40 enhancer was analyzed by point mutations within a viral genome deleted of one 72-bp sequence. In Herr's laboratory, three regions — A, B, and C — were defined in the viral enhancer (Figure 6). Mutations in one of them are sufficient to block enhancer function and viral expression. Functional revertants can be isolated that restore enhancer activity by duplicating the intact region(s) left. A double mutation affecting regions A and B can be overcome by duplication of region C. In the intact enhancer, multiple elements coexist and can compensate for one another. Region A (nt 250 to 271) is located outside the 72-bp sequence. Regions B (nt 199 to 220) and C (nt 234 to 248) are included in the 72-bp sequence. Each domain exhibits enhancer activity in both orientations when multimerized in front of a test gene.[70] A triple mutation does not allow the isolation of any revertant. The sizes of the duplications are variable, depending on the revertants analyzed, perhaps because some flexibility in the definition of the domains is introduced by the existence of auxiliary elements.[71]

In Chambon's laboratory, a different definition of the SV40 enhancer domains was given.[72] We will refer to this definition in our review. The minimal enhancer extends from nt 186 to nt 279 and is subdivided into two domains, A and B (nt 186 to 225 and 226 to 279, respectively). Each domain is poorly active separately, but both domains or two copies of one of them display high enhancer activity, somewhat independent of their relative distance and orientation. A mutation in the A or in the B domain has the same deleterious effect, suggesting that each domain has its own function. The A + B association provides much greater activity than A or B alone, whereas further multimerization of A + B induces only a slow, linear increase in enhancer activity. Thus, distinct mechanisms seem to be involved in the two phenomena. A third domain called C (nt 298 to 347) can be detected in the absence of A and B. When duplicated, this domain also shows enhancer activity. Mutation of the entire enhancer, 3 nt after 3 nt, has allowed the definition of individual sequence motifs within each enhancer domain. Every mutation reduces enhancer activity no more

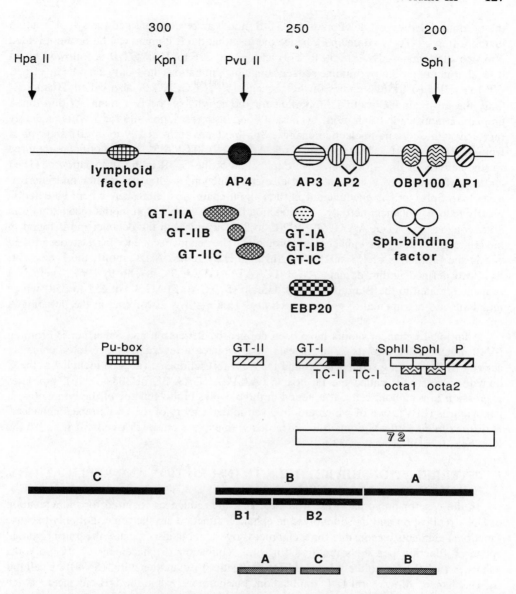

FIGURE 6.   DNA-protein interactions on the SV40 enhancer. The SV40 enhancer deleted of its origin-proximal 72-bp sequence is represented. Restriction sites for *Hpa*II, *Kpn*I, *Pvu*II, and *Sph*I are indicated by vertical arrows. The different factors interacting with the viral enhancer are schematized: AP1, AP2, AP3, and AP4 have been described by Tjian,[96] lymphoid factor by Pettersson and Schaffner,[74] OBP100 by Sturm et al.,[133] GT-II and GT-I binding proteins by Chambon,[125,129] Sph-binding factors by Herr,[96] and EBP20 by McKnight.[121] Below are indicated the functional enhancer motifs, the 72-bp sequence, the A, B, C (and B1, B2) domains defined in Chambon's laboratory (black boxes),[72] and the A, B, and C domains defined by Herr (shadowed boxes).[70.]

than eightfold, implying that multiple redundant elements are involved in total enhancer function.

Within the B domain, two sequence elements form a tandem repeat. The consensus sequence, called GT motif, is the following: 5' G(C/G)TGTGGAA(A/T)GT 3'. The more downstream motif is called GT-I (nt 239 to 250), the more upstream, GT-II (nt 262 to 273). The TGG sequence is of crucial importance in both motifs and mutation of other nucleotides are better tolerated in GT-I than in GT-II. Both GT motifs are necessary for the activity of the B domain. However, deletion of the GT-II motif can be compensated for by consecutive juxtaposition of GT-I to domain C, hence the idea that the B domain can be further divided into two subdomains, B1 (nt 259 to 279) and B2 (nt 226 to 258). GT-II is followed by an 8-bp alternating purine-pyrimidine sequence in which mutations have only a moderate effect. GT-I is followed by two copies of the sequence 5′ TCCCCAG 3′ also called TC motifs. Only the upstream TC motif is important for the activity of the B domain. Within the A domain, a motif called Sph motif because it contains a restriction site for *Sph*I is tandemly repeated with only one nucleotide change. Sph I extends from nt 199 to nt 207 and Sph II from nt 208 to nt 216. Their sequence is 5′ AAG(T/C)ATGCA 3′. Both motifs are required for the activity of the A domain. Like duplication of the GT-II motif, association of GT-II and Sph II constitutes a new enhancer element displaying high activity after multimerization.[73] An Sph motif is encountered in the κ-light chain gene enhancer, where its deletion greatly reduces enhancer activity. The junction between the two Sph motifs reconstitutes an 8-nt sequence (5′ ATGCAAAG 3′, nt 205 to 212) identical to the octamer motif found in the immunoglobulin gene enhancer and promoter. The octamer sequence has also been found within the promoters of U1 snRNA and histone H2B genes. A last motif, the P element, has been defined within domain A (5′ TCAATTAGTCA 3′, nt 186 to 196). There is a sequence related to the P motif in the B domain (5′ TCAGTTAG 3′, nt 251 to 258) where mutations are detrimental to enhancer activity. This motif is also found in the polyoma A enhancer.

Additional enhancer motifs have been defined by Pettersson and Schaffner.[74] From an SV40 mutant deleted of its 72-bp repeats, they isolated a revertant that restored enhancer activity by duplicating sequences around the *Nco*I restriction site (nt 333) included in the C domain. One motif, called the Pu box (5′ AAAGAGGAA 3′, nt 308 to 316), promotes expression in lymphoid cells. It is found in three copies in the genome of the lymphotropic papovavirus (LPV), two of which are involved in the activity of the 63-bp repeat enhancer. A distinct motif in the C domain is concerned with expression in CV1 cells. It is called the KID box, located around nt 330.

F. INTERRELATIONSHIP BETWEEN TRANSCRIPTION AND REPLICATION CONTROL

In the case of polyoma virus, it is clear that the sequences required for the activation of DNA replication and those involved in enhancer function are the same, not only because functional elements responsible for each process overlap, but also because the transcriptional enhancer function per se is required for viral replication. Replacement of the polyoma enhancer (which is absolutely required for replication) by another viral (SV40) or cellular (IgH) enhancer allows viral DNA replication. Furthermore, when the IgH enhancer, which is transcriptionally active only in lymphoid cells, replaces the polyoma enhancer, the recombinant viral genome replicates in myeloma cells, not in fibroblasts. Thus, the cell-type specificity of the transcriptional enhancer affects DNA replication.[75] Although the mechanism of action of enhancers is unknown, such observations suggest that a common molecular process activates both transcription and replication: for instance, the enhancer could induce a conformational change of the template, making it competent for transcription as well as replication. Alternatively, replication could be activated by transcription through the origin, as has been reported for prokaryotic systems.[76,77] If the transcriptional enhancer function is sufficient to activate viral replication, there are some differences, however, in the control of replication and transcription by this same element. To activate DNA replication, the polyoma enhancer cannot be separated from the origin core by more than 200 bp, whereas it remains capable of stimulating RNA synthesis 1400 bp upstream of the transcription start sites. It was suggested that the enhancer favors local melting of the DNA and consecutive entry of the polymerase. RNA polymerase would then be able to slide and reach the region of transcription initiation, whereas the DNA polymerase-primase complex could not do so toward the origin of bidirectional replication.[78] Furthermore, there are quantitative differences between both activation processes. Whereas two copies of the A element (nt 5109 to 5130) are sufficient to activate DNA replication as efficiently as the complete enhancer, seven copies of this element are required for a similar effect on transcription.[35] In the same way, if the polyoma enhancer is replaced by SV40 sequences, it has been demonstrated that a truncated 72-bp sequence (deleted of 19 bp on its early boundary) activates DNA replication, while transcription stimulation requires an intact 72-bp sequence or two copies of the truncated element.[75]

In the case of SV40, the enhancer is not absolutely required for DNA replication. In this case, the early promoter is located closer to the origin and is stronger than its polyoma counterpart, making the enhancer dispensable for activation of the viral origin. The 72-bp repeats stimulate SV40 DNA replication *in vivo* in an orientation-independent manner only in the absence of the GC-rich region. However, the enhancer cannot be separated from the core of the origin by more than 98 bp and activation preferentially occurs when the AT-rich sequence of the core is proximal to the enhancer. The 21-bp repeats themselves activate replication in an orientation- but not distance-independent fashion.[79]

IV. DNA-PROTEIN INTERACTIONS IN THE REGULATORY REGIONS OF POLYOMA AND SV40 GENOMES

A. DNASE I HYPERSENSITIVITY OF ENHANCER SEQUENCES

In cells infected by SV40, 5 to 20% of the viral minichromosomes appear devoid of nucleosomes around the origin of replication. A good correlation can be established between the absence of nucleosomes and the sensitivity to DNase I observed *in vivo* in the same region of the DNA along 250 bp or so (from 0.67 to 0.74 map units). Despite this particularity, the mean number of nucleosomes per molecule remains unchanged, suggesting that this gap in viral chromatin could participate in keeping the minichromosomes in a compact structure characteristic of physiological ionic strengths.[80] Gapped minichromosomes appear to be involved in the formation of active transcription complexes.[81] Similar observations are made with polyoma minichromosomes, a fraction of which are devoid of nucleosomes and are sensitive to nucleases in the regulatory region of the viral genome. Polyoma variants that duplicate their origin region duplicate this gap. A more refined analysis of the sequences accessible to nucleases detected two hypersensitive sites in the polyoma DNA. They are called HS1 and HS2 (Figure 5). They are located around nt 5101 and 5210, within the cores of enhancers A and B, respectively. They seem to be determined by the DNA sequence directly since a polyoma mutant (Py EC PCC4-97) that contains a duplication around 5101 and a deletion around 5210 duplicates HS1 and is deleted of HS2.[82] These sites could be preferentially exposed to the action of the recombination machinery, as suggested by the frequency of rearrangements that occurred around them *in vivo* in Py NB mutants.[69,82a] Mutagenesis of the sequences surrounding HS1 and HS2 showed that the peculiar chromatin structure generated by these regions is dependent on the presence of AAGCAPuPuAAG sequences flanked by short inverted repeats. The polyoma CSP variant bears a duplication of nt 5096 to 5139 and duplicates HS1. However, the isolated 44-bp sequence 5096 to 5139 induces weak DNase I sensitivity when inserted close to the origin of replication. The context represented by their natural surrounding sequences seems to be necessary to the formation of hypersensitive sites HS1 and HS2.[83] Nuclease hypersensitive regions in polyoma and SV40 are not uniform: they show alternance of hypersensitive and very resistant sequences.[84] Thus, although it has been suggested that some sequences have an intrinsic weak capacity to bind histones,[85] it seems highly probable that DNase I hypersensitivity is due to nonhistone proteins that bind to DNA in the regulatory region, prevent nucleosome assembly,[85a] and generate DNA conformations with several very sensitive phosphodiester bonds. This is supported by the observation that the SV40 enhancer separated from the viral context is able to generate a site devoid of nucleosomes in chromatin *in vivo*.[86]

B. ENHANCER-PROTEIN INTERACTIONS IN THE POLYOMA GENOME

In intact cells infected by polyoma virus, protein-DNA contacts have been mapped along the viral noncoding sequences using DNase I or DMS, according to the genomic sequencing technique described by Church and Gilbert.[87] It could be shown that several proteins bind to the enhancer sequences on the four functional domains (A, B, C, and D) defined previously, suggesting that the viral enhancer exists as a large nucleoprotein complex during productive infection.[88,89] DNase I footprinting *in vitro*, using soluble nuclear extracts prepared from uninfected cells, showed that cellular proteins that can bind to the four domains of the polyoma enhancer exist prior to viral infection.[89-91] Their interactions create three DNase I-hypersensitive sites, two of which correspond to the HS1 and HS2 sites detected *in vivo*. Several enhancer-binding factors were further characterized by this technique used in combination with gel retardation assays.

1. DNA-Protein Interactions in the Polyoma A Enhancer

DNase I footprinting experiments have shown strong protection of the A domain of the enhancer by proteins extracted from fibroblast nuclei at 0.4 *M* NaCl. The short protected domain encompasses three overlapping DNA motifs homologous to other viral and cellular enhancers: a first one homologous to the adenovirus 5-E1A gene enhancer (nt 5108 to 5116),[92] a second one homologous to the SV40 enhancer (nt 5114 to 5124),[72] and a third one homologous to the c-*fos* gene enhancer (nt 5121 to 5130).[93] Competition experiments using DNase I footprinting and gel retardation assays have demonstrated that three different cellular factors bind to the core sequence of the polyoma A enhancer (Figure 5). PEA1 binds to the SV40 motif; PEA2 binds to the c-*fos* homology. The protections provided by these two interactions slightly overlap, but both factors can bind to the A domain simultaneously, with only a weak cooperative effect.[90,94] Binding of PEA1 and PEA2 creates a hypersensitive site at nt 5108. In the same region, a second hypersensitive site is present at nt 5111; it is not induced by binding of PEA1 and PEA2 and can be specifically competed away by an oligonucleotide covering the adenovirus homology, without displacing the two factors. This hypersensitive site results from the binding of a third protein, PEA3, which recognizes the sequence 5' AGGAAG 3' (nt 5108 to 5113). However, the interaction does not confer protection against DNase I digestion; it can be detected either by the induction of a hypersensitive site or by gel retardation of a specific oligonucleotide. This motif is also found in the polyoma B enhancer (nt 5203 to 5208), suggesting that PEA3 recognizes both enhancers.[94] PEA1 also binds to the SV40 72-bp repeats (nt 110 to 128 in the origin-proximal 72-bp sequence) and to the c-*fos* enhancer, which lacks one nucleotide of the PEA2 sequence (5' TGACGCA 3' instead of 5' TGACCGCA 3'). The sequence homology between PEA1 and PEA2 binding sites in the polyoma A enhancer suggests that both factors could have evolved from a common ancestor.[90] The functional importance of these interactions is reinforced by the fact that the target domain corresponds to the core of the A enhancer, which is the major transcriptional enhancer of the virus in fibroblasts[52] and the minimal activator of the origin of replication. This domain, after multimerization, is a strong enhancer.[35] It is duplicated in a number of polyoma variants.[55] Furthermore, mutants defective for transcription and replication have been constructed by chemical mutagenesis. They bear several nucleotide substitutions. One of them is mutated in the PEA1 binding site and another in the PEA3 binding site. Viable revertants restore the wild-type site in each case.[95] Besides these three factors, Tjian described the binding of factor AP4 to the *Pvu* II site (nt5128) of the polyoma A enhancer.[96]

EC cells, which are nonpermissive to polyoma virus infection, have been tested for the presence of these A-binding activities since the A domain of the enhancer is deficient in such cells.[52,97] It was clearly demonstrated that no protection of the A domain could be obtained using EC cell extracts, that is, PEA1 and PEA2 were absent or inactive in EC cells.[98] After treatment with retinoic acid in the presence or absence of dibutyryl-cAMP, EC F9 cells differentiate *in vitro* into endoderm[99,100] and become permissive to polyoma virus.[101] Upon differentiation, at least PEA1 activity is induced, although it is less abundant in the endoderm-like cells than in fibroblasts.[98] The crucial role of PEA1 in polyoma virus biological activity is further supported by several observations. In transient expression and stable transformation experiments, it was shown that the B enhancer of polyoma is equally active in fibroblasts and in EC cells, whereas the A enhancer is 3.5-fold less efficient in undifferentiated than in differentiated cells.[52,102] The polyoma enhancer activity increases in the presence of the product of the c-Ha-*ras* oncogene in EC cells, but not in fibroblasts. In differentiated EC cells, an intermediate level of stimulation is obtained.[103] Depending on the amount of PEA1 activity initially present in each cell type, the Ras protein stimulates synthesis or activation of the factor until it reaches a plateau corresponding to the optimal activity observed in fibroblasts. The origin of replication is activated by the A domain of the enhancer in differentiated cells, but not in two-cell embryos.[97] A polyoma mutant (B1), defective for viral transcription and replication, bears several point mutations in its A domain. These mutations prevent the binding of PEA1 *in vitro*. Reversion of one of these mutations (at nt 5115 in mutant B1-5140) restores both PEA1 binding *in vitro* and viral expression *in vivo*.[95,104] PEA1 binds to polyoma and SV40 enhancers, which are similarly weak in EC cells and are together repressed by the product of the adenovirus 2-E1A gene.[105] PEA1 binds to the enhancer-like sequences of the mouse major histocompatibility complex H2-K gene,[106] which is developmentally regulated and expressed only after differentiation of F9 cells.[107] PEA1 binds to the SV40 and c-*fos* enhancers.[90] SV40 and c-*fos* gene expression is stimulated by serum and phorbol esters like TPA.[108,109] PEA1 itself is activated by serum or TPA in NIH3T3 cells.[110]

At present, we do not know whether the level of PEA1 synthesis increases upon differentiation or whether the factor preexists in an inactive form and is modified to become capable of efficient binding to the polyoma enhancer. In the case of NF-κB, which recognizes the immunoglobulin κ light chain gene enhancer, a posttranslational mechanism has been proposed to explain specific induction of the enhancer binding activity upon B cell differentiation.[111] Transcription factor AP1,[112] which is probably the human homolog of PEA1, is activated during treatment of the cells by TPA. In this case, a posttranslational modification of the factor is responsible for its induction.[113] One means used by Py EC mutants to compensate for the lack of PEA1 activity (hence, A enhancer activity) in EC cells is to duplicate their A domain to optimize the initially weak interaction.

2. DNA-Protein Interactions on the Polyoma B Enhancer

Gel retardation assays, with a fragment encompassing the polyoma B enhancer, allowed the isolation of a specific protein-DNA complex which is sensitive to EDTA and EGTA. The cellular protein involved is called PEB1 (Figure 5). Despite sequence homologies between the polyoma B enhancer and the SV40 and IgH enhancers, PEB1 recognizes neither SV40 enhancer nor IgH enhancer and appears quite specific for polyoma.[114] The protein has a sedimentation coefficient of 3.7 S, corresponding to an approximate molecular weight of 50 kDa[115] DNase I, DMS, and ethylnitrosourea (ENU) were used to perform footprinting experiments in order to define more precisely the interaction between PEB1 and its target DNA sequence in the polyoma B enhancer. On the early strand, protection extends along 30 nt, including the GC-rich palindrome and the SV40 core homology (domain B of the polyoma enhancer). On the late strand, protection is more extensive and includes the IgH homology. PEB1 interaction induces hypersensitive sites on the early side of the protected domain. Within the protected sequence, a strong interaction on the late strand involves the early part of the GC-rich palindrome and the first two G residues of the SV homology. On the opposite strand, only increased methylation of two G residues can be observed. PEB1 closely interacts with the major and minor grooves of the DNA along one helix turn. Bal 31 nuclease was used to define the minimal sequence required for binding of PEB1. These experiments demonstrate that complex formation primarily occurs on the GC-rich palindrome and is further stabilized by nonspecific interactions with the SV and IgH homologies. Furthermore, binding of the protein induces a bend in the DNA on the early side of the site. The asymmetry of PEB1 interaction is similar to the interaction of TFIIIA with the *Xenopus* 5S gene. TFIIIA is a zinc-finger protein. Comparison of the interactions of PEB1 on polyoma and of TFIIIA on the 5S gene shows that, in the absence of any real sequence homology, a similar interaction is displayed, with strong binding to one strand of a core DNA sequence, stabilized by nonspecific contacts.[116] The primary binding site of PEB1 is the core of the polyoma B enhancer, which has minor transcriptional enhancer activity in fibroblasts, but coincides with the β element involved in replication activation. This suggests a functional role of this protein in the control of polyoma DNA replication. PEB1 was also studied by Böhnlein and Gruss.[117]

A second factor binding to the polyoma B enhancer has been described by other groups.[118,119] It recognizes the small palindrome included in the C domain of the enhancer, between nt 5158 and 5172, and is called EF-C. We could detect this factor only after fractionation of the nuclear extracts on a heparin-agarose column. It is probably masked by PEB1 interaction in our assays with crude nuclear extracts. Both interactions may be mutually exclusive since the target sites overlap. We observed that EF-C concentration increases in slowly growing cells, suggesting that EF-C could be a repressor, the balance between PEB1 and EF-C allowing a fine regulation of enhancer activity during the cell cycle.[106] Supporting this hypothesis, one of the multiple negative elements contributing the "dehancer" function of the −424 to −1140 region of the murine c-*myc* gene binds EF-C nuclear factor, and a dimer of the polyoma EF-C binding site functions as a dehancer *in vivo*.[120]

Tijan and his group showed that several other factors can bind to the polyoma B enhancer.[96] AP2 and AP3, which recognize the SV40 enhancer core sequence (see below), also interact with the homologous motif found in polyoma. AP4 binds to the Pvu II site (nt 5265) and CTF (CCAAT transcription factor) binds upstream of EF-C.

Johnson et al.[121] described a protein from rat liver which binds to the SV40 enhancer (see below), to the HSV-*tk* gene, and to the MSV LTR. This protein, called EBP20, also recognizes three sequence motifs in the polyoma B enhancer. It binds to polyoma DNA between nt 5185 and 5201, a sequence including the SV core homology, and interacts at higher concentrations with two other regions: nt 5159 to 5175 and nt 5215 to 5228, respectively.

The polyoma wild-type enhancer is incapable of activating viral early expression and replication in EC cells. This restriction is due to the fact that enhancer A is deficient in these cells devoid of sufficient amounts of PEA1 and PEA2. The B enhancer retains a comparable activity in EC cells and in differentiated cells, but this is not sufficient to ensure viral expression. Py EC PCC4 mutants compensate for the weak activity of their A domain by duplicating this sequence. This duplication is accompanied by a deletion in the B enhancer, confirming the limited contribution of this element to the enhancer function in undifferentiated cells. Py EC F9 mutants exhibit a different strategy to escape the block imposed by EC cells. A single point mutation (A to G transition at nt 5233) is sufficient to restore viral enhancer activity in the case of mutant Py EC F9-1. This mutation affects the early region of enhancer B, located between the B domain and the replication origin and including the repeats homologous to the BPV enhancer. Two copies of this mutated sequence are found in several Py EC F9 mutants. It was suggested that the mutation creates a novel enhancer motif recognized by a transcription factor present in F9 cells and responsible for enhancer reactivation in undifferentiated cells.

In DNase I footprinting experiments, it was shown that the mutated region exhibits alternating hypersensitive and protected sequences. Two factors bind to this region, one between nt 5248 and 5264, the other between nt 5230 and 5238. The latter protein, called PEB2, has a very reduced affinity for the corresponding wild-type DNA sequence. It is therefore a good candidate for being responsible for Py EC F9-1 B enhancer activity in F9 cells. The other factor, PEB3, binds independently of the presence of the mutation and has not yet been assigned a biological role.[122] Two other laboratories have described the factor binding to the Py EC F9-1 B enhancer and not to its wild-type counterpart.[123,124] In both cases, the interaction site includes the sequence 5' TGGAATGT 3' (nt 5232 to 5239). An additional protection was observed on the late-proximal BPV repeat, which depends on prior binding to the mutated sequence.[123] F9 tk^- cells stably transformed by the HSV-tk gene, under control of its own promoter linked to the polyoma B enhancer, revealed that mutation at nt 5233 increases polyoma enhancer activity approximately tenfold.[102] The E domain (nt 5217 to 5249) of Py EC F9-1, associated with a 72-bp sequence of SV40, acts as an enhancer in F9 cells, whereas the SV40 sequence alone is inactive.[124] Xiao et al.[125] described a protein, GT-IIC, which binds to the GT-II motif of the SV40 enhancer (see below) and specifically recognizes the mutated sequence of the Py EC F9-1 enhancer. This protein binds neither to the polyoma wild type sequence (5' TAGAATGT 3') nor to the GT-I motif of SV40 (5' GTGGAAAG 3'). Its high sequence specificity and binding properties suggest that it is identical to PEB2.

According to Tjian, the mutated sequence of Py EC F9-1 is recognized by CTF.[96] However, attempts in our laboratory to detect NF1/CTF binding to this sequence failed.[106]

C. DNA-PROTEIN INTERACTIONS IN THE ENHANCER-PROMOTER REGION OF SV40

The enhancer-promoter region of SV40 appears protected *in vivo* against DNase I digestion along several DNA sequences. The 21-bp repeat region is protected by the binding of Sp1 (see below) and three proteins interact with the enhancer: one sequence (nt 255 to 275) is protected on the late side of the 72-bp repeats, near the *Pvu* II site, and two domains are protected in each 72-bp sequence, one around the *Eco*RII site (nt 150 to 169) and one around the *Sph*I site (nt 153 to 162).[88]

Protein-DNA interactions in this region were also characterized *in vitro* and several factors involved have now been purified (Figure 6).

1. 21-bp Repeats

The first DNA-binding factor to be purified was Sp1, which recognizes the sequence 5' GGGCGG 3' (GC box) represented six times within the 21-bp repeats of the SV40 promoter. Five Sp1 molecules can bind together to the SV40 promoter; they interact with the major groove of the DNA and align on the same side of the helix.[126] *In vitro*, SV40 early RNA synthesis is mediated by the interaction of Sp1 with GC boxes I, II, and III, whereas transcription in the late direction is mediated by binding to GC boxes III, V, and VI.[127] It was recently demonstrated that Sp1, a protein with a molecular weight of over 100 kDa and a binding site of only 10 nt, is a zinc-finger protein. It is ubiquitous and binds to several other viral and cellular regulatory sequences.[128]

2. Pu Box

This sequence included in the C domain of the SV40 enhancer binds a lymphoid-specific protein. One copy of the Pu box is sufficient to bind this factor but two copies are required for enhancer activity, suggesting the importance of protein-protein interactions.[74]

3. GT-II Motif

This motif can be subdivided into three overlapping sequences— GT-IIA, GT-IIB, and GT-IIC— each of which binds at least one distinct protein. The first sequence (A) is recognized by a protein called GT-IIA. The second one (B) interacts with protein GT-IIBα or with protein GT-IIBβ. GT-IIBα also binds to the μE3 motif of the IgH enhancer, despite the absence of any sequence homology to SV40. These three proteins are detected in a large number of cell types. The third motif (C) binds protein GT-IIC, which is absent from lymphoid cells.[125] Mutations in the GT-IIA and GT-IIB motifs do not affect enhancer function. These sequences may be active in other enhancers or here be involved in late promoter function (deletion of nt 270 to 300 reduces late transcription threefold). Mutations in the GT-IIC motif impair enhancer activity in all cells except lymphoid cells, supporting the possible role of protein GT-IIC as a *trans*-acting factor in the function of the B enhancer domain. Factor GT-IIC also binds to the mutated motif of the Py EC F9-1 enhancer, but not to the corresponding wild-type sequence, which differs only by one base pair. It does not recognize the GT-I motif present downstream of GT-II in the SV40 enhancer, although both sequences fit the core consensus defined by Weiher et al.[53]

consensus:	GTGG(A/T)(A/T)(A/T)G
GT-II:	GTGGAATG
GT-I:	GTGGAAAG

Converting a T:A bp to an A:T bp abolishes protein binding. Thus, the notion of core consensus sequence has little predictive value of which protein will bind to the related sequences.

4. GT-I Motif

This sequence is recognized by three proteins called GT-IA, GT-IB, and GT-IC. GT-IA and GT-IB are ubiquitous; GT-IC is absent from MPC-11 cells (a mouse plasmacytoma cell line). It binds to an upstream promoter element in the mouse β-major globin gene. The minimal sequence recognized, distinct from the Weiher core, is 5′ GGGTGTGG 3′ (nt 244 to 251). The core is not sufficient either for protein binding or for enhancer activity. Mutations of the A residues do not affect the binding of GT-IA, GT-IB, and GT-IC. GT-IB is distinct from GT-IA and is found in different forms, depending on the cell type. GT-IC also exists in different forms. Mutations in GT-I are compensated for by the presence of the TC-II motif *in vivo*, but they affect GT-IC binding *in vitro* as well as promoter activity *in vivo* in the case of the β-globin gene.[129] Protein EBP20 described by Johnson et al.[121] binds to the SV40 GT-I motif as well as to the HSV-*tk* gene, the MSV-LTR, and the polyoma B enhancer. The common sequence recognized is 5′ TGTGG(A/T)(A/T)(A/T) 3′, but no competition is exerted by HSV-*tk* on GT-IA, GT-IB, and GT-IC binding to the SV40 enhancer. Tjian reported the purification of a cellular factor, AP3, which binds to the sequence 5′ GGGTGTGGAAAG 3′ overlapping the core consensus. The protein is 57 kDa and does not bind to the core-like sequence corresponding to GT-II. Chiu et al.[130] demonstrated that TPA stimulation of the SV40 enhancer is mediated by AP3 as well as by AP1, but through different mechanisms. The isolated AP1 site functions as a TPA-responsive element (TRE). TPA induces post-translational modification of AP1 in HeLa cells, which leads to increased binding of this factor to the TRE and expression of the genes controlled by this sequence. Upon TPA induction, increased binding of AP3 to the SV40 enhancer is observed in HeLa *tk⁻* cells, but not in HepG2 or HeLa S3 cells. In these latter cell types, AP3 binding activity is constitutive and TPA induction via AP3 must involve a different regulatory step: it probably affects the transcriptional activation capacity of AP3, binding to DNA alone being insufficient to promote gene expression at the induced level.

5. TC Motifs

TC-I and TC-II motifs are both recognized by the HeLa cell protein AP2.[131,132] Whereas mutations in the TC-II motif have little effect on enhancer activity, one mutation in TC-I prevents AP2 binding and reduces enhancer activity in HeLa cells. One binding site for AP2 and one binding site for AP3 constitute the functional core domain of the enhancer, which is active in a wide range of cells after multimerization. AP2 binds with low affinity to the 21-bp region. It also recognizes the hMT-IIA, hGH, c-*myc*, H-2K, and BPV enhancers. It is a 50-kDa protein which is absent from HepG2 cells. All the genes recognized by AP2 are regulated by TPA. Some of them are induced by agents that increase the level of cAMP in the cell. TPA and forskolin increase the capacity of AP2 to activate transcription without affecting its affinity for the DNA. Thus, AP2 seems to be situated at a crossroad between two regulation cascades (protein kinase C and cAMP dependent kinase).

AP2 binding to DNA is inhibited by T Ag independently of the presence of T Ag binding sites (for instance, in the hMT-IIA enhancer). Protein-protein interactions prevent AP2 binding to DNA. AP2 is different from p53, which also forms a complex with SV40 T Ag, but it is not ruled out that both proteins could share a related functional domain. This could provide a model for the regulation of transcription by T Ag: the viral protein would displace AP2 from the enhancer, thus reducing early transcription and allowing other factors to bind to the DNA and promote late transcription.[131] Besides, AP2 binds to the 21-bp repeats, which are involved in the stimulation of early transcription and in the down-regulation of late transcription. T Ag could suppress the inhibitory effect of the 21-bp repeats on late transcription by preventing AP2 from binding to this region.[29]

6. Sph Motifs

Both Sph motifs (5′ AAG(T/C)ATGCA 3′) in the A domain bind the same factor in a very cooperative fashion.[96] GT-IIC and Sph motifs are bound by the same protein called TEF-1.[132a]

7. Octamer Motifs

An octamer motif is created in SV40 at the junction between *Sph* I and *Sph* II. This sequence, which is also found in the IgH enhancer, is known to bind several factors, some of them being ubiquitous, others being cell specific and involved in lymphoid-specific expression of the immunoglobulin genes.[96] Sturm et al.[133] described a 100-kDa HeLa cell protein, called OBP100, which binds to the SV40 octamer motif as well as to a related sequence overlapping the Sph I motif. The two sequences were called octa1 (5′ ATGCAAAG 3′) and octa2 (5 ′ ATGCATCT 3′). Although the same protein recognizes both octamer sites, different DNA-protein contacts are established. OBP100 would be identical to NF-A1, a ubiquitous octamer-binding factor described previously. Two genes oct-1 and oct-2 coding for the ubiquitous and the lympoid specific octamer binding factors were recently cloned and shown to contain an homeobox related domain (133a and references within).

8. P Motif

This motif binds human factor AP1, whose recognition site corresponds to the functional TPA-responsive element defined by the study of different inducible genes.[112,113] In the case of SV40, however, mutations in the P element do not affect the whole enhancer basal or induced activity. A factor present on the intact enhancer could prevent AP1 from binding to SV40 DNA and mediating TPA stimulation.[130] AP1 has been extensively studied and it has become clear now that this transcription factor is the product of a cellular proto-oncogene of the *jun* family. The C-terminal domain of the viral oncogene v-*jun* can substitute for the C-terminal domain of the yeast GCN4 protein, which is responsible for binding to a DNA sequence identical to the AP1 binding site (5′ TGACTCA 3′). Antibodies against v-*jun*

TABLE 1
Review of the Different Factors That Bind to the Polyoma and SV40 Regulatory Sequences

Factor	Recognition sequence	Enhancer	Factor	Recognition sequence	Enhancer
PEA1	T$_G^T$AGTCA	Consensus	PEB2	TGGAATGT	Consensus
AP1			EF-E		
			GT-IIC		
PEA2	TGACCGCA	Polyoma	PEB3	AATCATTACTATG	Polyoma
PEA3	AGGAAG	Polyoma	Sp1	GGGCGG	Consensus
PEB1	AGAGGGCAGTC	Polyoma	Lymphoid factor	AAAGAGGA	SV40
EF-C	GTTGCNNGGCAAC	Consensus	GT-IIA	GCAAAGAACCAGCT	SV40
AP2	TCCCCAG	Consensus	GT-IIB	CAGCTGT	SV40
AP3	GGGTGTGGAAAG	SV40	GT-IA	GGGTGTGG	SV40
			GT-IB		
			GT-IC		
AP4	CAGCTGTGG	Consensus	Sph factor	AAG$_C^T$ATGCA	SV40
CTF	TGGCTNNNAGCCAA	Consensus	OBP100	ATGCAAAT	Consensus
EBP20	TGTGG$_{TTT}^{AAA}$	Consensus			

recognize AP1. A β-gal/human c-*jun* fusion protein, recognized by antibodies directed against AP1, binds to the AP1 site in an indistinguishable manner, and the amino acid sequence of peptides isolated from HeLa cell AP1 preparations corresponds to the nucleotide sequence of c-*jun*.[134] The v-*jun* protein was shown to be located in the nucleus and to bind to the SV40 AP1 site with the same specificity as AP1 or GCN4.[135] Cotransfection of a v-*jun* expression vector with constructs that contain functional AP1 sites controlling CAT expression leads to a 20- to 100-fold relative increase in CAT activity.[136] The polyoma enhancer binding factor PEA1, which is the murine homolog of AP1, is also induced by TPA (but by a mechanism dependent on protein synthesis, in contrast to human AP1), by agents that increase cAMP levels, and after transformation by SV40 (suggesting a role of c-*jun* activation in SV40 transformation).[110] Pure PEA1 contains several distinct polypeptide chains coded by at least three related jun genes.[136a,b] These polypeptides form heterodimers with FOS, an association that increases their affinity for DNA.[136c]

D. FORMATION OF AN ACTIVE ENHANCER-PROMOTER COMPLEX WITHIN POLYOMA OR SV40 MINICHROMOSOMES

The mechanism of action of enhancers is still unknown, but several hypotheses have been proposed. For instance, enhancers could act as chromatin organizers, as entry sites for RNA polymerase, as regulators of local DNA superhelicity or as elements that directly interact with the promoters by DNA binding. At present, more and more data prove that these sequences are the targets of specific cellular factors: several enhancers exhibit a marked cell-type specificity, competitions have been observed *in vivo* between different expression vectors bearing identical or related enhancer sequences, and enhancers are especially sensitive to nucleases. Furthermore, numerous specific enhancer-protein interactions have been detected *in vitro* and *in vivo*, the case of polyoma and SV40 being quite demonstrative.

In a series of experiments performed by Takahashi et al.,[137] the distance between the SV40 early promoter or enhancer elements has been altered by inserting either odd or even multiples of half a turn of the DNA helix. There are marked differences in the *in vivo* effects of these two types of insertions on transcription initiation, which strongly suggests that protein-protein interactions, as well as DNA-protein interactions, are involved in the enhancer-promoter function.

Polyoma and SV40 enhancers display a modular organization with distinct redundant functional domains, each capable of binding one or several factor(s). Both viral enhancers appear as the sites of assembly of large multiprotein complexes within the nuclei of infected cells. Multiple copies of one enhancer domain or the association of several distinct domains, can be as active as the entire enhancer. Besides, Ondek et al.[73] have shown that enhancer domains in SV40 can be further divided into elementary subunits called enhansons. Each enhanson corresponds to an individual protein binding site which can be found in other enhancers. Two copies of an enhanson, or a combination of two distinct enhansons, behave as an enhancer element displaying high activity after multimerization. The spacing between both enhansons is crucial, whereas that between enhancer elements can vary. Redundancy is required for enhancer activity at the two levels of binary organization exhibited by the SV40 enhancer. Reiteration of a single DNA-protein complex, or juxtaposition of distinct complexes, is functional. The different enhancer-binding proteins already identified could belong to the same protein family, with different DNA-binding domains recognizing distinct specific sequences and sharing a common activation domain involved in transcription stimulation through interaction with general transcription factors or RNA polymerase itself. Alternatively, all the enhancer-binding proteins could interact with various components of the transcription machinery, the action of these factors being additive.

SV40 and polyoma chromosomes also provide models for the mechanisms of chromatin replication. At present, both conservative and distributive mechanisms of nucleosome propagation have gained experimental support, and inheritance of nuclease-hypersensitive sites during development has been demonstrated. The study of SV40 chromosome replication *in vitro* has shown that nucleosome assembly is concomitant with DNA replication.[138] Nuclease hypersensitivity of the viral origin region is transiently abrogated by the progression of the replication forks and *de novo* induction of this particular chromatin structure rapidly occurs thereafter as a consequence of the binding of nonhistone proteins to the viral noncoding sequences.[139]

Most probably, the study of polyoma and SV40 chromosomes will continue to contribute to a better understanding of the nature and dynamics of active chromatin.

REFERENCES

1. **Tooze, J., Ed.,** *Molecular Biology of Tumor Viruses: DNA Tumor Viruses,* 2nd ed., Cold Spring Harbor Laboratory, Cold Spring Harbor, NY, 1981.
2. **Fried, M. and Prives, C.,** The biology of simian virus 40 and polyomavirus, in *Cancer Cells,* Vol. 4, Botchan, M., Grodzicker, T., and Sharp, P. A., Eds., Cold Spring Harbor Laboratory, Cold Spring Harbor, NY, 1986,1.
3. **Kamen, R. I., Favaloro, J. M., Parker, J. T., Treisman, R. H., Lania, L., Fried, M., and Mellor, A.,** Comparison of polyoma virus transcription in productively infected mouse cells and transformed rodent cell lines, *Cold Spring Harbor Symp. Quant. Biol.,* 44, 63, 1979.
4. **Lewis, E. D., Chen, S., Kumar, A., Blanck, G., Pollack, R. E., and Manley, J. L.,** A frameshift mutation affecting the carboxylterminus of the simian virus 40 large tumor antigen results in a replication- and transformation-defective virus, *Proc. Natl. Acad. Sci. U.S.A.,* 80, 7065, 1983.
5. **Khalili, K., Brady, J., and Khoury, G.,** Translational regulation of SV40 early mRNA defines a new viral protein, *Cell,* 48, 639, 1987.
6. **Treisman, R.,** Characterization of polyoma late mRNA leader sequences by molecular cloning and DNA sequence analysis, *Nucleic Acids Res.,* 8, 4867, 1980.
7. **Acheson, N. H.,** Polyoma virus giant RNAs contain tandem repeats of the nucleotide sequence of the entire viral genome, *Proc. Natl. Acad. Sci. U.S.A.,* 75, 4754, 1978.
8. **Kamen, R., Favaloro, J., and Parker, J.,** Topography of the three late mRNAs of polyoma virus which encode the virion proteins, *J. Virol.,* 33, 637, 1980.
9. **Good, P. J., Welch, R. C., Ryu, W.-S., and Mertz, J. E.,** The late spliced 19 S and 16 S mRNAs of simian virus 40 can be synthesized from a common pool of transcripts, *J. Virol.,* 62, 563, 1988.
10. **Grass, D. S. and Manley, J. L.,** Selective translation initiation on bicistronic simian virus 40 late mRNA, *J. Virol.,* 61, 2331, 1987.

11. **Sedman, S. A. and Mertz, J. E.,** Mechanisms of synthesis of virion proteins from the functionally bigenic late mRNAs of simian virus 40, *J. Virol.,* 62, 954, 1988.

12. **Good, P. J., Welch, R. C., Barkan, A., Somasekhar, M. B., and Mertz, J. E.,** Both VP2 and VP3 are synthesized from each of the alternatively spliced late 19 S RNA species of simian virus 40, *J. Virol.,* 62, 944, 1988.

13. **Kamen, R., Jat, P., Treisman, R., and Favaloro, J.,** 5′ termini of polyoma virus early region transcripts synthesized in vivo by wild-type virus and viable deletion mutants, *J. Mol. Biol.,* 159, 189, 1982.

14. **Cowie, A., Jat, P., and Kamen, R.,** Determination of sequences at the capped 5′ ends of polyoma virus early region transcripts synthesized in vivo and in vitro demonstrates an unusual microheterogeneity, *J. Mol. Biol.,* 159, 225, 1982.

15. **Jat, P., Novak, U., Cowie, A., Tyndall, C., and Kamen, R.,** DNA sequences required for specific and efficient initiation of transcription at the polyoma virus early promoter, *Mol. Cell. Biol.,* 2, 737, 1982.

15a. **Katinka, M. and Yaniv, M.,** Deletions of N-terminal sequences of polyoma virus T-antigens reduce but do not abolish transformation of rat fibroblasts, *Mol. Cell. Biol.,* 2, 1238, 1982.

16. **Mueller, C. R., Mes-Masson, A.-M., Bouvier, M., and Hassell, J. A.,** Location of sequences in polyomavirus DNA that are required for early gene expression in vivo and in vitro, *Mol. Cell. Biol.,* 4, 2594, 1984.

17. **Dailey, L. and Basilico, C.,** Sequences in the polyomavirus DNA regulatory region involved in viral DNA replication and early gene expression, *J. Virol.,* 54, 739, 1985.

18. **Farmerie, W. G. and Folk, W. R.,** The polyomavirus early promoter: role of proximal promoter elements in the formation of 5′ termini in vivo, *Virology,* 150, 518, 1986.

19. **Wasylyk, B., Wasylyk, C., Matthes, H., Wintzerith, M., and Chambon, P.,** Transcription from the SV40 early-early and late-early overlapping promoters in the absence of DNA replication, *EMBO J.,* 2, 1605, 1983.

20. **Cowie, A., Tyndall, C., and Kamen, R.,** Sequences at the capped 5′-ends of polyoma virus late region mRNAs: an example of extreme terminal heterogeneity, *Nucleic Acids Res.,* 9, 6305, 1981.

21. **Bourachot, B., Yaniv, M., and Herbomel, P.,** Control elements situated downstream of the major transcriptional start site are sufficient for highly efficient polyomavirus late transcription, *J. Virol.,* 63, 2567, 1989.

22. **Jat, P., Roberts, J. W., Cowie, A., and Kamen, R.,** Comparison of the polyoma virus early and late promoters by transcription in vitro, *Nucleic Acids Res.,* 10, 871, 1982.

23. **Kern, F. G., Dailey, L., and Basilico, C.,** Common regulatory elements control gene expression from polyoma early and late promoters in cells transformed by chimeric plasmids, *Mol. Cell. Biol.,* 5, 2070, 1985.

24. **Piatak, M., Ghosh, P. K., Norkin, L. C., and Weissman, S. M.,** Sequences locating the 5′ ends of the major simian virus 40 late mRNA forms, *J. Virol.,* 48, 503, 1983.

25. **Ernoult-Lange, M., Omilli, F., O'Reilly, D., and May, E.,** Characterization of the simian virus 40 late promoter: relative importance of sequences within the 72-base-pair repeats differs before and after viral DNA replication, *J. Virol.,* 61, 167, 1984.

26. **Keller, J. M. and Alwine, J. C.,** Activation of the SV40 late promoter: direct effects of T antigen in the absence of viral DNA replication, *Cell,* 36, 381, 1984.

27. **Brady, J. and Khoury, G.,** Trans activation of the simian virus 40 late transcription unit by T-antigen, *Mol. Cell. Biol.,* 5, 1391, 1985.

28. **Keller, J. M. and Alwine, J. C.,** Analysis of an activatable promoter: sequences in the simian virus 40 late promoter required for T-antigen-mediated trans activation, *Mol. Cell. Biol.,* 5, 1859, 1985.

29. **May, E.,** personal communication, 1988.

30. **Kern, F. G., Pellegrini, S., and Basilico, C.,** Cis- and trans-acting factors regulating gene expression from the polyomavirus late promoter, in *Cancer cells,* Vol. 4, Botchan, M., Grodzicker, T. and Sharp, P. A., Eds., Cold Spring Harbor Laboratory, Cold Spring Harbor, NY, 1986, 115.

31. **Katinka, M. and Yaniv, M.,** DNA replication origin of polyoma virus: early proximal boundary, *J. Virol.,* 47, 244, 1983.

32. **Triezenberg, S. J. and Folk, W. R.,** Essential nucleotides in the polyomavirus origin region, *J. Virol.,* 51, 437, 1984.

33. **Muller, W. J., Mueller, C. R., Mes, A.-M., and Hassell, J. A.,** Polyomavirus origin for DNA replication comprises multiple genetic elements, *J. Virol.,* 47, 586, 1983.

34. **Hassell, J. A., Mueller, C. R., and Muller, W. J.,** The polyoma virus enhancer: multiple sequence elements required for transcription and DNA replication, in *Current Communications in Molecular Biology, Eukaryotic Transcription,* Cold Spring Harbor Laboratory, Cold Spring Harbor, NY, 1985, 33.

35. **Veldman, G. M., Lupton, S., and Kamen, R.,** Polyomavirus enhancer contains multiple redundant sequence elements that activate both DNA replication and gene expression, *Mol. Cell. Biol.,* 5, 649, 1985.

36. **Hendrickson, E. A., Fritze, C. E., Folk, W. R., and De Pamphlis, M. E.,** The origin of bidirectional replication in polyoma virus, *EMBO J.,* 6, 2011, 1987.

37. **Li, J. J., Peden, K. W. C., Dixon, R. A. F., and Kelly, T.,** Functional organization of the simian virus 40 origin of DNA replication, *Mol. Cell. Biol.,* 6, 1117, 1986.
38. **Hertz, G. Z., Young, M. R., and Mertz, J. E.,** The A + T-rich sequence of the simian virus 40 origin is essential for replication and is involved in bending of the viral DNA, *J. Virol.,* 61, 2322, 1987.
39. **Hay, R. T. and De Pamphilis, M. L.,** Initiation of SV40 DNA replication in vivo: location and structure of 5' ends of DNA synthesized in the ori region, *Cell,* 28, 767, 1982.
40. **Cowie, A. and Kamen, R.,** Multiple binding sites for polyomavirus large T antigen within regulatory sequences of polyomavirus DNA, *J. Virol.,* 52, 750, 1984.
41. **Cowie, A. and Kamen, R.,** Guanine nucleotide contacts within viral DNA sequences bound by poly-omavirus large T antigen, *J. Virol.,* 57, 505, 1986.
42. **De Lucia, A. L., Lewton, B. A., Tjian, R., and Tegtmeyer, P.,** Topography of simian virus 40 A protein-DNA complexes: arrangement of pentanucleotide interaction sites at the origin of replication, *J. Virol.,* 46, 143, 1983.
43. **Tegtmeyer, P., Lewton, B. A., De Lucia, A. L., Milson, V. G., and Ryder, K.,** Topography of simian virus 40 A protein-DNA complexes: arrangement of protein bound to the origin of replication, *J. Virol.,* 46, 151, 1983.
44. **Deb, S., Tsui, S., Koff, A., De Lucia, A. L., Parsons, R., and Tegtmeyer, P.,** The T-antigen-binding domain of the simian virus 40 core origin of replication, *J. Virol.,* 61, 2143, 1987.
45. **Dean, F. B., Dodson, M., Echols, H., and Hurwitz, J.,** ATP-dependent formation of a specialized nucleoprotein structure by simian virus 40 (SV40) large tumor antigen at the SV40 replication origin, *Proc. Natl. Acad. Sci. U.S.A.,* 84, 8981, 1987.
46. **Farmerie, W. G. and Folk, W. R.,** Regulation of polyomavirus transcription by large tumor antigen, *Proc. Natl. Acad. Sci. U.S.A.,* 81, 6919, 1984.
47. **Banerji, J., Rusconi, S., and Schaffner, W.,** Expression of a β-globin gene is enhanced by remote SV40 DNA sequences, *Cell,* 29, 299, 1981.
48. **De Villiers, J. and Schaffner, W.,** A small segment of polyoma virus DNA enhances the expression of a cloned β-globin gene over a distance of 1400 base pairs, *Nucleic Acids Res.,* 9, 6251, 1981.
49. **Treisman, R. and Maniatis, T.,** Simian virus 40 enhancer increases the number of RNA polymerase II molecules on linked DNA, *Nature (London),* 315, 72, 1985.
50. **Weber, F. and Schaffner, W.,** Simian virus 40 enhancer increases RNA polymerase density within the linked gene, *Nature (London),* 315, 75, 1985.
51. **Wasylyk, B., Wasylyk, C., Augereau, P., and Chambon, P.,** The SV40 72 bp repeat preferentially potentiates transcription starting from proximal natural or substitute promoter elements, *Cell,* 32, 503, 1983.
52. **Herbomel, P., Bourachot, B., and Yaniv, M.,** Two distinct enhancers with different cell specificities coexist in the regulatory region of polyoma, *Cell,* 39, 653, 1984.
53. **Weiher, H., König, M., and Gruss, P.,** Multiple point mutations affecting the simian virus 40 enhancer, *Science,* 219, 626, 1983.
54. **Mueller, C. R., Muller, W. J., and Hassell, J. A.,** The polyomavirus enhancer comprises multiple functional elements, *J. Virol.,* 62, 1667, 1988.
55. **Ruley, H. E. and Fried, M.,** Sequence repeats in a polyoma virus DNA region important for gene expression, *J. Virol.,* 47, 233, 1983.
56. **Swartzendruber, D. E. and Lehman, J. M.,** Neoplastic differentiation: interaction of simian virus 40 and polyoma virus with the murine teratocarcinoma cells in vitro, *J. Cell. Physiol.,* 65, 179, 1975.
57. **Kelly, F. and Condamine, H.,** Tumor viruses and early mouse embryos, *Biochim. Biophys. Acta,* 651, 105, 1982.
58. **Martin, G. R.,** Cell lines derived from teratocarcinomas, in *Teratocarcinoma Stem Cells,* Silver, L. M., Martin, G. R., and Strickland, S., Eds., Cold Spring Harbor Laboratory, Cold Spring Harbor, NY, 1983, 690.
59. **Katinka, M., Yaniv, M., Vasseur, M., and Blangy, D.,** Expression of polyoma early functions in mouse embryonal carcinoma cells depends on sequence rearrangements in the beginning of the late region, *Cell,* 20, 393, 1980.
60. **Katinka, M., Vasseur, M., Montreau, N., Yaniv, M., and Blangy, D.,** Polyoma DNA sequences involved in control of viral gene expression in murine embryonal carcinoma cells, *Nature (London),* 290, 720, 1981.
61. **Vasseur, M., Kress, C., Montreau, N., and Blangy, D.,** Isolation and characterization of polyoma virus mutants able to develop in multipotential murine embryonal carcinoma cells, *Proc. Natl. Acad. Sci. U.S.A.,* 77, 1068, 1980.
62. **Vasseur, M., Katinka, M., Herbomel, P., Yaniv, M., and Blangy, D.,** Physical and biological features of polyoma virus mutants able to infect embryonal carcinoma cell lines, *J. Virol.,* 43, 800, 1982.
63. **Fujimura, F. K., Deininger, P. L., Friedmann, T., and Linney, E.,** Mutation near the polyoma DNA replication origin permits productive infection of F9 embryonal carcinoma cells, *Cell,* 23, 809, 1981.

64. **Sekikawa, K. and Levine, A. J.,** Isolation and characterization of polyoma host range mutants that replicate in nullipotential embryonal carcinoma cells, *Proc. Natl. Acad. Sci. U.S.A.,* 78, 1100, 1981.
65. **Mélin, F., Pinon, H., Reiss, C., Kress, C., Montreau, N., and Blangy, D.,** Common features of polyomavirus mutants selected on PCC4 embryonal carcinoma cells, *EMBO J.,* 4, 1799, 1985.
66. **Mélin, F., Pinon, H., Kress, C., and Blangy, D.,** Isolation of polyomavirus mutants multiadapted to murine embryonal carcinoma cells, *J. Virol.,* 53, 862, 1985.
67. **Tanaka, K., Chowdhury, K., Chang, K. S. S., Israel, M., and Ito, Y.,** Isolation and characterization of polyoma virus mutants which grow in murine embryonal carcinoma and trophoblast cells, *EMBO J.,* 1, 1521, 1982.
68. **De Simone, V., La Mantia, G., Lania, L., and Amati, P.,** Polyomavirus mutation that confers a cell-specific cis-advantage for viral DNA replication, *Mol. Cell. Biol.,* 5, 2142, 1985.
69. **Maione, R., Passananti, C., De Simone, V., Delli-Bovi, P., Augusti-Tocco, G., and Amati, P.,** Selection of mouse neuroblastoma cell-specific polyoma virus mutants with stage differentiative advantages of replication, *EMBO J.,* 4, 3215, 1985.
70. **Ondek, B., Shepard, A., and Herr, W.,** Discrete elements within the SV40 enhancer region display different cell-specific enhancer activities, *EMBO J.,* 6, 1017, 1987.
71. **Clarke, J. and Herr, W.,** Activation of mutated simian virus 40 enhancers by amplification of wild-type enhancer elements, *J. Virol.,* 61, 3536, 1987.
72. **Zenke, M., Grundström, T., Matthes, H., Wintzerith, M., Schatz, C., Wildeman, A., and Chambon, P.,** Multiple sequence motifs are involved in SV40 enhancer function, *EMBO J.,* 5, 387, 1986.
73. **Ondek, B., Gloss, L., and Herr, W.,** The SV40 enhancer contains two distinct levels of organization, *Nature (London),* 333, 40, 1988.
74. **Pettersson, M. and Schaffner, W.,** A purine-rich DNA sequence motif present in SV40 and lymphotropic papovavirus binds a lymphoid-specific factor and contributes to enhancer activity in lymphoid cells, *Genes Dev.,* 1, 962, 1987.
75. **De Villiers, J., Schaffner, W., Tyndall, C., Lupton, S., and Kamen, R.,** Polyoma virus DNA replication requires an enhancer, *Nature (London),* 312, 242, 1984.
76. **Dove, W. F., Inokuchi, H., and Stevens, W. F.,** Replication control in phage lambda, in *The Bacteriophage Lambda,* Hershey, A. D., Ed., Cold Spring Harbor Laboratory, Cold Spring Harbor, NY, 1971, 747.
77. **Panayotatos, N.,** DNA replication regulated by the priming promoter, *Nucleic Acids Res.,* 12, 2641, 1984.
78. **Hassell, J. A., Muller, W. J., and Mueller, C. R.,** The dual role of the polyomavirus enhancer in transcription and DNA replication, in *Cancer Cells,* Vol. 4, Botchan, M., Grodzicker, T., and Sharp, P. A. Eds., Cold Spring Harbor Laboratory, Cold Spring Harbor, NY, 1986, 561.
79. **Chandrasekharappa, S. C. and Subramanian, K. N.,** Effects of position of orientation of the 72-base-pair-repeat transcriptional enhancer on replication from the simian virus 40 core origin, *J. Virol.,* 61, 2973, 1987.
80. **Saragosti, S., Moyne, G., and Yaniv, M.,** Absence of nucleosomes in a fraction of SV40 chromatin between the origin of replication and the region coding for the late leader RNA, *Cell,* 20, 65, 1980.
81. **Choder, M., Bratosin, S., and Aloni, Y.,** A direct analysis of transcribed minichromosomes: all transcribed SV40 minichromosomes have a nuclease-hypersensitive region within a nucleosome-free domain, *EMBO J.,* 3, 2929, 1984.
82. **Herbomel, P., Saragosti, S., Blangy, D., and Yaniv, M.,** Fine structure of the origin-proximal DNase I-hypersensitive region in wild-type and EC mutant polyoma, *Cell,* 25, 651, 1981.
82a. **Caruso, M., Felsani, A., and Amati, P.,** Sites hypersensitive to, and protected from, nuclease digestion in the regulatory region of wild-type and mutant polyoma chromatin, *EMBO J.,* 5, 3539, 1986.
83. **Bryan, P. N. and Folk, W. R.,** Enhancer sequences responsible for DNase I hypersensitivity in polyomavirus chromatin, *Mol. Cell. Biol.,* 6, 2249, 1986.
84. **Cereghini, S. and Yaniv, M.,** Assembly of transfected DNA into chromatin: structural changes in the origin-promoter-enhancer region upon replication, *EMBO J.,* 3, 1243, 1984.
85. **Wasylyk, B., Oudet, P., and Chambon, P.,** Preferential in vitro assembly of nucleosome cores on some AT-rich regions of SV40 DNA, *Nucleic Acids Res.,* 7, 705, 1979.
85a. **Ambrose, C., Blasquez, V., and Bina, M.,** A block in initiation of simian virus 40 assembly results in the accumulation of minichromosomes containing an exposed regulatory region, *Proc. Natl. Acad. Sci. U.S.A.,* 83, 3287, 1986.
86. **Jongstra, J., Reudelhuber, T. L., Oudet, P., Benoist, C., Chae, C.-B., Jeltsch, J.-M., Mathis, D. J., and Chambon, P.,** Induction of altered chromatin structures by simian virus 40 enhancer and promoter elements, *Nature (London),* 307, 708, 1984.
87. **Church, G. M. and Gilbert, W.,** Genomic sequencing, *Proc. Natl. Acad. Sci. U.S.A.,* 81, 1991, 1984.
88. **Piette, J., Cereghini, S., Kryszke, M.-H., and Yaniv, M.,** Identification of cellular proteins that interact with polyomavirus or simian virus 40 enhancers, in *Cancer Cells,* Vol. 4, Botchan, M., Grodzicker, T., and Sharp, P. A., Eds., Cold Spring Harbor Laboratory, Cold Spring Harbor, NY, 1986, 103.

89. **Piette, J., Kryszke, M.-H., and Yaniv, M.,** The polyoma enhancer, in *Molecular Aspects of the Papo-viruses,* Aloni, Y., Ed., Martinus Nijhoff, Boston, 1987, 85.

90. **Piette, J. and Yaniv, M.,** Two different factors bind to the α-domain of the polyoma virus enhancer, one of which also interacts with the SV40 and c-fos enhancers, *EMBO J.,* 6, 1331, 1987.

91. **Kryszke, M.-H., Piette, J., and Yaniv, M.,** Identification of cellular proteins that interact with the polyoma virus enhancer, in *Viral Carcinogenesis,* Alfred Benzon Symposium 24, Kjeldgaard, N. O. and Forchhammer, J., Eds., Munksgaard, Copenhagen, 1987, 68.

92. **Hearing, P. and Shenk, T.,** The adenovirus type 5 E1A transcription control region contains a duplicated enhancer element, *Cell,* 33, 695, 1983.

93. **Treisman, R.,** Transient accumulation of c-fos RNA following serum stimulation requires a conserved 5′ element and c-*fos* 3′ sequences, *Cell,* 42, 889, 1985.

94. **Martin, M. E., Piette, J., Yaniv, M., Tang, W.-J., and Folk, W. R.,** Activation of the polyomavirus enhancer by a murine activator protein 1 (AP1) homolog and two contiguous proteins, *Proc. Natl. Acad. Sci. U.S.A.,* 85, 5839, 1988.

95. **Tang, W. J., Berger, S. L., Triezenberg, S. J., and Folk, W. R.,** Nucleotides in the polyomavirus enhancer that control viral transcription and DNA replication, *Mol. Cell. Biol.,* 7, 1681, 1987.

96. **Jones, N. C., Rigby, P. W. J., and Ziff, E. B.,** Trans-acting protein factors and the regulation of eukaryotic transcription: lessons from studies on DNA tumour viruses, *Genes Dev.,* 2, 267, 1988.

97. **Wirak, D. O., Chalifour, L. E., Wassarman, P. M., Muller, W. J., Hassell, J. A., and De Pamphilis, M. L.,** Sequence-dependent DNA replication in preimplantation mouse embryos, *Mol. Cell. Biol.,* 5, 2924, 1985.

98. **Kryszke, M.-H., Piette, J., and Yaniv, M.,** Induction of a factor that binds to the polyoma virus A enhancer on differentiation of embryonal carcinoma cells, *Nature (London),* 328, 254, 1987.

99. **Strickland, S. and Mahdavi, V.,** The induction of differentiation in teratocarcinoma stem cells by retinoic acid, *Cell,* 15, 393, 1978.

100. **Strickland, S., Smith, K. K., and Marotti, K. R.,** Hormonal induction of differentiation in teratocarcinoma stem cells: generation of parietal endoderm by retinoic acid and dibutyryl-cAMP, *Cell,* 21, 347, 1980.

101. **Fujimura, F. K., Silbert, P. E., Eckhart, W., and Linney, E.,** Polyoma virus infection of retinoic acid-induced differentiated teratocarcinoma cells, *J. Virol.,* 39, 306, 1981.

102. **Linney, E. and Donerly, S.,** DNA fragments from F9 Py EC mutants increase expression of heterologous genes in transfected F9 cells, *Cell,* 35, 693, 1983.

103. **Wasylyk, C., Imler, J. L., Perez-Mutul, J., and Wasylyk, B.,** The c-Ha-ras oncogene and a tumor promoter activate the polyoma virus enhancer, *Cell,* 48, 525, 1987.

104. **Piette, J. and Folk, W. R.,** unpublished data, 1987.

105. **Borrelli, E., Hen, R., and Chambon, P.,** Adenovirus-2 E1A products repress enhancer-induced stimulation of transcription, *Nature (London),* 312, 608, 1984.

106. **Piette, J. and Yaniv, M.,** unpublished observations, 1987.

107. **Gachelin, G.,** The cell-surface antigens of mouse embryonal carcinoma cell, *Biochim. Biophys. Acta,* 516, 27, 1978.

108. **Imbra, R. J. and Karin, M.,** Phorbol ester induces the transcriptional stimulatory activity of the SV40 enhancer, *Nature (London),* 323, 555, 1986.

109. **Greenberg, M. E. and Ziff, E. B.,** Stimulation of 3T3 cells induces transcription of the c-fos proto-oncogene, *Nature (London),* 311, 433, 1984.

110. **Piette, J., Hirai, S.-H., and Yaniv, M.,** Constitutive synthesis of activator protein 1 transcription factor after viral transformation of mouse fibroblasts, *Proc. Natl. Acad. Sci. U.S.A.,* 85, 3401, 1988.

111. **Sen, R. and Baltimore, D.,** Inducibility of κ immunoglobulin enhancer-binding protein NF-κB by a posttranslational mechanism, *Cell,* 47, 921, 1986.

112. **Lee, W., Haslinger, A., Karin, M., and Tjian, R.,** Activation of transcription by two factors that bind promoter and enhancer sequences of the human metallothionein gene and SV40, *Nature (London),* 325, 368, 1987.

113. **Angel, P., Imagawa, M., Chiu, R., Stein, B., Imbra, R. J., Rahmsdorf, H. J., Jonat, C., Herrlich, P., and Karin, M.,** Phorbol ester-inducible genes contain a common cis element recognized by a TPA-modulated trans-acting factor, *Cell,* 49, 729, 1987.

114. **Piette, J., Kryszke, M.-H., and Yaniv, M.,** Specific interaction of cellular factors with the B enhancer of polyoma virus, *EMBO J.,* 4, 2675, 1985.

115. **Kryszke, M.-H. and Yaniv, M.,** unpublished data, 1985.

116. **Piette, J. and Yaniv, M.,** Molecular analysis of the interaction between an enhancer binding factor and its DNA target, *Nucleic Acids Res.,* 14, 9595, 1986.

117. **Böhnlein, E. and Gruss, P.,** Interaction of distinct nuclear proteins with sequences controlling the expression of polyomavirus early genes, *Mol. Cell. Biol.,* 6, 1404, 1986.

118. **Fujimura, F. K.,** Nuclear activity from F9 embryonal carcinoma cells binding specifically to the enhancers of wild-type polyomavirus and Py EC mutant DNAs, *Nucleic Acids Res.,* 14, 2845, 1986.

119. **Ostapchuk, P., Diffley, J. F. X., Bruder, J. T., Stillman, B., Levine, A. J., and Hearing, P.,** Interaction of a nuclear factor with the polyomavirus enhancer region, *Proc. Natl. Acad. Sci. U.S.A.,* 83, 8550, 1986.
120. **Weisinger, G., Remmers, E. F., Hearing, P., and Marcu, K. B.,** Multiple negative elements upstream of the murine c-myc gene share nuclear factor binding sites with SV40 and polyoma enhancers, *Oncogene.* 3, 635, 1988.
121. **Johnson, P. F., Landschulz, W. H., Graves, B. J., and McKnight, S. L.,** Identification of a rat liver nuclear protein that binds to the enhancer core element of three animal viruses, *Genes Dev.,* 1, 133, 1987.
122. **Kryszke, M.-H. and Yaniv, M.,** unpublished data, 1987.
123. **Kovesdi, I., Satake, M., Furukawa, K., Reichel, R., Ito, Y., and Nevins, J. R.,** A factor discriminating between the wild-type and a mutant polyomavirus enhancer, *Nature (London),* 328, 87, 1987.
124. **Hearing, P.,** personal communication, 1988.
125. **Xiao, J. H., Davidson, I., Ferrandon, D., Rosales, R., Vigneron, M., Macchi, M., Ruffenach, F., and Chambon, P.,** One cell-specific and three ubiquitous nuclear proteins bind in vitro to overlapping motifs in the domain B1 of the SV40 enhancer, *EMBO J.,* 6, 3005, 1987.
126. **Gidoni, D., Dynan, W. S., and Tjian, R.,** Multiple specific contacts between a mammalian transcription factor and its cognate promoters, *Nature (London),* 312, 409, 1984.
127. **Gidoni, D., Kadonaga, J. T., Barrera-Saldana, H., Takahashi, K., Chambon, P., and Tjian, R.,** Bidirectional SV40 transcription mediated by tandem Sp1 binding interactions, *Science,* 230, 511, 1985.
128. **Kadonaga, J. T., Carner, K. R., Masiarz, F. R., and Tjian, R.,** Isolation of cDNA encoding transcription factor Sp1 and functional analysis of the DNA binding domain, *Cell,* 51, 1079, 1987.
129. **Xiao, J.-H., Davidson, I., Macchi, M., Rosales, R., Vigneron, M., Staub, A., and Chambon, P.,** In vitro binding of several cell-specific and ubiquitous nuclear proteins to the GT-I motif of the SV40 enhancer, *Genes Dev.,* 1, 794, 1987.
130. **Chiu, R., Imagawa, M., Imbra, R. J., Bockoven, J. R., and Karin, M.,** Multiple cis- and trans-acting elements mediate the transcriptional response to phorbol esters, *Nature (London),* 329, 648, 1987.
131. **Mitchell, P. J., Wang, C., and Tjian, R.,** Positive and negative regulation of transcription in vitro: enhancer-binding protein AP-2 is inhibited by SV40 T antigen, *Cell,* 50, 847, 1987.
132. **Imagawa, M., Chiu, R., and Karin, M.,** Transcription factor AP-2 mediates induction of two different signal-transduction pathways: protein kinase C and cAMP, *Cell,* 51, 251, 1987.
132a. **Davidson, I., Xiao, J.-H., Rosales, R., Staub, A., and Chambon, P.,** The Hela cell protein TEF-1 binds specifically and cooperatively to two SV40 enhancer motifs of unrelated sequence, *Cell,* 54, 931, 1988.
133. **Sturm, R., Baumruker, T., Franza, B. R., Jr., and Herr, W.,** A 100-kd HeLa cell octamer binding protein (OBP100) interacts differently with two separate octamer-related sequences within the SV40 enhancer, *Genes Dev.,* 1, 1147, 1987.
133a. **Herr, W., Sturm, R. A., Clerc, R. G., Corcoran, L. M., Baltimore, D., Sharp, P. A., Ingraham, H. A., Rosenfeld, M. G., Finney, M., Ruvkun, G., and Horvitz, H. R.,** The POU domain: a large conserved region in the mammalian *pit-1, oct-1, oct-2,* and *Caenorhabditis elegans unc-86* gene products, *Genes Dev.,* 2, 1513. 1988.
134. **Bohmann, D., Bos, T. J., Admon, A., Nishimura, T., Vogt, P. K., and Tjian, R.,** Human proto-oncogene c-jun encodes a DNA-binding protein with structural and functional properties of transcription factor AP-1, *Science,* 238, 1386, 1987.
135. **Bos, T. J., Bohmann, D., Tsuchie, H., Tjian, R., and Vogt, P. K.,** V-jun encodes a nuclear protein with enhancer binding properties of AP-1, *Cell,* 52, 705, 1988.
136. **Angel, P., Allegretto, E. A., Okino, S. T., Hattori, K., Boyle, W. J., Hunter, T., and Karin, M.,** Oncogene jun encodes a sequence-specific trans-activator similar to AP-1, *Nature (London),* 332, 166, 1988.
136a. **Nakabeppu, Y., Ryder, K., and Nathans, D.,** DNA binding activities of three murine Jun proteins: stimulation by *Fos. Cell,* 55, 907, 1988.
136b. **Ryseck, R. P., Hirai, S. I., Yaniv, M., and Bravo, R.,** Transcriptional activation of *c-jun* during the G0/G1 transition in mouse fibroblasts, *Nature,* 334, 535, 1988.
136c. **Curran, T. and França, B. R., Jr.,** Fos and Jun: the AP1 connection, *Cell,* 55, 395, 1988.
137. **Takahashi, K., Vigneron, M., Matthes, H., Wildeman, A., Zenke, M., and Chambon, P.,** Requirement of stereospecific alignments for initiation from the simian virus 40 early promoter, *Nature (London),* 319, 121, 1986.
138. **Stillman, B.,** Chromatin assembly during SV40 DNA replication in vitro, *Cell,* 45, 555, 1986.
139. **Solomon, M. J. and Varshavsky, A.,** A nuclease-hypersensitive region forms de novo after chromosome replication, *Mol. Cell. Biol.,* 7, 3822, 1987.
140. **Soeda, E., Arrand, J. R., Smolar, N., and Griffin, B. E.,** Sequence from early region of polyoma virus DNA containing viral replication origin and encoding small, middle and (part of) large T antigens, *Cell,* 17, 357, 1979.

Chapter 6

THE GENOME OF CAULIFLOWER MOSAIC VIRUS: ORGANIZATION AND GENERAL CHARACTERISTICS

J.-M. Mesnard, G. Lebeurier, and L. Hirth

TABLE OF CONTENTS

I. INTRODUCTION

Cauliflower mosaic virus[1] (CaMV) is a plant virus which possesses some unexpected characteristics. Having remained in relative obscurity until the discovery of the sequence of its genome, this virus has aroused considerable interest resulting in numerous publications (for reviews, see References 2 to 4). The prototype of the caulimoviruses, CaMV was described as early as 1961[5] and recognized in 1970 as being the only plant virus to carry its genetic information in a double-stranded DNA molecule.[6] The discovery of geminiviruses,[7] single-stranded DNA viruses, has shown that the existence of DNA viral genomes was not limited to caulimoviruses. Nevertheless, the large majority of plant viruses consists of RNA viruses, including viroids. The small concentration of virions in infected turnip cells (10 to 20 mg/kg of freshly infected leaves) slowed down research concerning the protein structure of the virus. Although the proteins were initially considered to be numerous in the virions, studies by Al Ani et al.[9] showed that the viral capsid was composed of a single major protein of 42 kDa. The other viral proteins described corresponded to dimeric associations or to degradation products. This observation suggested the existence of a relatively simple genome and was confirmed by the determination of the sequence of the viral DNA from isolate Cabb-S of CaMV.[1] The discovery of the replication process of the viral genome, the process which relates the caulimoviruses to the large group of retroids[10,11] and whose characteristics will be described later, represents a second phase of research relative to the understanding of a biological system unique in the viral plant world. Compared to the replication mechanism of the viral DNA which involves a reverse transcriptase, the expression of the viral genes and the function of their products of expression pose problems not yet understood. We have decided to avoid discussion of these known problems in order to devote ourselves to those which are presently the object of extensive research.

II. THE STRUCTURE OF THE VIRION

CaMV is the only representative of the caulimovirus family whose structure has been the object of considerable research, but the structures of the other members of the caulimovirus family are probably analogous to CaMV. The structure of this virus has mostly been studied by means of neutron diffusion.[12] The results show that the relaxed, double-stranded viral DNA is associated with the internal region of the protein capsid, leaving a region of this protein oriented toward the interior of the capsid: the center of this capsid is apparently empty. The structure of the particle poses other problems; the aforecited study shows a low compaction in the assembly of the protein subunits, suggesting a weak stability of the virions; however, the virus particles are very stable. It is, in fact, possible to extract the viral DNA, at a high pH level, without destroying or even modifying the capsid.[13] These results are consistent with the fact that the molecule of extruded DNA can reintegrate the capsid when the pH value of the reactive environment nears neutrality.[13] The observations were explainable in view of the great stability of the virion. There exists, therefore, a particuliar structure of the capsid of CaMV which is most likely linked to the amino acid composition of the protein subunits and to their distribution in the polypeptide chain. The exact structure of CaMV can only be determined if the crystals form correctly in highly purified virion preparations, which has not been the case so far. Only bidimensional crystals have been obtained, thanks to the use of the Pasquali-Ronchetti method.[127] The organization of the virions is probably a structure of the icosahedral type $T = 7$.[14]

III. STRUCTURE OF VIRAL DNA

Viral DNA purified from virions is circular, double-stranded, and presents discontinuities corresponding to regions sensitive to nuclease S1.[15,16] The number of discontinuities varies

$$T \; T \; T \; T \; T \; T \; T^{rA}_{dA} \text{-5'}$$
$$T$$
$$A$$

3'-G G G C G A * T T T T T T T A A C C A T A-5'
5'-C C C G C T T A A A A A A A T T G G T A T-3'

FIGURE 1.   Nucleotide structure at the sequence discontinuity situated on the α strand of CaMV, Cabb-S strain. The interruption, represented by the * sign, is due to the absence of a phosphodiester link, between the A + T nucleotides in this case. At the interruption, a triple-stranded structure is created.

according to the caulimovirus studied: the CM4-184 strain of CaMV has only two,[16] whereas the figwort mosaic virus (FMV) can have up to four.[17] The isolate Cabb-S of CaMV has three.[1] However, in all cases, one of the two strands of DNA has only one discontinuity: this strand is called the α strand. The other discontinuities located on the other strand delimit the single-strand DNA fragments called β, γ, and δ.

An extensive study of these discontinuities[18] revealed a sequence interruption due to the absence of a phosphodiester link between two nucleotides, the 5′ end of the region following the interruption being the repetition of the 3′ end of the same strand (Figure 1). Consequently, at this gap the DNA presents a triple-stranded structure which can extend to 40 nucleotides or more. The 5′ end of the third strand is generally associated with a short polyribonucleotide[19] involved in the replication of the viral genome (see Section VII). The existence of sequence interruptions is not necessary to express the pathogenicity of the genome since viral DNA obtained by cloning in *Escherichia coli,* and thus devoid of its interruptions, is always infectious[20,21] after inoculation of sensitive plants.

Besides this relaxed structure, viral DNA can present a supercoiled form,[22] unencapsidated and localized in the nucleus of infected cells.[23] This supercoiled DNA is associated with nuclear proteins, probably histones, and is present in the form of a minichromosome.[24] The observation of this minichromosome with the electron microscope shows that the number of nucleosomes corresponding to the value of the length of the viral genome is 41 ± 2.[24] Whereas on the SV40 minichromosome there exists a region not associated with histones and certainly corresponding to the origin of replication of the viral genome, no equivalent has been described for the CaMV minichromosome. These observations must be linked to the fact that the replication system of CaMV DNA is closer to that of the retroviruses than that of the animal DNA viruses (see Section VII).

Besides complete genomes, there exists in infected plants free supercoiled forms which are shorter than that of a whole viral genome.[25,26] They can be associated, or not, with proteins and their origin is not clearly known. In addition to these DNAs which are smaller than that of a whole genome, there exists a small DNA, 8S DNA, single-stranded and encapsidated with the viral genome.[19] The origin of this DNA is linked to the replication of the viral genome (see Section VIII).

Finally, complete molecules of viral DNA purified from virions possess knots, most likely resulting from the activity of a topoisomerase during the replication process of viral DNA. The existence of knots is not particular to CaMV DNA since these forms exist for P4 DNA, a satellite of the P2 bacteriophage.[24] It is possible that the existence of knotted molecules is linked to viral morphogenesis, but the significance and the role of such molecules are only a matter of speculation.

IV. CHARACTERIZATION OF CaMV GENES AND THEIR EXPRESSION PRODUCT

The genomes of different caulimoviruses have been sequenced: the CaMV genome, with

TABLE 1

**Coordinates of Possible Genes Deduced from the Nucleotide Sequences of
CaMV Isolates CABB-S, CM1841, D/H, and Xinjiang**

	Cabb-S	CM1841	D/H	Xinjiang
Genome size (base pairs)	8024	8031	8016	8060
Gene VII	13—303	13—303	13—303	13—303
Gene I	364—1347	364—1347	365—1348	365—1348
Gene II	1349—1828	1349—1828	1345—1824	1345—1824
Gene III	1830—2219	1830—2219	1826—2215	1826—2215
Gene IV	2201—3670	2201—3667	2197—3669	2197—3708
Gene V	3633—5672	3633—5669	3623—5650	3665—5689
Noncoding region (base pairs)	104	105	104	104
Gene VI	5776—7338	5774—7336	5754—7322	5793—7364
Noncoding region (base pairs)	698	707	706	708

TABLE 2

**Detection of CaMV Gene I Product in
Infected Turnips**

Protein molecular weight (theoretical molecular weight: 36.9 kDa)	Detected in		Ref.
	Viral inclusion bodies	Replication complexes	
46, 42, and 38 kDa	No	Yes	32
45 and 36 kDa	Yes	Not tested	33
41 kDa	Yes	Not tested	34
40 kDa	No	Yes	35

the Cabb-S,[1] CM1841,[27] D/H,[28] and Xinjiang[29] strains, and that of the carnation etched ring virus (CERV)[30] and of FMV.[31] The size of the genome varies from 7743 base pairs for the FMV to 8060 base pairs for the Xinjiang strain of CaMV. The study of the distribution of nonsense codons allowed the definition of six open reading frames and two intergenic regions common to all these caulimoviruses, the transcription of a single α-strand permitting the translation of six genes (genes I to VI). To simplify the presentation of the genomic organization of the caulimoviruses, CaMV was chosen as the model for the group. The interruption of the sequence localized on the α strand determined the point of reference for numbering the nucleotide sequence.

In comparing the four strains of CaMV for which the genome was sequenced (Table 1), in addition to the six genes common to all the caulimoviruses, a seventh open reading frame (gene VII) was localized 60 nucleotides upstream of gene I. An eighth open reading frame (gene VIII) overlapping the C-terminal region of gene IV was defined for the Cabb-S, CM1841, and D/H strains. However, it is likely that this gene is not required for multiplication of the virus since the Xinjiang strain contains two in-phase termination codons in this potential gene, rendering impossible the synthesis of a functional protein. Moreover, whether it be the case of gene VIII or gene VII, no viral product coded by these genes has been detected in extracts of infected plants, which is not the case for the six other potential genes of CaMV.

Gene I and its product of expression have only recently been studied. As one can observe in Table 2, the different authors who have studied the product of gene I are not all in agreement.[32-35] (Hereafter, the gene I product will be called P1 and this nomenclature will be applied to the other viral products of CaMV referred to in this article.) In comparing the sequence of P1 with other viral proteins, a homology of sequence with P30 of tobacco mosaic virus (TMV) was obvious.[30] However, P30 of TMV is involved in the diffusion of the virus from cell to cell in the infected plant[36] and was localized in the plasmodesmata of leaves infected by TMV.[37] Thus, P1 could also play a role in cell-to-cell diffusion, especially as this protein was recently detected exclusively in preparations enriched in the cell walls of turnip leaves 26 d after infection with CaMV.[35] The process by which the P1 intervenes at the diffusion level is not yet known. However, its absence in the viral particle,[33-35] its inability to interact with the CaMV genome,[128] and especially its probable localization in the cell wall indicate that P1 would act more on a cellular structure than directly on a viral element such as the genome or the virion. In the case of the dahlia mosaic virus (DaMV) which belongs to the caulimovirus group, it has been established that after infection of zinnia plants, the plasmodesmata of foliar tissues increase in diameter from 25—35 to 60—80 nm and elongate as well.[38] P1 could be directly or indirectly responsible for the modifications of the plasmodesmata, but this is only a hypothesis which remains to be confirmed.

The transmission of CaMV from plant to plant is carried out by aphids and depends upon a factor induced by the virus.[39] Aphids are capable of transmitting nontransmissible strains, provided they coinfect these strains with a transmissible viral strain. Indeed, transmission by aphids can be blocked by introducing mutants in gene II of CaMV, either by deletion[40,41] or by insertion[42] of nucleotides. Such mutants remain infectious, however. Likewise, a natural strain of CaMV (CM4-184) containing a deletion of 421 base pairs in gene II[43] is always infectious without being transmitted by aphids. Gene II codes for a protein, P2, of 18 kDa. However, this protein is also expressed for nontransmissible strains having a whole gene II such as the Campbell and CM1841 strains.[44] Thanks to the construction of recombinants between the Campbell strain and a transmissible strain, Cabb B-JI, the region of gene II involved in the diffusion of virus by insects was localized to a sequence of 129 base pairs.[45] By comparing this sequence among various strains of CaMV,[45,46] only one difference was detected, localized at amino acid 94 of the protein coded by gene II. Transmissible strains (Cabb-S, Cabb B-JI, D/H, Xinjiang, PV147) possess a glycine at this position, whereas the nontransmissible strains (Campbell, CM1841) possess an arginine. The transmission of CaMV by aphids is a negative transmission, i.e., insects feeding upon infected plants can acquire the viral agent responsible for the transmission of the virus on their stylet, this being possible only minutes after prehension. As purified virions are not transmissible,[47] whereas viroplasm preparations are,[48] the viral agent transmitted is certainly the viroplasm. This idea is reinforced by the conspicuous role played by P2 in the viroplasm structure,[49] whereas this protein is absent from the purified virions.[50]

Of all the CaMV proteins, the protein coded by gene III has been studied the least. P3 is a protein of low molecular weight: 15 kDa for CaMV,[1] 14 kDa for CERV,[30] and 13 kDa for FMV.[31] When one compares the percentage of homology between these different products, one observes 41.4% direct homology between P3 of CaMV and CERV,[30] 37% for FMV and CaMV, and 31% for CERV and FMV.[31] Compared to other viral proteins (Table 3), P3 shows a homology ratio in amino acids similar to that of the capsid protein coded by gene IV, but very far from that of P1 or the reverse transcriptase coded by gene V. However, these last two proteins have an exclusively intracellular function. Intracellular conditions of eukaryote cells change very slowly and thus viral proteins which function in the interior of the cell are generally better conserved during evolution. Apparently, this is not the case for P3. This viewpoint is reinforced by the fact that the protein was localized at the surface of the virion by immunocytochemistry.[51] By this same approach, P3 was localized in situ exclusively at the viroplasm level associated with the viral particles.[51] P3 purified from viroplasms has a molecular weight of 15 kDa,[52] whereas when it is associated with the virion, the C-terminal region degrades and P3 has a mass of only 11 kDa.[51] Moreover, not only does the structure of P3 change according to its localization, but it is capable of

TABLE 3
Direct Amino Acid Homologies Between the Proteins
Encoded by the Genes of CaMV, CERV, and FMV

	Percent		
	CaMV/CERV	CaMV/FMV	FMV/CERV
Gene I	48.3	54	49
Gene II	53.6	42	38
Gene III	41.4	37	31
Gene IV	41.3	38	33
Gene V	66.9	64	64
Gene VI	33.1	26	21

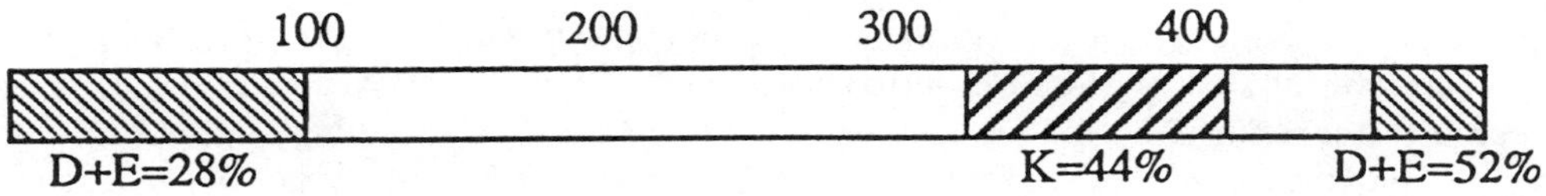

FIGURE 2.   Distribution of acidic (D + E) and basic (K) amino acid residues in the CaMV gene IV product.

interacting strongly with double-stranded DNA when it is expressed in its complete form in *E. coli*,[51,53] which is no longer the case when it is degraded in the C-terminal region.[51] In view of these results, it would seem that P3 is a component of the viral capsid, but probably has a very specific function. One can suppose that P3 would play a role in the decapsidation of the virion or, considering its localization, it could intervene in the movement of the virus in the cell by interacting with microtubules.

Gene IV codes for the 57-kDa protein P4, which corresponds to the major capsid protein.[54] After purification and dissociation of the viral particles, 57-kDa protein is detected in a polyacrylamide-SDS gel, but in small quantities, whereas polypeptides with molecular weights of 42 and 37 kDa appear in large amounts,[9,55] Following these results, it seemed reasonable to suppose that the primary product of gene IV of 57 kDa was processed to the polypeptides of 42 and 37 kDa which join together to form the viral capsid. However, it was recently established that the complete product of gene IV in a polyacrylamide-SDS gel did not migrate at 57 kDa, but, rather, at 76 to 80 kDa,[50,56] suggesting that the 57-kDa protein previously detected did not correspond to the complete product of gene IV, but, rather, to an already processed product. This hypothesis was confirmed by showing that, in fact, 57-kDa polypeptide detected *in vivo* did not possess the N-terminal end of P4.[57] The abnormal migration of P4 in polyacrylamide-SDS gels can be explained by the structure of this protein: a strong base region framed by two acid regions in the N- and C-terminal positions (Figure 2). The 57- and 42-kDa polypeptides can be phosphorylated[58] by a kinase protein associated with the virion.[59,60] Moreover, the 42-kDa protein can also be glycosylated.[61] Following experiments showing the structural modifications of the virion after phosphorylation,[59] it was suggested that this phosphorylation altered the capacity of the structural proteins to interact with the DNA of CaMV. One can also assume that the phosphorylation of these polypeptides is coupled with their maturation, with the phosphorylated proteins being cleaved by a specific protease.

The protein of gene V, P5, has a molecular weight of 80 kDa.[62,63] Strong sequence homologies between the different domains of the pol protein of the retroviruses, such as the Moloney murine leukemia virus (MoMLV) and the P5 protein of CaMV,[64-66] allowed the localization of two enzymatic sites on P5, one site characteristic of proteases specific to aspartic acid such as pepsin and one site corresponding to an activity such as reverse

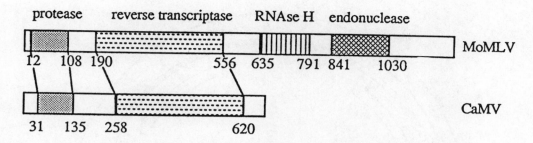

FIGURE 3.   Homology profiles along the length of the MoMLV polymerase and the CaMV gene V product. Homologous regions are indicated by the connecting lines. The sequence of MoMLV polymerase is divided into four regions: protease region,[66] reverse transcriptase region,[65,67,68] RNase H region[67-69] and endonuclease region.[66]

transcriptase. On the other hand, enzyme activity sites of the RNase H type and endonuclease coded by the pol protein of MoMLV (Figure 3) were not revealed. The absence of an endonuclease activity is not unusual since, as opposed to retroviruses, the double-stranded DNA of CaMV does not integrate into the genome of the infected cell. On the other hand, the nonexistence of an RNase site can seem surprising as such an activity is indispensible to the replication of a genome by a reverse transcriptase. However, an activity of the RNase type was detected in healthy turnips and was stimulated after infection by CaMV and even by the RNA viruses.[50] For the moment, only the reverse transcriptase activity has been associated without ambiguity with the P5 of CaMV.[70,71]

Protein P6 (66 kDa), coded by gene VI of CaMV, is the prime constituent of viroplasms,[72] cytoplasmic inclusions present in cells infected by caulimoviruses. Most of the viral particles are localized in these cytoplasmic inclusions, except in the case of CERV and DaMV, where the virions are generally free in the cytoplasm. The construction *in vitro* of recombinants between two strains of CaMV (D4 and CM1841) showed that the first half of gene VI determined the ability of CaMV to give a systemic infection on a solanaceous, *Datura stramonium*.[73] This experiment and others carried out with different strains of CaMV and different host plants[74,75] suggest that, in certain cases, the systemic diffusion of CaMV in the host plant depends on the N-terminal region of P6, but that, in other cases, P6 would act in association with other viral proteins. Moreover, P6 is capable of inducing viral symptoms in transgenic plants of *Nicotiana tabacum* which only express the product of gene VI of CaMV,[76] whereas *N. tabacum* is a nonsensitive host of CaMV.

V. TRANSCRIPTION OF THE VIRAL GENOME

Of all the CaMV genes previously described, only the mRNA of gene VI was able to be characterized with certainty.[77-79] This RNA, called 19S RNA (Figure 4), is transcribed by RNA polymerase II of the host cell from the α-strand of the genome,[80] capped, polyadenylated, and nonspliced. The 5' end is localized at nucleotide 5764 and the 3' end at nucleotide 7615 for the Cabb-S strain of CaMV.[81] Minor RNAs able to code for the proteins of gene I[82] and gene V[83] were described, but these results were never confirmed.

Besides the 19S RNA, another major capped, polyadenylated, and nonspliced RNA was characterized in cells infected by CaMV, the 35S RNA.[84] This RNA corresponds to the complete transcript of the genomic DNA (Figure 4). For the Cabb-S strain, its 5' end begins at nucleotide 7435 and its 3' end terminates at nucleotide 7615, as for the 19S RNA, thus presenting a terminal redundancy of 180 nucleotides.[81] This major 35S RNA is transcribed from the α strand of the genomic DNA, which is present in a supercoiled form without sequence interruption and is associated with nuclear proteins such as histones.[23] In addition to this major 35S RNA synthesized by RNA polymerase II from the host cell, a series of

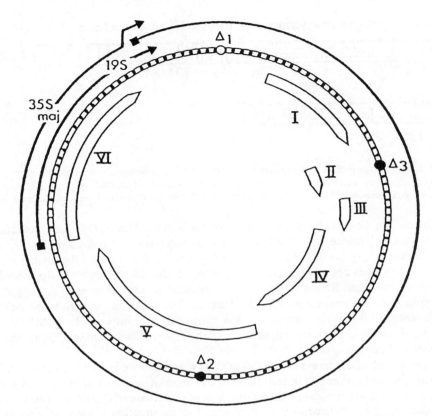

FIGURE 4.   Genetic map of CaMV, with the two transcripts indicated that are discussed in the text. Stippled circle represents the virion DNA with the discontinuities ($\Delta 1$, $\Delta 2$, and $\Delta 3$) indicated by a circle. Open arrows inside the circle represent open reading frames, and arrows outside the circle represent the transcripts.

       -143   -125   -104   -85
agcatcGTGGAAAaagaagacgttCCAACCACgtcttcaaagcaaGTGGATTG atgtgatatctCCACTga

      -64   -57   -30   +1
cgtaagggatgacgCACAATcCCACTatccttcgcaagacccttcctcTATATAAggaagttcatttcatttggagagg

FIGURE 5.   Structure of the CaMV 35S RNA promoter. Sequences homologous to the TATA box, CAAT box, and the SV40 core enhancer (GTGGWWWG) are written in capital letters.

minor 35S RNAs were characterized.[81,85] These minor RNAs overlap the whole viral genome, but present nonredundant 5' and 3' ends different from those of the major 35S RNA.

When one compares the activity of the promoter of the major 35S RNA to that of the promoter of gene VI of CaMV[86] or to that of the promoter of the nopaline synthase gene of the Ti plasmid of *Agrobacterium tumefaciens*,[87] one observes a clearly higher activity in the CaMV promoter. In addition to the classical sequences of the "TATA box" and "CAAT box" type, situated upstream of the promoter of the major 35S RNA, homologous sequences exist at the sequence enhancer of SV40 at positions -86 to -148 (Figure 5), position $+1$ corresponding to the initiation site of transcription.[86,88,89] This nucleotide region is capable of acting independently of its orientation and of stimulating a heterologous promoter.[86,89] The addition of several sequences of this type upstream of the promoter has an additive effect upon the stimulation of transcription.[86,90] However, unlike enhancers, the activity of this nucleotide region is influenced by its position: the further this sequence is moved from the initiation site of transcription, the more its effectiveness diminishes.[86,89] As these stimulation experiments were performed with plant cells insensitive to CaMV (tobacco, carrot, soya, maize), it is probable that the factors responsible for this stimulation were not of viral origin and are present in most plant cells.

The two major transcripts of CaMV possess the same termination site, localized at about 20 nucleotides downstream of the AATAAA sequence,[81] involved in the polyadenylation of eukaryote mRNA. However, for the major 35S RNA, the termination of transcription apparently takes place only during the second passage of RNA polymerase II at the termination site. It is possible that a weak termination signal allows the synthesis of large-sized RNA, which is then matured by a signal involving the polyadenylation sequence. In fact, it is more probable that a particular structure present on the RNA molecule at the end of synthesis interferes with the termination of transcription.

VI. TRANSLATION OF THE VIRAL GENOME

With the exception of gene VI, no individual mRNA was able to be characterized with certitude for the other genes of CaMV (see Section V). The absence of such molecules led us to consider the 35S RNAs as potential mRNAs. Among the different 35S RNAs described, two minor RNAs having their 5' end upstream of either gene II or gene IV[85] could allow the translation of these two genes. However, the existence of these two molecules was never confirmed and the authors could not explain how the synthesis of these RNAs began or ended. On the other hand, the existence of a third, minor 35S RNA was confirmed by several groups.[81,85] This RNA could be synthesized by RNA polymerase II, which could recognize the interrruption of the sequence situated on the α-strand of a viral DNA as the initiation and the termination site, possessing once again a circular, relaxed form upon its arrival at the nucleus of the infected cell. This idea of RNA polymerase II recognizing the sequence interruption as an initiation or termination site is reinforced by the detection of an 8S RNA.[81] In fact, this RNA begins at nucleotide 7435, a model initiation site recognized by RNA polymerase II (see Section V) and ends at nucleotide 8024, where the sequence interruption of the α-strand is located. And thus two 35S RNAs coexist in the nucleus of an infected cell, one minor RNA, synthesized from a circular, relaxed DNA and the other major RNA synthesized from a circular, supercoiled DNA. In both cases, these 35S RNAs, if they were messengers, would be polycistronic.

With such a hypothesis, different models are possible for explaining the translation of different genes. As for the RNA bacteriophages such as MS2, a nucleotide sequence located upstream of the AUG of each gene of CaMV, equivalent to the Shine-Dalgarno sequence of prokaryotes, would allow the binding of ribosomes. Such sequences have never been described for eukaryotes, for which the 40S ribosome subunit attaches itself to the cap of the mRNA[91] and moves along this molecule until it recognizes an AUG.[92] If this AUG presents itself in the optimal context of ANNAUGG,[93] the 40S subunit is coupled to the 60S subunit and thus the translation in initiated at this site. If the AUG is not in a favorable context, the 40S subunit continues until it finds a more favorable AUG.[93] In certain cases, the mRNA can present two AUGs recognized as an initiation site. Thus, when the AUG is not in a favorable context, initiation takes place either with this AUG or with the next, more effective AUG; this is the leaky scanning mechanism.[93] If the two initiation codons are in phase, we have a short and a long version of the same polypeptide.[94] If both AUGs are in a different reading frame, we have a dicistronic mRNA, giving two totally different proteins.[95] In the case of the leaky scanning mechanism, the translation of the second cistron can be more efficient than the first.[94] Lastly, the termination-reinitiation model applies when an in-frame terminator codon exists between two AUG initiators. In this case, the 40S ribosome subunit initiates the translation at the AUG by coupling with the 60S subunit, with protein synthesis occurring until the recognition of a codon stop. The 60S subunit then dissociates from the mRNA, whereas the 40S subunit remains attached to the mRNA and migrates to a favorable AUG, which will reinitiate translation.[96] As the initiation of translation of two cistrons is realized by the same 40S ribosome subunit, the quantity of protein synthesized from the second cistron is dependent upon the effectiveness of the reinitiation. The second cistron will never translate more effectively than the first cistron,[96-98] its rate of translation being inversely proportional to the distance between the codon stop of the first cistron and the initiation codon of the second cistron.

By modifying the nucleotide sequence of gene II, it is possible to increase the number of nucleotides between the codon stop of gene II and the codon AUG of gene III. Although the information of gene II is dispensable in the infection of CaMV (see Section IV), symptoms are slow to appear in plants infected by these mutants. However, if one reinfects new plants with crushed leaves coming from the first infection, no delay is noted. Deletions appeared at gene II of the mutants in order to position the codon stop once more in the proximity of the AUG of gene III,[99,100] thus allowing a cycle of normal infection. The translation of gene III therefore seems coupled to that of gene II, responding to the termination-reinitiation model previously described.

Whichever 35S RNA is used as a messenger, the first AUG of a potential gene is that of gene VII. Although the nucleotide context of the initiation codon is not very favorable at first sight (GCCAUGA[101]), a mutation of the AUG to ACG causes a delay in the appearance of symptoms.[102] As this mutation often reverses in infected plants, either the gene VII product is indispensable in multiplication or the AUG of gene VII is indispensable in the translation of the viral genome. However, the first hypothesis can be excluded as mutants deprived of a functional gene VII show no delay in the appearance of symptoms.[103] In view of its localization, gene VII can influence the expression of gene I, as in the agnogene of SV40 situated upstream of the genes coding for coat proteins.[104] In fact, it has been shown *in vitro* that, on one hand, the translation of gene I could depend upon that of gene VII, according to the termination-reinitiation model, but that, on the other hand, an internal initiation exists at the AUG level of gene I; in this case, the 40S ribosome subunit does not recognize the AUG of gene VII, which is consistent with the leaky scanning model.[50]

In view of these results, it is possible that the translation of genes VII, I, II, and III is intimately linked. However, in the case of gene IV, responsible for the synthesis of the coat protein, it is difficult to conclude that the translation of this protein is influenced by that of genes located upstream of gene IV. In fact this protein, compared to other viral proteins, seems to be present in a much larger quantity, which is also the case for the major protein of viroplasms, the only difference being that this major protein possesses its own mRNA.

Finally, the translation of gene V could depend upon that of gene IV. Whether for retroviruses,[105-107] hepadnaviruses,[108] or retroposons,[109] the translation of the reverse transcriptase gene is always linked to that of the coat protein gene, since in all cases the coat protein-reverse transcriptase polypeptide precursor is synthesized and then matured due to an aspartic acid protease of viral origin. As CaMV potentially possesses such a protease (see Section IV), a similar translation can be envisaged, although in certain cases it would seem that the reverse transcriptase could be synthesized independently of the coat protein.[110]

VII. REPLICATION OF THE VIRAL GENOME

The first step in replication is localized in the nucleus, where sequence interruptions in the relaxed DNA are eliminated and replaced by covalent links by a process as yet unknown. The modified DNA thus possesses a supercoiled form associated with nuclear proteins (see Section III) and in this way serves as the matrix for the synthesis of the major 35S RNA by RNA polymerase II of the host cell (for more details, see Section V). This RNA, due to its redundant ends of 180 nucleotides, serves as a template for the reverse transcriptase step localized in the viroplasms.[111,112] The synthesis of the α-strand of the viral DNA is probably primed by a tRNA^{Met} [113] (Figure 6.1) which is positioned 600 nucleotides from the 5′ end of the 35S RNA[19] (Figure 6.2). This initiation site is responsible for the formation of the sequence interruption present in the α strand. The reverse transcriptase, coded by gene V of CaMV (see Section IV), begins to copy the RNA into single-stranded DNA.[64,114] When the enzyme reaches the 5′ end of the 35S RNA (Figure 6.3), the synthesis can terminate, producing a single-stranded DNA of approximately 600 nucleotides.[19,115] However, in general, the synthesized DNA reassociates with the free 3′ end of the 35S RNA in a manner similar to its 5′ end[116-118] (Figure 6.4). After this passage from one end of the molecule to the other, the reverse transcriptase continues to the initiation site, moving the primer tRNA^{Met} and several nucleotides 5′ of the α-strand and thus creating a sequence interruption (Figure 6.5). The synthesis of the complementary strand of the α strand is certainly carried out by reverse transcriptase,[114] caused by the presence of ribonucleotide primers localized at future sequence interruptions[19] (Figure 6.6). In addition to the reverse transcriptase activity, replication of the viral genome requires an RNase H activity necessary for the removal of the 35S RNA matrix (Figure 6.3) and the tRNA^{Met} (Figure 6.6) as well as for the formation of the ribonucleotide primers involved in the initiation of the synthesis of the second strand of DNA.[116-118] The origin of the RNase H is not yet known (see Section IV). Besides the reverse transcriptase and RNase H, the coat protein is also associated with the DNA in replication,[119,120] suggesting that encapsidation is intimately linked to replication. This idea is reinforced by the presence in the coat protein of a sequence formed of three cysteines (n, n + 3, n + 13) and one histidine (n + 8) known for its capacity to interact with the RNA.[121] The encapsidation of the DNA matrix offers several advantages: (1) to protect the RNA from cellular RNases, (2) to separate the RNA involved in replication from the RNA involved in translation, and (3) to compartmentalize replication of the viral genome and thus avoid the synthesis of DNA from RNA of cellular origin.

VIII. RECOMBINATION OF THE VIRAL DNA

The DNA of CaMV was cloned in a bacterial vector which allows rapid replication and facilitates its purification, as well as *in vitro* and *in vivo* manipulations. It was thus shown that the DNA of CaMV, amplified in a bacterial vector, purified, and excised from the vector by the action of the restriction enzyme before being inoculated into sensitive plants, is infectious.[20,21] These experiments show that the whole viral genome, but inoculated in a linear form, penetrates the cell nucleus where it is circularized by nuclear enzymes and then maintained in the minichromosome state before being transcribed by RNA polymerase II. Furthermore, turnips inoculated with two populations of viral genomes, each integrated at a different restriction site (homologous for the vector and the viral genome) and maintained in association with the vector, are infected. Study of the encapsidated viral DNA isolated from such infected plants shows that the genome is identical to that of the cloned viral strain. On the other hand, if the inoculum contains only one population of viral genomes, all integrated at the same restriction site of the vector, this infection is not successful.[122] This process corresponds to a classical method of recombination inducing the formation of heteroduplexes between viral DNA molecules and by crossing-over generating infectious viral DNA molecules. Similarly, viral DNA molecules rendered noninfectious by deletion can, by this method, yield infectious molecules, provided that two mutants deleted from different regions of the viral genome are coinoculated.[122-125] The detection of virions in plants treated in this manner was significant in demonstrating the existence of a nuclear stage in the infection process of the virus. Nevertheless, some questions remain unanswered. The struc-

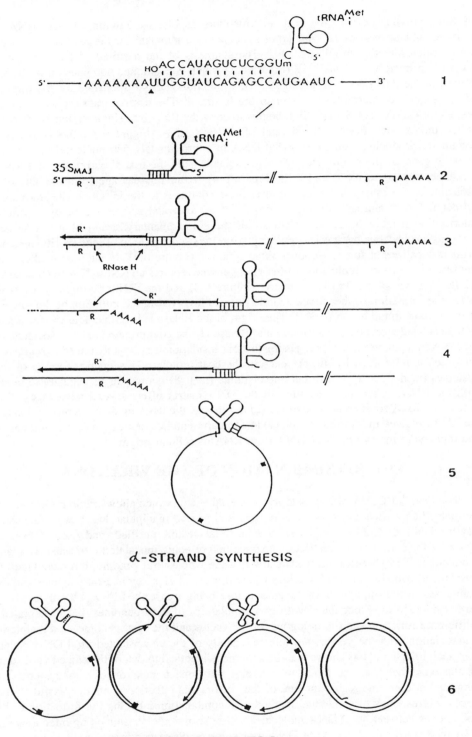

FIGURE 6.   The model for synthesis of the CaMV α, β and γ strands by reverse transcription of the major 35 S transcript using tRNA[Met] as primer. R designates the repeated sequences at the 5′ and 3′ ends of the 35 S transcript. cDNA is indicated by thicker lines. Small sequences associated with the complete α strand represent primers for initiation of complementary strand synthesis (Courtesy of Dr. K. E. Richards, Institut de Biologie Moléculaire des Plantes, Strasbourg, France).

ture of the minichromosome is such that apparently no replication origin analogous to that of SV40, for example, exists; one must therefore assume that inoculated viral DNA molecules associated with nuclear histones will participate in the recombination process.

As previously described (see Section VII), the amplification of encapsidated DNA molecules is brought about by the transcription of one of the viral DNA strands into a genomic RNA molecule (35 S RNA.) After migration toward the cytoplasm, this RNA molecule is used as a template by reverse transcriptase to generate circular, relaxed DNA viral molecules. Considering this stage of replication, it was possible to show that a process other than the one just described could be the origin of the infectious heterologous DNA molecules. This process depends upon the structure of the viral DNA molecules contained in the inoculum. Thus, when the inoculum contains a heterologous pair of viral genomes — one infectious, the other deleted (gene VI is replaced by the fragment of the transposon Tn5 capable of expressing the resistance factor to kanamycin) — lined up at the same cloning site and coinoculated, it induces the multiplication of viral DNA molecules, recombinant or not, in the infected cells.[126] The analysis of the genome of the heterologous strain demonstrates that the initial template must be a 35S RNA molecule generated by the transcription of a dimer (Figure 7). The viral DNA molecules present in the nucleus associate as dimers due to their identical cohesive ends. Each dimer organizes itself in a large minichromosome upon which RNA polymerase II finds the transcription signal and synthesizes a 35S RNA molecule heterologous in sequence information. This population of heterologous 35S RNA molecules migrates toward the cytoplasm, where it follows the viral replication cycle. If the inoculum contains two heterologous infectious genomes, with linear molecules having the same cohesive ends, the infected plants contain the two viral populations. Last, when the inoculum corresponds to a homologous viral DNA population — where one fraction is infectious and the other contains deletions, and with the ends lined up by the same restriction enzyme — there follows an inhibition in the multiplication of the infectious strain as if a ligation polarity existed between the viral DNA molecules, a polarity favoring the formation of dimers carrying the information of the transposon and, therefore, that of the transcription of the defective 35S RNA molecules. The formation of a dimer used as a matrix by RNA polymerase II to give birth to heterologous molecules was shown by using an inoculum containing heterologous DNA molecules truncated by enzymatic treatment, either at the 5' end of the infectious DNA molecules or at the 3' end of the defective ones. The plants inoculated with such an inoculum showed no symptoms of virosis. This dimerization process was considered to be at the origin of recombinants studied by Penswick et al.[110] The inoculum is represented by two heterologous parents, one defective in the intergenic region (about 422), and the other having a linker in the ORFV (about 169). Thus, the viral DNA molecules, having reached the nucleus of the cells of the inoculated plants, can, according to their initial structure, either separate into single strands and then form a new double-stranded structure of the minichromosome type, which is the origin of the infectious genomes described, or form dimers which will be transcribed into genomic RNA. The possibility of obtaining infectious genomes with hybrid information *in vivo* is an open possibility in the study of interactions existing between certain viral genes.

IX. REMARKS

Cauliflower mosaic virus (CaMV) is, without doubt, a very interesting plant virus whose study has probably raised more questions that it has answered. Concerning its structure, the fact that it is type $T = 7$ (unique for plant viruses) and that its genetic information is DNA have prompted many questions, the least difficult to resolve being those concerning the structure of the viral DNA, its mode of replication, and the determination of its products expressed by different genes as well as their role. This does not mean that we know their

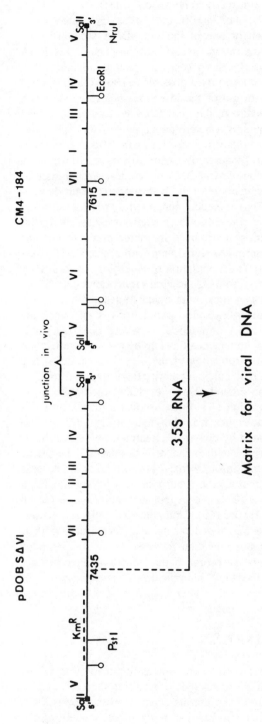

FIGURE 7.   Formation of the hybrid 35S RNA used as a template to generate recombinants. This dimer, represented in a linear form, would be circular in infected cells. Roman numerals indicate the localization of genes on each DNA molecule. *Pst*I, *Nru*I, and *Bgl*I indicate the localization of DNA sequences recognized by these enzynes. *Eco*RI sites are represented by downward lines ending with circles.

tertiary structure and their properties. Remember that the structure of viral proteins and their mode of association are not yet known. Progress has been made in this domain, but the organization of the capsid remains a mystery. This is due to the difficulty of obtaining tridimensional crystals. We believe that, among the problems to solve in the near future, those concerning DNA-protein and protein-protein relationships are among the most interesting, but also the most difficult to resolve.

The role of a reverse transcriptase in the multiplication of a plant virus is without doubt one of the most important discoveries in molecular biology. This has confirmed that plants and animals, as well as their viruses, probably have a common origin. Several authors have suggested that the RNA molecule was the molecule from which various primitive, information-packed molecules were formed. However, the relative instability or fragility of the RNA molecule was not favorable for the rapid development of the information necessary for the evolution of organisms. Stability occurred by acquiring stronger DNA structures. One can then ask the question, why are plant viruses mainly RNA and animal viruses mainly DNA? We cannot answer this question satisfactorily at present. We can only assume (by carefully avoiding all exceptions to this rule) that, "in general", plants furnish large quantities of virus, which is rarely the case in animal cells; the necessity of protecting genetic information is perhaps less important for plants than for animals.

Are animal retroviruses an intermediate form between RNA viruses and DNA viruses? If we accept this hypothesis, then, in this case, are the caulimoviruses linked to the animal retroviruses? This is not evident, of course, but one must remember that many analogies exist between RNA plant and animal viruses, in particular concerning the genomic regions corresponding to the expression of the functional genes in certain RNA viruses. The discovery of the structure of CaMV[1] and of its functional genes has reinforced the concept of an initial unity of life on earth.

REFERENCES

1. **Franck, A., Guilley, H., Jonard, G., Richards, K., and Hirth, L.,** Nucleotide sequence of cauliflower mosaic virus DNA, *Cell,* 21, 285, 1980.
2. **Hirth, L.,** Organisation des génomes viraux de plantes, *Ann. Sci. Nat. Bot. Paris,* 5, 163, 1983.
3. **Hirth, L.,** The molecular biology of caulimoviruses, *Microbiol. Sci.,* 3, 260, 1986.
4. **Lebeurier, G.,** Cauliflower mosaic virus potential vector gene. Genome organization and replication, *Symbiosis,* 2, 19, 1986.
5. **Pirone, T. P., Pound, G. S., and Shepherd, R. J.,** Properties and serology of purified cauliflower mosaic virus, *Phytopathology,* 51, 541, 1961.
6. **Shepherd, R. J., Bruening, G. E., and Wakeman, R. J.,** Double-stranded DNA from cauliflower mosaic virus, *Virology,* 41, 339, 1970.
7. **Goodman, R. M.,** Geminiviruses, *J. Gen. Virol.,* 54, 9, 1981.
8. **Hull, R. and Shepherd, R. J.,** The coat proteins of cauliflower mosaic virus, *Virology,* 70, 217, 1976.
9. **Al Ani, R., Pfeiffer, P., and Lebeurier, G.,** The structure of cauliflower mosaic virus, *Virology,* 93, 188, 1979.
10. **Hohn, T., Hohn, B., and Pfeiffer, P.,** Reverse transcription in CaMV, *Trends Biochem. Sci.,* 10, 205, 1985.
11. **Hull, R. and Covey, S. N.,** Genome organization and expression of reverse transcribing elements: variations and a theme, *J. Gen. Virol.,* 67, 1751, 1986.
12. **Chauvin, C., Jacrot, B., Lebeurier, G., and Hirth, L.,** The structure of cauliflower mosaic virus. A neutron diffraction study, *Virology,* 96, 640, 1979.
13. **Al Ani, R., Pfeiffer, P., Lebeurier, G., and Hirth, L.,** The structure of cauliflower mosaic virus, *Virology,* 93, 175, 1979.
14. **Krüse, J., Timmins, P., and Witz, J.,** The spherically averaged structure of a DNA isometric plant virus: cauliflower mosaic virus, *Virology,* 159, 166, 1987.

15. **Volovitch, M., Drugeon, G., and Yot, P.,** Studies on the single-stranded discontinuities of the cauliflower mosaic virus genome, *Nucleic Acids Res.,* 5, 2913, 1978.
16. **Hull, R. and Howell, S. H.,** Structure of the cauliflower mosaic virus genome, *Virology,* 86, 482, 1978.
17. **Hull, R. and Donson, J.,** Physical mapping of the DNAs of carnation etched ring and figwort mosaic viruses, *J. Gen. Virol.,* 60, 125, 1982.
18. **Richards, K. E., Guilley, H., and Jonard, G.,** Further characterization of the discontinuities in cauliflower mosaic virus DNA, *FEBS Lett.,* 134, 67, 1981.
19. **Guilley, H., Richards, K. E., and Jonard, G.,** Observations concerning the discontinuous DNAs of cauliflower mosaic virus, *EMBO J.,* 2, 277, 1983.
20. **Lebeurier, G., Hirth, L., Hohn, T., and Hohn, B.,** Infectivities of native and cloned cauliflower mosaic virus DNA, *Gene,* 12, 139, 1980.
21. **Howell, S. H., Walker, L. L., and Dudley, R. K.,** Cloned cauliflower mosaic virus DNA infects turnips, *Science,* 208, 1265, 1980.
22. **Ménissier, J., Lebeurier, G., and Hirth, L.,** Free cauliflower mosaic virus supercoiled DNA in infected leaves, *Virology,* 117, 322, 1982.
23. **Olszewski, N., Hagen, G., and Guilfoyle, T.,** A transcriptionally active, covalently closed minichromosome of cauliflower mosaic virus DNA isolated from infected turnip leaves, *Cell,* 29, 395, 1982.
24. **Ménissier, J., De Murcia, G., Lebeurier, G., and Hirth, L.,** Electron microscopic studies of the different topological forms of the cauliflower mosaic virus DNA: knotted encapsidated DNA and nuclear minichromosome, *EMBO J.,* 2, 1067, 1983.
25. **Olszewski, N. E. and Guilfoyle, T. J.,** Nuclei purified from cauliflower mosaic virus-infected turnip leaves contain subgenomic, covalently closed circular cauliflower mosaic virus DNAs, *Nucleic Acids Res.,* 11, 8961, 1983.
26. **Rollo, F. and Covey, S. N.,** Cauliflower mosaic virus DNA persists as supercoiled forms in cultured turnip cells, *J. Gen. Virol.,* 66, 603, 1985.
27. **Gardner, R. C., Howarth, A. J., Hahn, P., Brown-Luedi, M., Shepherd, R. J., and Messing, J.,** The complete nucleotide sequence of an infectious clone of cauliflower mosaic virus by M13mp7 shotgun sequencing, *Nucleic Acids Res.,* 9, 2871, 1981.
28. **Balàzs, E., Guilley, H., Jonard, G., and Richards, K.,** Nucleotide sequence of DNA from an altered-virulence isolate D/H of the cauliflower mosaic virus, *Gene,* 19, 239, 1982.
29. **Rongxiang, F., Xiaojun, W., Ming, B., Yingchuan, T., Faxing, C., and Keqiang, M.,** Complete nucleotide sequence of cauliflower mosaic virus (Xinjiang isolate) genomic DNA, *Chinese J. Virol.,* 1, 247, 1985.
30. **Hull, R., Sadler, J., and Longstaff, M.,** The sequence of carnation etched ring virus DNA. Comparison with cauliflower mosaic virus and retroviruses, *EMBO J.,* 5, 3083, 1986.
31. **Richins, R. D., Scholthof, H. B., and Shepherd, R. J.,** Sequence of figwort mosaic virus DNA (caulimovirus group), *Nucleic Acids Res.,* 15, 8451, 1987.
32. **Harker, L. C., Mullineaux, P. M., Bryant, J. A., and Maule, A. J.,** Detection of CaMV gene I and gene VI protein products in vivo using antisera raised to COOH-terminal β-galactosidase fusion proteins, *Plant Mol. Biol.,* 8, 275, 1987.
33. **Young, M. J., Daubert, S. D., and Shepherd, R. J.,** Gene I products of cauliflower mosaic virus detected in extracts of infected tissue, *Virology,* 158, 444, 1987.
34. **Martinez-Izquierdo, J. A., Fütterer, J., and Hohn, T.,** Protein encoded by ORF I of cauliflower mosaic virus is part of the viral inclusion body, *Virology,* 160, 527, 1987.
35. **Albrecht, H., Geldreich, A., Ménissier de Murcia, J., Kirchherr, D., Mesnard, J. M., and Lebeurier, G.,** Cauliflower mosaic virus gene I product detected in a cell-wall-enriched fraction, *Virology,* 163, 503, 1988.
36. **Atabekov, J. G. and Dorokhov, Y. L.,** Plant virus-specific transport function and resistance of plants to viruses, *Advan. Virus Res.,* 29, 313, 1984.
37. **Tomenius, K., Clapham, D., and Meshi, T.,** Localization by immunogold cytochemistry of the virus-coded 30 K protein in plasmodesmata of leaves infected with tobacco mosaic virus, *Virology,* 160, 363, 1987.
38. **Kitajima, E. W. and Lauritis, J. A.,** Plant virions in plasmodesmata, *Virology,* 37, 681, 1969.
39. **Lung, M. C. Y. and Pirone, T. P.,** Studies on the reason for differential transmission of cauliflower mosaic virus isolates by aphids, *Phytopathology,* 63, 910, 1973.
40. **Armour, S. L., Melcher, U., Pirone, T. P., Lyttle, D. J., and Essenberg, R. G.,** Helper component for aphid transmission encoded by region II of cauliflower mosaic virus DNA, *Virology,* 129, 25, 1983.
41. **Woolston, C. J., Covey, S. N., Penswick, J. R., and Davies, J. W.,** Aphid transmission and a polypeptide are specified by a defined region of the cauliflower mosaic virus genome, *Gene,* 23, 15, 1983.
42. **Daubert, S., Shepherd, R. J., and Gardner, R. C.,** Insertional mutagenesis of the cauliflower mosaic virus genome, *Gene,* 25, 201, 1983.

43. **Howarth, A. J., Gardner, R. C., Messing, J., and Shepherd, R. J.,** Nucleotide sequence of naturally occurring deletion mutants of cauliflower mosaic virus, *Virology,* 112, 678, 1981.

44. **Harker, C. L., Woolston, C. J., Markham, P. G., and Maule, A. J.,** Cauliflower mosaic virus aphid transmission factor protein is expressed in cells infected with some aphid non transmissible isolates, *Virology,* 160, 252, 1987.

45. **Woolston, C. J., Czaplewski, L. G., Markham, P. G., Goad, A. S., Hull, R., and Davies, J. W.,** Location and sequence of a region of cauliflower mosaic virus gene 2 responsible for aphid transmissibility, *Virology,* 160, 246, 1987.

46. **Modjtahedi, N., Volovitch, M., Mazzolini, L., and Yot, P.,** Comparison of the predicted secondary structure of aphid transmission factor for transmissible and non-transmissible cauliflower mosaic virus strains, *FEBS Lett.,* 181, 223, 1985.

47. **Pirone, T. P. and Megahed, E.,** Aphid transmissibility of some purified viruses and viral RNAs, *Virology,* 30, 631, 1966.

48. **Rodriguez, D., Lopez-Abella, D., and Diaz-Ruiz, J. R.,** Viroplasms of an aphid-transmissible isolate of cauliflower mosaic virus contain helper component activity, *J. Gen. Virol.,* 68, 2063, 1987.

49. **Givord, L., Xiong, C., Giband, M., Koenig, I., Hohn, T., Lebeurier, G., and Hirth, L.,** A second cauliflower mosaic virus gene product influences the structure of the viral inclusion body, *EMBO J.,* 3, 1423, 1984.

50. **Gordon, K., Pfeiffer, P., Fütterer, J., and Hohn, T.,** In vitro expression of cauliflower mosaic virus genes, *EMBO J.,* 7, 309, 1988.

51. **Giband, M., Mesnard, J. M., and Lebeurier, G.,** The gene III product (P15) of cauliflower mosaic virus is a DNA-binding protein while an immunologically related P11 polypeptide is association with virions, *EMBO J.,* 5, 2433, 1986.

52. **Xiong, C., Lebeurier, G., and Hirth, L.,** Detection in vivo of a new gene product (gene III) of cauliflower mosaic virus, *Proc. Nat. Acad. Sci. U.S.A.,* 81, 6608, 1984.

53. **Mesnard, J. M., Geldreich, A., Xiong, C., Lebeurier, G., and Hirth, L.,** Expression of a putative plant viral gene in *Escherichia coli, Gene,* 31, 39, 1984.

54. **Daubert, S., Richins, R., Shepherd, R. J., and Gardner, R. C.,** Mapping of the coat protein gene of cauliflower mosaic virus by its expression in a prokaryotic system, *Virology,* 122, 444, 1982.

55. **Hahn, P. and Shepherd, R. J.,** Evidence for a 58-kilodalton polypeptide as precursor of the coat protein of cauliflower mosaic virus, *Virology,* 116, 480, 1982.

56. **Albrecht, H. and Lebeurier, G.,** Expression of CaMV ORF IV in *E. coli, Ann. Virol. (Inst. Pasteur),* 139, 263, 1988.

57. **Kirchherr, D., Albrecht, H., Mesnard, J. M., and Lebeurier, G.,** Expression of the cauliflower mosaic virus capsid gene in vivo, *Plant Mol. Biol.,* 11, 271, 1988.

58. **Hahn, P. and Shepherd, R. J.,** Phosphorylated proteins in cauliflower mosaic virus, *Virology,* 107, 295, 1980.

59. **Ménissier-de Murcia, J., Geldreich, A., and Lebeurier, G.,** Evidence for a protein kinase activity associated with purified particles of cauliflower mosaic virus, *J. Gen. Virol.,* 67, 1885, 1986.

60. **Martinez-Izquierdo, J. and Hohn, T.,** Cauliflower mosaic virus coat protein is phosphorylated in vitro by a virion-associated protein kinase, *Proc. Natl. Acad. Sci. U.S.A.,* 84, 1824, 1987.

61. **Du Plessis, D. H. and Smith, P.,** Glycosylation of the cauliflower mosaic virus capsid polypeptide, *Virology,* 109, 403, 181.

62. **Ziegler, V., Laquel, P., Guilley, H., Richards, K., and Jonard, G.,** Immunological detection of cauliflower mosaic virus gene V protein produced in engineered bacteria or infected plants, *Gene,* 36, 271, 1985.

63. **Pietrazak, M. and Hohn, T.,** Translation products of cauliflower mosaic virus ORF V, the coding region corresponding to the retrovirus pol gene, *Virus Genes,* 1, 83, 1987.

64. **Volovitch, M., Modjtahedi, N., Yot, P., and Brun, G.,** RNA-dependent DNA polymerase activity in cauliflower mosaic virus-infected plant leaves, *EMBO J.,* 3, 309, 1984.

65. **Toh, H., Kikuno, R., Hayashida, H., Miyata, T., Kugimiya, W., Inouye, S., Yuki, S., and Saigo, K.,** Close structural resemblance between putative polymerase of a *Drosophila* transposable genetic element 17.6 and pol gene product of Moloney murine leukaemia virus, *EMBO J.,* 4, 1267, 1985.

66. **Toh, H., Ono, M., Saigo, K., and Miyata, T.,** Retroviral protease-like sequence in the yeast transposon Ty1, *Nature (London),* 315, 691, 1985.

67. **Kotewicz, M. L., Sampson, C. M., D'Alessio, J. M., and Gerard, G. F.,** Isolation of cloned Moloney murine leukemia virus reverse transcriptase lacking ribonuclease H activity, *Nucleic Acids Res.,* 16, 265, 1988.

68. **Tanesse, N. and Goff, S. P.,** Domain structure of the Moloney murine leukemia virus reverse transcriptase: mutational analysis and separate expression of the DNA polymerase and RNase H activities, *Proc. Natl. Acad. Sci. U.S.A.,* 85, 1777, 1988.

69. **Johnson, M. S., Mc Clure, M. A., Feng, D.-F., Gray, J., and Doolittle, R. F.,** Computer analysis of retroviral pol genes: assignment of enzymatic functions to specific sequences and homologies with nonviral enzymes, *Proc. Nat. Acad. Sci. U.S.A.,* 83, 7648, 1986.

70. **Laquel, P., Ziegler, V., and Hirth, L.,** The 80 K polypeptide associated with the replication complexes of cauliflower mosaic virus is recognized by antibodies to gene V translation product, *J. Gen. Virol.,* 67, 197, 1986.

71. **Takatsuji, H., Hirochika, H., Fukushi, T., and Ikeda, J.,** Expression of cauliflower mosaic virus reverse transcriptase in yeast, *Nature (London),* 319, 240, 1986.

72. **Al Ani, R., Pfeiffer, P., Whitechurch, O., Lesot, A., Lebeurier, G., and Hirth, L.,** A virus-specified protein produced upon infection by cauliflower mosaic virus (CaMV), *Ann. Virol. (Inst. Pasteur),* 131E, 33, 1980.

73. **Schoelz, J., Shepherd, R. J., and Daubert, S.,** Region VI of cauliflower mosaic virus encodes a host range determinant, *Mol. Cell. Biol.,* 6, 2632, 1986.

74. **Daubert, S., Schoelz, J., Debao, L., and Shepherd, R. J.,** Expression of disease symptoms in cauliflower mosaic virus genomic hybrids, *J. Mol. Appl. Genet.,* 2, 537, 1984.

75. **Schoelz, J. and Shepherd, R. J.,** Host range control of cauliflower mosaic virus, *Virology,* 162, 30, 1988.

76. **Baughman, G. A., Jacobs, J. D., and Howell, S. H.,** Cauliflower mosaic virus gene VI produces a symptomatic phenotype in transgenic tobacco plants, *Proc. Natl. Acad. Sci. U.S.A.,* 85, 733, 1988.

77. **Odell, J. T. and Howell, S. H.,** The identification, mapping, and characterization of mRNA for P66, a cauliflower mosaic virus-coded protein, *Virology,* 102, 349, 1980.

78. **Covey, S. N. and Hull, R.,** Transcription of cauliflower mosaic virus DNA. Detection of transcripts, properties, and location of the gene encoding the virus inclusion body protein, *Virology,* 111, 463, 1981.

79. **Xiong, C., Muller, S., Lebeurier, G., and Hirth, L.,** Identification by immunoprecipitation of cauliflower mosaic virus in vitro major translation product with a specific serum against viroplasm protein, *EMBO J.,* 1, 971, 1982.

80. **Guilfoyle, T. J.,** Transcription of the cauliflower mosaic virus genome in isolated nuclei from turnip leaves, *Virology,* 107, 71, 1980.

81. **Guilley, H., Dudley, R. K., Jonard, G., Balàzs, E., and Richards, K. E.,** Transcription of cauliflower mosaic virus DNA: detection of promoter sequences, and characterization of transcripts, *Cell,* 30, 763, 1982.

82. **Condit, C., Hagen, T. J., McKnight, T. D., and Meagher, R. B.,** Characterization and preliminary mapping of cauliflower mosaic virus transcripts, *Gene,* 25, 101, 1983.

83. **Plant, A. L., Covey, S. N., and Grierson, D.,** Detection of a subgenomic mRNA for gene V, the putative reverse transcriptase gene of cauliflower mosaic virus, *Nucleic Acids Res.,* 13, 8305, 1985.

84. **Howell, S. H. and Hull, R.,** Replication of cauliflower mosaic virus and transcription of its genome in turnip leaf protoplasts, *Virology,* 86, 468, 1978.

85. **Condit, C. and Meahger, R. B.,** Multiple, discrete 35 S transcripts of cauliflower mosaic virus, *J. Mol. Appl. Genet.,* 2, 301, 1983.

86. **Ow, D. W., Jacobs, J. D., and Howell, S. H.,** Functional regions of the cauliflower mosaic virus 35 S RNA promoter determined by use of the firefly luciferase gene as a reporter of promoter activity, *Proc. Nat. Acad. Sci. U.S.A.,* 84, 4870, 1987.

87. **Sanders, P. R., Winter, J. A., Barnason, A. R., Rogers, S. G., and Fraley, R. T.,** Comparison of cauliflower mosaic virus 35 S nopaline synthase promoter in transgenic plants, *Nucleic Acids Res.,* 15, 1543, 1987.

88. **Odell, J. T., Nagy, F., and Chua, N.-H.,** Identification of DNA sequences required for activity of the cauliflower mosaic virus 35 S promoter, *Nature (London),* 313, 810, 1985.

89. **Odell, J. T., Knowlton, S., Lin, W., and Mauvais, C. J.,** Properties of an isolated transcription stimulating sequence derived from the cauliflower mosaic virus 35 S promoter, *Plant Mol. Biol.,* 10, 263, 1988.

90. **Kay, R., Chan, A., Daly, M., and McPherson, J.,** Duplication of CaMV 35 S promoter sequences creates a strong enhancer for plant genes, *Science,* 236, 1299, 1987.

91. **Shatkin, A. Y.,** mRNA cap binding proteins: essential factors for initiating translation, *Cell,* 40, 223, 1985.

92. **Kozak, M.,** Influences of mRNA secondary structure on initiation by eukaryotic ribosomes, *Proc. Natl. Acad. Sci. U.S.A.,* 83, 2850, 1986.

93. **Kozak, M.,** Bifunctional messenger RNAs in eukaryotes, *Cell,* 47, 481, 1986.

94. **Sedman, S. A. and Mertz, J. E.,** Mechanisms of synthesis of virion proteins from the functionally bigenic late mRNAs of Simian Virus 40, *J. Virol.,* 62, 954, 1988.

95. **Giorgi, C., Blumberg, B. M., and Kolakofsky, D.,** Sendai virus contains overlapping genes expressed from a single mRNA, *Cell,* 35, 829, 1983.

96. **Peabody, D. S. and Berg, P.,** Termination-reinitiation in the translation of mammalian cell mRNAs, *Mol. Cell. Biol.,* 6, 2695, 1986.

97. **Peabody, D. S., Subramani, S., and Berg, P.,** Effect of upstream reading frames on translation efficiency in simian virus 40 recombinants, *Mol. Cell. Biol.,* 6, 2704, 1986.

98. **Kaufman, R. J., Murtha, P., and Davies, M. V.,** Translational efficiency of polycistronic mRNAs and their utilization to express heterologous genes in mammalian cells, *EMBO J.,* 6, 187, 1987.

99. **Sieg, K. and Gronenborn, B.,** Introduction and propagation of foreign DNA in plants using cauliflower mosaic virus as a vector, Abstr. NATO/FEBS Course Str. Funct. Plant Genome, 1982, 154.

100. **Melcher, U., Steffens, D. L., Lyttle, D. J., Lebeurier, G., Lin, H., Choe, I. S., and Essenberg, R. C.,** Infectious and non-infectious mutants of cauliflower mosaic virus DNA, *J. Gen. Virol.,* 67, 1491, 1986.

101. **Hull, R.,** A model for the expression of CaMV nucleic acid, *Plant Mol. Biol.,* 3, 121, 1984.

102. **Dixon, L., Jiricny, J., and Hohn, T.,** Oligonucleotide directed mutagenesis of cauliflower mosaic virus DNA using a repair-resistant nucleoside analogue: identification of an agnogene initiation codon, *Gene,* 41, 1986.

103. **Dixon, L. K. and Hohn, T.,** Initiation of translation of the cauliflower mosaic virus genome from a polycistronic mRNA: evidence from deletion mutagenesis, *EMBO J.,* 3, 2731, 1984.

104. **Good, P. J., Welch, R. C., Barkan, A., Somasekhar, M. B., and Mertz, J. E.,** Both VP2 and VP3 are synthesized from each of the alternatively spliced late 19 S RNA species of simian virus 40, *J. Gen. Virol.,* 62, 944, 1988.

105. **Opperman, H., Bishop, J. M., Varmus, H. E., and Levintow, L.,** A joint product of the genes gag and pol of avian sarcoma virus: a possible precursor of reverse transcriptase, *Cell,* 12, 993, 1977.

106. **Yoshinaka, T., Katoh, I., Copeland, T. D., and Oroszlan, S.,** Murine leukemia virus protease is encoded by the gag-pol gene and is synthesized through suppression of an amber termination codon, *Proc. Natl. Acad. Sci. U.S.A.,* 83, 1618, 1985.

107. **Moore, R., Dixon, M., Smith, R., Peters, G., and Dickson, C.,** Complete nucleotide sequence of a milk-transmitted mouse mammary tumor virus: two frameshift suppression events are required for translation of gag and pol, *J. Virol.,* 61, 480, 1987.

108. **Will, H., Salfeld, J., Pfaff, E., Manso, C., Theilmann, L., and Schaler, H.,** Putative reverse transcriptase intermediates of human hepatitis B virus in primary liver carcinomas, *Science,* 231, 594, 1986.

109. **Wilson, W., Malim, M. H., Mellor, J., Kingsman, A. J., and Kingsman, S. M.,** Expression strategies of the yeast retrotransposon Ty: a short sequence directs ribosomal frameshifting, *Nucleic Acids Res.,* 14, 7001, 1986.

110. **Penswick, J., Hübler, R., and Hohn, T.,** A viable mutation in cauliflower mosaic virus, a retroviruslike plant virus, separates its capsid protein and polymerase genes, *J. Virol.,* 62, 1460, 1988.

111. **Mazzolini, L., Bonneville, J. M., Volovitch, M., Magazin, M., and Yot, P.,** Strand-specific viral DNA synthesis in purified viroplasms isolated from turnip leaves infected with cauliflower mosaic virus, *Virology,* 145, 293, 1985.

112. **Thomas, C. M., Hull, R., Bryant, J. A., and Maule, A. J.,** Isolation of a fraction from cauliflower mosaic virus-infected protoplasts which is active in the synthesis of ($+$) and ($-$) strand viral DNA and reverse transcription of primed RNA templates, *Nucleic Acids Res.,* 13, 4557, 1985.

113. **Turner, D. S. and Covey, S. N.,** A putative primer for the replication of cauliflower mosaic virus by reverse transcription is virion-associated, *FEBS Lett.,* 165, 285, 1984.

114. **Pfeiffer, P., Laquel, P., and Hohn, T.,** Cauliflower mosaic virus replication complexes: characterization of the associated enzymes and of the polarity of the DNA synthesized in vitro, *Plant Mol. Biol.,* 3, 261, 1984.

115. **Covey, S. N., Turner, D., and Mulder, G.,** A small DNA molecule containing covalently-linked ribonucleotides originates from the large intergenic region of the cauliflower mosaic virus genome, *Nucleic Acids Res.,* 11, 251, 1983.

116. **Guilley, H., Jonard, G., and Richards, K.,** Structure and expression of cauliflower mosaic virus DNA, in *Current Communications in Molecular Biology,* Robertson, H. D., Howell, S. H., Zaitlin, M., and Malmberg, R. L., Eds., Cold Spring Harbor, Laboratory, Cold Spring Harbor, NY, 1983, 17.

117. **Hull, R. and Covey, S. N.,** Does cauliflower mosaic virus replicate by reverse transcription? *Trends Biochem. Sci.,* 8, 119, 1983.

118. **Pfeiffer, P. and Hohn, T.,** Involvement of reverse transcription in the replication of cauliflower mosaic virus: a detailed model and test of some aspects, *Cell,* 33, 781, 1983.

119. **Marsh, L., Kuzj, A., and Guilfoyle, T. J.,** Identification and characterization of cauliflower mosaic virus replication complexes, *Virology,* 143, 212, 1985.

120. **Marsh, L. and Guilfoyle, T. J.,** Cauliflower mosaic virus replication intermediates are encapsidated into virion-like particles, *Virology,* 161, 129, 1987.

121. **Covey, S. N.,** Amino acid sequence homology in gag region of reverse transcribing elements and the coat protein gene of cauliflower mosaic virus, *Nucleic Acids Res.,* 14, 623, 1986.

122. **Lebeurier, G., Hirth, L., Hohn, B., and Hohn, T.,** In vivo recombination of cauliflower mosaic virus DNA, *Proc. Natl. Acad. Sci. U.S.A.,* 79, 2932, 1982.

123. **Howell, S. H., Walker, L. L., and Walden, R. M.,** Rescue of in vitro generated mutants of cloned cauliflower mosaic virus genome in infected plants, *Nature (London),* 293, 483, 1981.

124. **Walden R. M. and Howell, S. H.,** Intergenomic recombination events among pairs of defective cauliflower mosaic virus genomes in plants, *J. Mol. Appl. Genet.,* 1, 447, 1982.
125. **Choe, I. S., Melcher, U., Richards, K., Lebeurier, G., and Essenberg, R. C.,** Recombination between mutant cauliflower mosaic virus DNAs, *Plant Mol. Biol.,* 5, 281, 1985.
126. **Geldreich, A., Lebeurier, G., and Hirth, L.,** In vivo dimerization of cauliflower mosaic virus DNA can explain recombination, *Gene,* 48, 277, 1986.
127. **Ménissier-de Murcia, J.,** unpublished data.
128. **Mesnard, J. M.,** unpublished data.

Chapter 7

REINITIATION OF DNA REPLICATION IN BACTERIOPHAGE λ

Ross B. Inman

TABLE OF CONTENTS

I. INTRODUCTION

The growth cycle of bacteriophage λ begins when the phage adsorbs to *Escherichia coli*. The distal end of the tail attaches to the bacterium and DNA is injected into the host through the tail. Thereafter, events proceed rapidly and lead eventually to the production of about 100 progeny phage per bacteria, which are released by lysis of the host. The complete growth cycle is accomplished in about 50 min at 37°C. The mechanism whereby the λ genome is replicated has been studied in great detail and the following general outline of its replicative strategy has emerged. The infecting DNA molecule is linear and contains protruding single-stranded ends which are each 12 nucleotides long[1] and complementary (Figure 1a). Once inside the host, these complementary ends cohere to make a circle and ligase then produces a covalently closed structure which is now able to be supercoiled (Figure 1b and c). Young and Sinsheimer[2] were able to demonstrate the existence of such structures in intracellular DNA after infection of *E. coli* with λ phage. DNA replication now commences and at early times proceeds by a bidirectional mechanism leading to so-called theta structures (Figure 1d) these will be discussed in greater detail below. At later times, replication continues by a rolling circle mechanism and replicative intermediates have a sigma structure, which generates a DNA tail containing many copies of the λ genome (Figure 1e). The concatenated genome is then cut into monomeric units and packaged within heads, and phage tails are attached to produce mature phage particles. Complete details of the λ growth cycle may be found in two books edited by Hershey[3] and Hendrix et al.,[4] respectively.

As already mentioned, the very first round of λ replication involves theta-type replication, initiating from a negatively supercoiled template. This chapter will be concerned with the events that surround this early phase of the λ growth cycle. We will specifically address the following questions. How does replication initiate on a covalently closed circular template? How are these initiation events controlled? What are the consequences of disturbing the control mechanism?

II. FIRST ROUND OF DNA REPLICATION

After the infecting DNA is circularized, an initiation event occurs and replication commences. During the first round, the circular replicative intermediate possesses two branch points and has the appearance of a θ structure in the electron microscope.[5] These replicators initiate from a unique position on the circular template and both branch points move away from this origin.[6] Replication is therefore bidirectional during this initial phase of genome duplication. Growing forks consist of sites of continuous leading strand and discontinuous lagging strand synthesis. The arm of the fork containing the site of lagging strand synthesis can therefore possess one or more gaps and these are observed as single-stranded regions in the electron microscope. The gaps are always deployed in a *trans* configuration about the two growing points, confirming that replication is bidirectional.[7]

Although the majority of first-round replicative intermediates are bidirectional, a minor fraction is found to be unidirectional and replicates either to the left or to the right. Furthermore, bidirectional replication is by no means exact; left and rightward growing points can propagate unequal distances from the origin of replication.[6] The round of replication terminates when the bidirectional growing points collide. Because bidirectional propagation is not exact, termination is not necessarily at the exact antipode to the origin; there is apparently no special terminating position in the λ genome.[8]

The sequence of events discussed above is driven by much of the *E. coli* host replicative machinery (about 20 different proteins) and also requires several phage-encoded proteins. Lambda mutants that are defective in O and P proteins do not replicate[9] and several other gene products are known to alter the efficiency of phage production.[10]

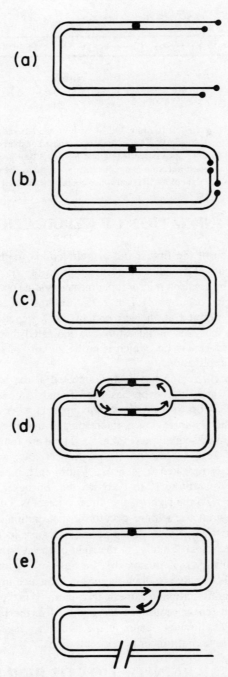

FIGURE 1.   Species present during λ DNA replication. (a) Linear λ DNA injected from the phage. This molecule is 48,514 bp long and at each end has protruding complementary single strands, each 12 nucleotides long. (b) The complementary ends cohere by formation of a short helical segment to produce a circular structure containing two discontinuities. (c) Ligase seals the nicks to produce a covalently closed molecule. DNA gyrase can now subject this structure to negative superhelical tension. (d) Replication is initiated from the origin sequence and proceeds bidirectionally around the circle (generating a θ-like structure). At each growing point, one template strand is copied by a continuous mechanism (leading strand synthesis), while the other is copied discontinuously (lagging strand synthesis). (e) Later rounds of replication proceed by a different mechanism. A single growing point repeatedly traverses the circle to produce a long tail composed of concatenated λ genomes. This mechanism (a rolling circle) leads to σ-like structures. Single λ genomes are derived from the concatenated tail and packaged into λ head structures. The position of the origin of replication is shown by the filled-in area.

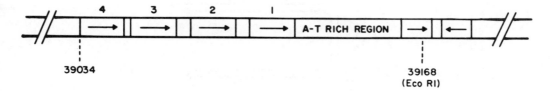

FIGURE 2.   λ origin region. From the left (starting 39,034 bp from the left end of the λ map) are four 18- to 19-bp direct repeated sequences (4, 3, 2, and 1). Each of these is composed of short and imperfect inverted repeats. The direct repeats are followed by a 42-bp AT-rich region and then by 12-bp inverted repeated sequences. The minimal region that is known to support initiation and replication is delineated by the sequences between 39,034 to 39,168; however, sequences to the right of 39,168 (and which contain the inverted repeats) may play an important part in efficient λ replication.

III. INITIATION OF REPLICATION

The details of initiation of the first round of bidirectional replication in λ are still not understood. Initiation occurs at a well-defined position on the superhelical parental circle. This was first established by electron microscopic measurements on partially denatured theta structures.[6] The origin region occurred 81.7 ± 2.9% from the left end of the infecting linear molecule (the left end is defined as the left end of the λ genetic map). The λ genome contains 48,514 bp[11] and thus this initial estimation indicated that the origin was situated within the region 38,229 to 41,043 bp, which is in the vicinity of genes O and P which are known to be required for replication.

More recently, the origin region has been delineated at the base sequence level. The sequence between 39,034 and 39,168 bp has been found to contain the essential origin functions;[12] an additional 34 bp to the right of this appears to enhance replication.[13] The actual base sequence of the origin region is interesting in that it contains a complex array of direct and inverted repeated sequences (Figure 2). There are four direct-repeat sequences (4, 3, 2, 1) composed of 18, 18, 18, and 19 bp, respectively; within these, 11% of the bases deviate from perfectly repeated sequences. Within each repeat there is some degree of inverted base symmetry involving 10 to 12 bases and, of these, 13% are mismatched in terms of perfect symmetry. To the right of the direct repeats is a 42-bp region of high AT content (72%) which is asymmetric with respect to the distribution of purine and pyrimidine bases on the l and r DNA strands; 69% of the purine bases are located in the l strand (and include a run of 18 purine bases). Finally, to the right of the AT-rich region are two 12-bp segments of inverted symmetry (8% mismatch). Because of the inverted symmetry within the four direct repeats, it is possible to draw the origin sequence in a number of alternative configurations; one such configuration, a cloverleaf structure,[14] makes full use of this inverted symmetry. The four direct-repeat sequences are involved in the binding of the λ O gene product; at low levels of O, binding occurs at the two inner repeats (3 and 2, Figure 2) while all four repeats are bound at higher O concentrations.[15]

IV. UNUSUAL REINITIATION OF REPLICATION

A. NORMALLY, REINITIATION IS A RARE EVENT

Normally, the first round of replication proceeds in an orderly fashion; once initiation has taken place, the two growing points move bidirectionally around the circle. After a round is in progress, two daughter origin sequences must be generated in the theta structure, but normally they rarely reinitiate (Figure 1d). Not until termination of the first round are they able to support an initiation leading to a further round of replication. Although the apparent inactivity of daughter origins leads to an orderly progression of replication, one must ask why this is so. The daughter origin sequences must be the same and they can be assumed to be in the same environment as the parental origin. Why is it, then, that they do not reinitiate during a round of replication? Perhaps there is a lack of replicative machinery to service so many origins. This does not appear to be the case because it has been shown that a λ phage containing two separate origins can initiate replication from either or both origins with significant frequency.[16] The simplest explanation for the normal lack of reinitiation rests on the notion that there must be some type of regulation system whose purpose is to ensure that daughter origins cannot reinitiate while a round of replication is in progress.

B. ABNORMAL REINITIATION EVENTS

When λ infects *E. coli* growing in the presence of 2 m*M* caffeine, a striking departure from the normal orderly sequence of events is observed. A significant number of replicative intermediates (up to 20%) have more than the normal two growing points.[17] An example of such a molecule is shown in Figure 3. Electron microscopic studies show that the extra growing points are nested within the normal ones and that they involve reinitiation from the origin and move bidirectionally (Figure 4). Apparently, therefore, under these conditions reinitiation events occur frequently and presumably this is because the control process that normally prevents reinitiation has been disturbed. It is interesting to note that although reinitiation frequently occurs in the presence of 2 m*M* caffeine, the yield of phage is not significantly decreased. A number of other abnormal conditions also lead to reinitiation events; these include infection in the presence of the antitumor drug *cis*-Pt or when λ infects a P2 lysogenic strain.[18]

Under the abnormal conditions discussed above, it is possible to interfere with the control process that normally prevents reinitiation. One can therefore utilize this fact to gain some insight into the control process itself. The effect of *cis*-Pt has been revealing in this respect. It is known that the primary effect of *cis*-Pt is to cause intrastrand cross-linking between neighboring guanine bases.[19,20] As discussed above, reinitiations occur if λ replicates in a host growing in the presence of this drug, and under these conditions both host and phage DNA will contain these crosslinked bases. If the host is grown in *cis*-Pt and then the drug removed prior to infection, one still observes abnormal reinitiation during λ replication;[18] therefore, *cis*-Pt-induced lesions within the host DNA are sufficient to lead to this phenomenon. It is well known that lesions of the type produced by *cis*-Pt can lead to the SOS response in *E. coli*; under these conditions, a special error-prone mode of DNA replication takes place which allows modified replication machinery to pass over lesions.[21] UV irradiation of the host prior to λ infection also leads to reinitiation;[22] such treatment creates intrastrand cross-links between neighboring thymine bases and these, in turn, should again elicit the SOS response. In the above experiments involving lesions produced by *cis*-Pt or UV irradiation, it was shown by phage reactivation experiments that the host was in fact exhibiting SOS functions at the time of the abnormal λ reinitiation events. There is, therefore, a definite correlation between expression of SOS and reinitiation. Definitive evidence that SOS function is responsible for λ reinitiation was obtained by an examination of λ infection of two *E. coli* strains containing mutations that disturb the regulation of SOS. Strain GC3217 chronically expresses SOS functions at 42°C.[23] In this strain, a mutation in the recA gene (recA441) results in SOS expression in the absence of template damage at 42°C; however, at 32°C, SOS is not expressed unless the template is damaged. Lambda infection of this host, grown at 42°C, leads to λ reinitiation, whereas infection of this host, grown at 32°C, does not.[22] Another strain (DM1180) contains an additional mutation in the lexA gene which renders the host incapable of SOS expression:[24] in this case, there is a greatly reduced level of λ reinitiation (however, not as low as in an infection of wild-type *E. coli* without SOS functions being expressed). Hence, these experiments strongly imply that one or more SOS functions are responsible for the λ reinitiation events. One might therefore suspect that these SOS functions have the ability to interfere with the λ regulatory system that normally ensures

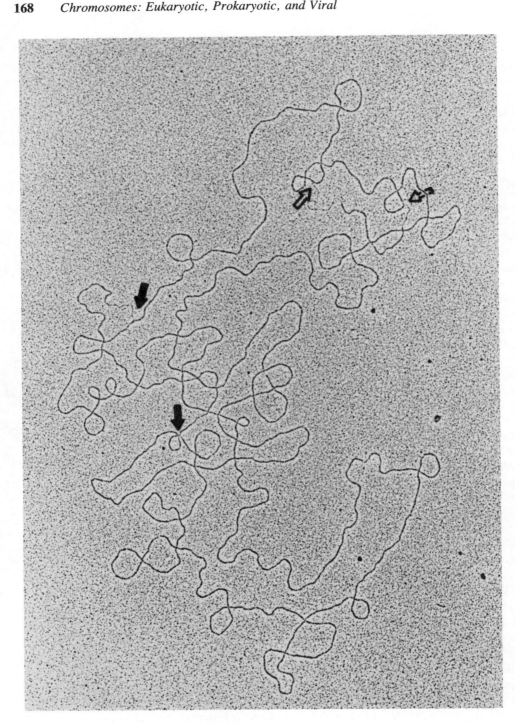

FIGURE 3. Electron micrograph showing a λ replicative intermediate that involves abnormal extra growing points. The growing points arising from the first wave of replication are indicated by solid arrows. Open arrows show the abnormal growing points resulting from reinitiation from one of the daughter origins.

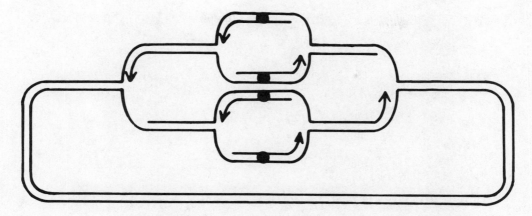

FIGURE 4.   Diagram of a replicative intermediate in which both daughter origins have reinitiated. The origin region is shown by the black area.

that reinitiation cannot take place during a round of replication. During SOS, it appears that λ origins can take part in initiation events regardless of the stage of the replicator.

V. REGULATION OF INITIATION OF REPLICATION

So far, we have seen that SOS functions can interfere with the regulation of λ initiation to the extent that daughter origins can reinitiate even when a round of replication is in progress. We should now discuss the λ regulatory mechanism itself to gain some insight into how it operates to normally prevent reinitiation during replication.

Once the first round begins and bidirectional replication has proceeded a very short distance, two daughter origin sequences will be produced. The regulation system must be such that normally these daughter origins remain inactive until the round terminates to produce mature daughter circles. Is there some differences between the daughter origins present in the replicative intermediate (Figure 1d), compared with the parent or mature daughters (Figure 2c)? The base sequence must be the same and we would suspect that any chemical changes that could be made to create a different biochemical "ori" activity would act similarly on parental and matured daughter origins. Is there a physical difference between these various types of origin sequences? The λ replicative intermediate (Figure 1d) has two growing points and two daughter arms, each containing a daughter origin. The topological properties of the origins in these daughter arms could be distinctly different from the parental or matured daughter origins (Figure 1c) because in the latter cases the origin sequences are part of a double-stranded and covalently continuous structure, whereas in the replicative intermediate, they are not. Because of the strand discontinuities, daughter arms (and therefore daughter origins) can never be subjected to the negative superhelical tension that can be induced in covalently closed structures. Thus, the daughter origins in a θ replicative intermediate must always be in a relaxed state, whereas parental or matured daughter origins could be under negative superhelical tension. If such tension were required for initiation, then this would provide an automatic control system that would ensure that daughter origins would not reinitiate during a round of replication. Once the round terminates and daughter circles become covalently continuous, they could be negatively supertwisted and thus become a substrate for the initiating machinery.

Support for these notions comes from several types of experiments. First, it is known that intracellular λ circles are supercoiled.[2] Second, it has also recently been shown that θ structures do have the expected topological domains discussed above; that is, superhelical parental sections and nonsuperhelical daughter arms. Figure 5 shows a very clear demon-

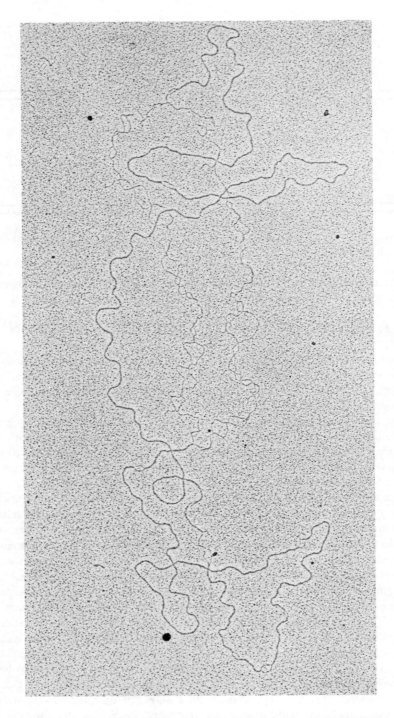

FIGURE 5.   Electron micrograph of a λ replicative intermediate after psoralen-UV-heat treatment. The parental section is essentially native, whereas the daughter sections are highly denatured. The single-stranded daughter regions are held together by psoralen-UV induced interstrand cross-links. (From Inman, R. and Schnos, M., *J. Mol. Biol.*, 193, 377, 1987. With permission.)

stration of this fact.[25] The λ replicative intermediate shown in this figure has been reacted with the drug psoralen, irradiated with UV light, and then heated to a temperature which would completely denature linear λ DNA. The reaction with psoralen and UV light simply produces interstrand crosslinks which prevent the strands from dissociating after thermal denaturation.[26] The daughter arms in the replicative intermediate show the effects of this cross-linking; although this material is completely denatured, the two strands cannot dissociate because of the cross-links and we see a consecutive series of single-stranded regions bounded by the interstrand cross-links. Although the daughter arms behave normally and denature after the heating step, this is certainly not true for the parental section. As can be seen in the figure, it is essentially native in spite of the heat treatment. A series of experiments with simple plasmid DNA clearly shows that this effect arises because the parental section is covalently continuous and therefore able to supercoil.[25] Figure 6a shows an electron micrograph of a covalently closed plasmid DNA which has been psoralen-UV treated and then heated to beyond the normal denaturation temperature for this DNA. As can be seen, the covalently continuous molecule is essentially native. However, if this same DNA is linearized before or immediately after the psoralen-UV treatment, the molecule becomes highly denatured after the heating step (Figure 6 b and c). Again, the psoralen-UV induced cross-links serve to hold the highly denatured DNA together. The fundamental reason for the native appearance of the parental section in the θ structure (Figure 5) and the covalently closed circle (Figure 6a) is that covalently closed DNA has a higher melting temperature than linear DNA or DNA that is not covalently continuous.[27] Thus, Figure 5 shows that in this molecule there is a global difference between daughter arms and the parental section; the daughter arms are not able to be subjected to superhelical tension. A study of many θ structures shows that this is true for most λ replicative intermediates.[25]

Because of this finding, the model for regulation of initiation already discussed is at least feasible. If this model is correct, it is of some interest to determine why negative superhelical tension at the origin is required by the initiating machinery. The λ origin sequence is composed of a complex array of repeated sequences which can be drawn in an alternative clover-leaf configuration[15] or in some more simple hairpin type of alternative state. It is well known that negative superhelical tension can provide a driving force that will tend to induce these otherwise less stable configurations. Perhaps, therefore, the origin region has to be in some type of alternative configuration before it becomes a suitable site for initiation of replication.

VI. REQUIREMENT FOR NEGATIVE SUPERHELICAL TENSION AT THE ORIGIN

Two phage-encoded genes, O and P, are known to be essential for λ DNA replication.[9] Complementation tests between phages producing various hybrid gene O products and a knowledge of the origin base sequence indicates that O protein interacts directly with the origin. This occurs within the amino terminal segment of O.[28-30] This protein has been purified and shown to bind specifically to the λ ori sequence. At low concentrations of O, binding takes place at the two inner sections of the four repeated sequences within ori, while at higher levels of O all four repeats are bound.[15] Binding of O protein to the origin has also been observed by electron microscopy. The size of the nucleoprotein structure is increased if λP and host DnaB proteins are also added. No binding of λP and DnaB is observed, however, if O is omitted.[31] Although O protein can bind to both linear and negatively supercoiled plasmids containing the λ origin sequence, there is a profound difference in the subsequent effect of binding on the structure of the ori region, depending on the presence or absence of superhelical tension. Negatively supercoiled ori-containing plasmids show greatly increased S1 or P1 nuclease sensitivity after O binding, whereas a similar linearized

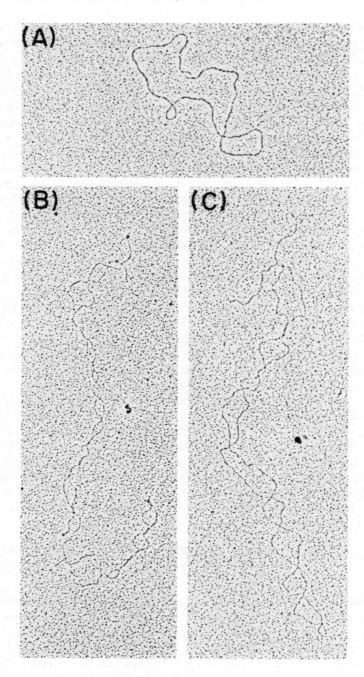

FIGURE 6.   Electron micrograph of simple plasmid DNA after psoralen-UV-heat treatment. (A) Covalently closed circle; (B) linearized before treatment; (C) linearized immediately after the psoralen-UV treatment but before heating. (From Inman, R. and Schnos, M., *J. Mol. Biol.*, 193, 377, 1987. With permission.)

plasmid does not.[32] The increased nuclease sensitivity occurs within the origin sequence. The actual nuclease-sensitive sites are located throughout an AT-rich region just to the right of the O binding domain (Figure 2). Certain origin mutations which greatly reduce origin function show a decreased S1 sensitivity when compared with the wild-type sequence. These studies clearly show that O binding to the origin causes an alteration in DNA structure to the immediate right of the site of O binding and that this alteration requires that the origin sequence be under superhelical tension. The altered state has not yet been definitively defined, but it has properties consistent with an unwound or destabilized region. Initiation of replication on a covalently closed circular template requires some type of unwinding event to allow access of the initiation and replication machinery. The alteration in origin structure, which is a consequence of O binding, appears to correspond to this prepriming step. A more complex experiment shows a much more extensive unwinding reaction that proceeds away from the O binding site. When the six proteins O, P, DnaB, DnaJ, and DnaK, and Ssb are incubated with a superhelical, origin-containing plasmid, an extensive (800-bp) unwound area can be observed by electron microscopy.[33]

The experiments discussed above clearly show that superhelical tension plays an important role in the prepriming event associated with O binding. Thus, the model presented earlier to explain why daughter origins in θ structures do not reinitiate during replication becomes more attractive. Once a round of replication is in progress, these daughter origins are not able to be subjected to superhelical tension and, as indicated above, the O-mediated unwinding step is not favored and, therefore, initiation cannot take place.

VII. EFFECT OF SOS ON THE REGULATION OF INITIATION

According to the arguments presented so far, we envision the following scenario for normal initiation of λ replication. The parental circle is covalently closed and, thus, the origin can be subjected to negative superhelical tension and is therefore able to undergo the unwinding modification that is induced by O protein interaction. This prepriming step then sets in motion the cascade of events that involve λP protein and *E. coli* proteins, which eventually lead to the initiation and creation of replicative growing points and bidirectional replication. At this stage, the replicative intermediate has a θ structure with two copies of the origin region (daughter origins).

As already shown, these daughter origins cannot be subjected to negative superhelical tension. Thus, O binding will not induce unwinding and, therefore, they will not be able to support initiation. Once the round terminates and the replicator is resolved into two daughter circles which are covalently continuous, it is possible for superhelical tension to be once again applied to the matured daughter origin sequences and O binding will induce the unwinding modification that is the critical prepriming step. Clearly, in this scenario superhelical tension provides the controlling mechanism that prevents reinitiation while replication is in progress.

We have already discussed the experiments that show how SOS functions can disturb this normal control mechanism: replication in the presence of these SOS functions allows reinitiation even when a round of replication is in progress. How does this fact fit in with the scenario formulated above for normal replication? There are two extreme possibilities to consider. First, perhaps an SOS function will allow daughter origins to become supercoiled. If SOS functions did, indeed, lead to this series of events, then one would expect to find superhelical daughter arms in molecules that had reinitiated. Figure 7a shows a λ replicative intermediate and we will assume that both growing points have encountered template lesions and consequently have stalled. If a similar situation existed at the host growing points, SOS functions would be induced. Perhaps one function of SOS that could enable reinitiation provides some way that allows the stalled leading strand site of DNA synthesis to now copy

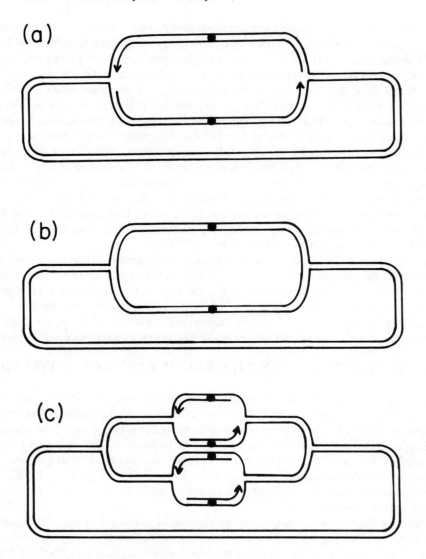

FIGURE 7. One possible sequence of events that could lead to superhelical tension at daughter origins during SOS. (a) A normal replicative intermediate whose growing points have stalled because of template lesions. (b) As a result of an SOS function, the leading strand site of synthesis switches to the complementary template and ligase then seals this strand to the trailing end of the lagging strand, thus creating covalent continuity within the daughter segment and subsequent negative superhelicity. (c) Reinitiation can now take place from the origin and leads to error-prone replication.

the complementary template. If this occurred, further synthesis would eventually extend to the trailing end of the lagging strand. Once this was accomplished at both growing points, daughter segments could be made covalently continuous by ligase and the origins subjected to superhelical tension (Figure 7b). This would then allow reinitiation and lead to subsequent bidirectional, error-prone replication. Figure 7c shows a molecule of this type. The expectation would be that both parental and daughter segments would be superhelical, whereas the granddaughter segments would not. This possibility has been tested using the psoralen-UV irradiation-denaturation technique already described.[22] It is found that under SOS conditions, daughter segments in reinitiated molecules are in fact relaxed and not able to be subjected to superhelical tension; therefore, the scenario shown in Figure 7 is unlikely. There are, however, other ways in which SOS might allow daughter origins to be subjected to superhelical tension. Perhaps a protein "glue" at growing points prevents rotation of DNA and thus causes the daughter arms to behave as if they are covalently closed. This possibility has not as yet been subjected to an experimental test.

The second extreme scenario which would explain how SOS functions allow reinitiation involves a model in which SOS provides an environment in which O binding will lead to the unwinding-prepriming event even in the absence of superhelical tension. Perhaps unwinding factors are provided that can replace the normally required superhelical tension. One important SOS function lowers the fidelity of replication; the ensuing error-prone mode of replication allows such a growing fork to traverse lesions. Perhaps another function provides a means to bypass the normally stringent initiation requirement for superhelical tension at the origin. In this way, a replicator that has been subjected to template damage resulting from hostile environmental conditions can survive because initiation can be accomplished during an aborted round of replication. This then allows the error-prone replication mechanism to complete a replicative cycle by making an imperfect copy of the template.

REFERENCES

1. **Wu, R. and Taylor, E.,** Nucleotide sequence analysis of DNA. II. Complete nucleotide sequence of the cohesive ends of bacteriophage λ DNA, *J. Mol. Biol.,* 57, 491, 1971.
2. **Young, E. T. and Sinsheimer, R. L.,** Novel intra-cellular forms of lambda DNA, *J. Mol. Biol.,* 10, 562, 1964.
3. **Hershey, A. D.,** *The Bacteriophage Lambda,* Cold Spring Harbor Laboratory, Cold Spring Harbor, NY, 1971.
4. **Hendrix. R. W., Roberts, J. W., Stahl. F. W., and Weisberg, R. A.,** *Lambda II,* Cold Spring Harbor Laboratory, Cold Spring Harbor, NY, 1983.
5. **Tomizawa, J. and Ogawa, T.,** Replication of phage lambda DNA, *Cold Spring Harbor Symp. Quant. Biol.,* 33, 533, 1968.
6. **Schnos, M. and Inman, R. B.,** Position of branch points in replicating λ DNA, *J. Mol. Biol.,* 51, 61, 1970.
7. **Inman, R. B. and Schnos, M.,** Structure of branch points in replicating DNA: presence of single-stranded connections in λ DNA branch points, *J. Mol. Biol.,* 56, 319, 1971.
8. **Valenzuela, M., Freifelder, D., and Inman, R. B.,** Lack of a unique termination site for the first round of λ DNA replication, *J. Mol. Biol.,* 102, 569, 1976.
9. **Ogawa, T. and Tomizawa, J.,** Replication of bacteriophage DNA. I. Replication of DNA of lambda phage defective in early functions, *J. Mol. Biol.,* 38, 217, 1968.
10. **Furth, M. E. and Wickner, S. H.,** Lambda DNA replication, in *Lambda II,* Hendrix, R. W., Roberts, J. W., Stahl, F.W., and Weisberg, R. A., Eds., Cold Spring Harbor Laboratory, Cold Spring Harbor, NY, 1983, 145.
11. **Daniels, D. L., Schroeder, J. L., Szybalski, W., Sanger, F., and Blattner, F. R.,** A molecular map of coliphage lambda, in *Lambda II,* Hendrix, R. W., Roberts, J. W., Stahl, F. W., and Weisberg, R. A., Eds., Cold Spring Harbor Laboratory, Cold Spring Harbor, NY, 1983, 469.
12. **Denniston-Thompson, K., Moore, D. D., Kruger, K. E., Furth, M. E., and Blattner, F. R..** Physical structure of the replication origin of bacteriophage lambda, *Science,* 198, 1051, 1977.
13. **Lusky, M. and Hobom, G.,** Inceptor and origin of DNA replication in lambdoid coliphages. I. The λ DNA minimal replication system, *Gene,* 6, 137, 1979.
14. **Grosschedl, R. and Hobom, G.,** DNA sequences and structural homologies of the replication origins of lambdoid bacteriophages, *Nature (London),* 277, 621, 1979.
15. **Tsurimoto, T. and Matsubara, K.,** Purified bacteriophage λO protein binds to four repeating sequences at the λ replication origin, *Nucleic Acids Res.,* 9, 1789, 1981.
16. **Schnos, M., Denniston, K. J., Blattner, F. R., and Inman, R. B.,** Replication of bacteriophage lambda DNA: examination of variants containing double origins and observations of a bias in directionality, *J. Mol. Biol.,* 159, 441, 1982.

17. **Schnos, M. and Inman, R. B.,** Caffeine-induced reinitiation of phage λ DNA replication, *J. Mol. Biol.,* 159, 457, 1982.

18. **Schnos, M. and Inman, R. B.,** Reinitiation during λ DNA replication resulting from either cis-Pt treatment or infection of a P2 lysogenic strain, *Virology,* 145, 304, 1985.

19. **Stone, P. J., Kelman, A. D., Sinex, F. M., Bhargava, M. M., and Halvorson, H. O.,** Resolution of α, β and γ DNA of *Saccharomyces cerevisiae* with the antitumor drug cis-Pt(NH₃)₂Cl₂. Evidence for preferential drug binding by GpG sequences of DNA, *J. Mol. Biol.,* 104, 793, 1976.

20. **Kelman, A D. and Buchbinder, M.,** Platinum-DNA crosslinking; platinum antitumor drug interactions with native lambda bacteriophage DNA studied using a restriction endonuclease, *Biochimie,* 60, 893, 1978.

21. **Witkin, E. M.,** Ultraviolet mutagenesis and inducible DNA repair in *Escherichia coli, Bacteriol. Rev.,* 40, 869, 1976.

22. **Schnos, M. and Inman, R. B.,** Reinitiation at the λ DNA origin accompanies the host SOS response, *Virology,* 158, 294, 1987.

23. **Castellazzi, M., George, J., and Buttin, G.,** Prophage induction and cell division in *E. coli.* I. Further characterization of the thermosensitive mutation *tif*-1 whose expression mimics the effect of UV irradiation, *Mol. Gen. Genet.,* 119, 139, 1972.

24. **Mount, D. W., Low, K. B., and Edmiston, S. J.,** Dominant mutations (lex) in *Escherichia coli* K-12 which affect radiation sensitivity and frequency of ultraviolet light-induced mutations, *J. Bacteriol.,* 112, 886, 1972.

25. **Inman, R. B. and Schnos, M.,** Electron microscopic identification of supercoiled regions in complex DNA structures, *J. Mol. Biol.,* 193, 377, 1987.

26. **Hanson, C. V., Shen, C. J., and Hearst, J. E.,** Cross-linking of DNA in situ as a probe for chromatin structure, *Science,* 193, 62, 1976.

27. **Vinograd, J., Lebowitz, J., and Watson, R.,** Early and late helix-coil transitions in closed circular DNA. The number of superhelical turns in polyoma DNA, *J. Mol. Biol.,* 33, 173, 1968.

28. **Furth, M. E., McLeester, C., and Dove, W. F.,** Specificity determinants for bacteriophage lambda DNA replication. I. A chain of interactions that controls the initiation of replication, *J. Mol. Biol.,* 126, 195, 1978.

29. **Furth, M. E. and Yates, J. L.,** Specificity determinants for bacteriophage lambda DNA replication. II. Structure of O proteins of λ-φ80 and λ-82 hybrid phages and of a λ mutant defective in the origin of replication, *J. Mol. Biol.,* 126, 227, 1978.

30. **Moore, D., Denniston, K., and Blattner, F.,** Sequence organization of the origins of DNA replication in lambdoid coliphages, *Gene,* 14, 91, 1981.

31. **Dodson, M., Roberts, J., McMacken, R., and Echols, H.,** Specialized nucleoprotein structures at the origin of replication of bacteriophage λ: complexes with λO protein and with λO, λP and *Escherichia coli* DnaB proteins, *Proc. Natl. Acad. Sci. U.S.A.,* 82, 4678, 1985.

32. **Schnos, M., Zahn, K., Inman, R. B., and Blattner, F. R.,** Initiation protein induced helix destabilization at the λ origin: a prepriming step in DNA replication, *Cell,* 52, 385, 1988.

33. **Dodson, M., Echols, H., Wickner, S., Alfano, C., Mensa-Wilmot, K., Gomes, B., LeBowitz, J., Roberts, J. D., and McMacken, R.,** Specialized nucleoprotein structures at the origin of replication of bacteriophage λ: localized unwinding of duplex DNA by a six-protein reaction, *Proc. Natl. Acad. Sci. U.S.A.,* 83, 7638, 1986.

Chapter 8

IN VIVO FATE OF BACTERIOPHAGE T4 DNA

Andrzej W. Kozinski

TABLE OF CONTENTS

I. INTRODUCTION

Over a number of years, research on the intracellular fate of T4 DNA was controversial and at times resulted in prolonged polemics. This chapter, will restrict itself mainly to descriptions of the progress of research in our laboratory, contending that disagreements with other groups have largely vanished, in our favor, and thus elaborations on the past results of others are not necessary.

With the model of T4 phage, we will show how a number of novel qualities of the intracellular fate of DNA were discovered — for instance, breakage, exchanges, and covalent joining as a molecular mechanism of recombination; multiple, bidirectional, and specific initiations of DNA replication; visualization of the leading end of progeny DNA strands; and formation of concatenated structures of DNA by the process of recombination.

Throughout, this chapter will try to draw attention to areas which will likely demand further study or reconsideration as perhaps they are oversimplified in available reviews and monographs.

II. EARLY RESEARCH: PROOFS OF DNA BREAKAGE, EXCHANGES, AND COVALENT JOINING AS A MOLECULAR MECHANISM OF RECOMBINATION

T4 phage is a large virus containing a molecule of DNA of 166 kilobases. Instead of cytosine, T4 phage DNA contains hydroxymethylcytosine which is modified by glycosylation. The DNA molecules are terminally redundant, having repetitive regions at the ends, and are circularly permutated. It was noted in the early era of phage genetics by Delbruck that each T4 phage undergoes six to eight rounds of mating (recombination) in the course of its life cycle. Because of its recombinational proficiency, T4 bacteriophage offered an ideal model for elucidating the molecular mechanism of recombination. There were two putative models of molecular recombination. The first was copy choice, a model in vogue during the late 1950s and early 1960s which assumed that recombination occurs as a result of a replicative switchover between two different molecules. The second model called for the breaking of DNA molecules and joining of the resulting fragments. Due to the simplistic view of the vulnerability of DNA to breakage which was popular at that time, the breakage and reunion mode of recombination was much less favored.

Using T4 phage as a model system, we provided the first proof of breakage, molecular exchanges, and covalent repair of recombinant molecules.[1,2] Even though not published promptly, the first experiment which provided clear proof of breakage and reunion of DNA had already been performed, in cooperation with Uchida, in 1957. We demonstrated at that time[3] that a very short pulse of high-specific-activity ^{32}P applied during T4 DNA replication leads to the dispersion of such label over the majority of the progeny phage, resulting in suicidal death due to ^{32}P decay. Of course, given a strict semiconservative mode of DNA replication not accompanied by molecular exchanges, only a small proportion of the progeny phage should become labeled and thus suicidal. We concluded that "exchange of homologous subunits of DNA with each other may occur during their sojourn at the pool." However, it was some 3 years later, after publication of the results showing molecular fragmentation of the parental DNAs, as proven by double-density experiments and CsCl equilibrium analysis,[1] that the paper with Uchida[3] was published.

I will describe the design of the double-density experiment in great detail as the construction of the experiment and mode of analysis later became widely used in a very large number of experiments and publications from many laboratories. Bacteria grown in heavy (containing 5-bromodeoxyuridine, 5-BrdUrd) media were infected with light,^{32}P-labeled T4 bacteriophage. After lysis of the bacteria, the progeny bacteriophage were purified and their

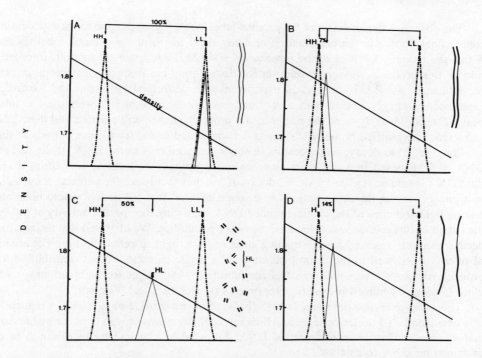

FIGURE 1. Evidence for recombination by breakage and joining, as provided by differential density labeling experiment. Idealized outcome of CsCl density gradient analysis compiled from Kozinski[1] and Kozinski and Kozinski.[2] 32P-labeled parental DNA is represented by a solid line and density references are represented by a broken line. (A) Parental phage DNA (32P-labeled) was bistranded and light (LL) and thereby cobands with LL reference DNA. The observed distance between the heavy and light location is normalized to 100. (B) Upon infection of bacteria in the presence of the density label 5-BrdUrd, progeny phage were isolated and their DNA extracted. CsCl analysis of the progeny DNA revealed that the parental 32P-labeled DNA (originally 100% light) banded close to the heavy location. The observed displacement indicated a 7% gross contribution of parental light DNA to progeny molecules, i.e., there was breakdown and reunion leading to dispersion of parental label within the length of the heavy, progeny DNA. (C) Progeny DNA was sheared, by sonication, to short fragments. Upon CsCl analysis, the parental label assumed an HL density halfway between the two references, i.e., parental DNA subunits underwent semiconservative replication. (D) Progeny DNA was codenatured with reference DNA. Upon CsCl analysis, the light parental label remained joined to the heavy progeny DNA. Therefore, discrete stretches of parental light strands must be covalently joined to the heavy progeny strands.

DNA analyzed in CsCl equilibrium gradients. Figure 1 idealizes the results of this experiment. The light, parental DNA contributed only 7% of the mass of the entire progeny molecule (Figure 1B). (Of course, semiconservative replication not followed by breakage and reunion would result in a 50% contribution of parental density.) When such recombinant molecules were sheared to 0.5 to 1 million Da by sonication, all the fragments containing the parental 32P contribution assumed an intermediate density at the hybrid (HL, one strand heavy and one strand light) location in the CsCl gradient (Figure 1C). Not only had these results provided the second proof of semiconservative DNA replication, after the classical experiment of Meselson and Stahl,[4] but more importantly, they had, for the first time, demonstrated breakage and reunion of DNA.* Importantly, upon denaturation (Figure 1D), the light parental label remained attached to the heavy progeny strands, shifting only slightly toward the lighter location in the CsCl gradient. This proved that molecular exchanges are followed by covalent repair (joining) of the participating subunits.[2] Thus, the process of recombination does not leave molecular scars.

* Breakage and exchanges of eukaryotic chromosomes were previously observed microscopically, but such observations did not provide a molecular explanation of recombination.

The observed displacement of the parental label, some 7% away from the heavy location, could either be due to the contribution of many short fragments of parental label residing within the progeny DNA or to the presence of a single, larger parental subunit. In order to resolve this problem, obviously relevant for determining the frequency of recombinations per molecule, a ^{32}P suicide-scission experiment was performed by Shahn and Kozinski.[5] Light bacteriophage, labeled with a very high specific activity of ^{32}P, were used to infect heavy *E. coli*. After lysis of the bacteria, the progeny phage were purified and their DNA extracted. The purified progeny DNA was denatured and stored for various periods of time to allow the ^{32}P to decay, which results in single-stranded cuts to the DNA. If the parental DNA were disposed in many short stretches in the progeny molecules, very little, if any, light DNA would be released upon ^{32}P decay. If, on the other hand, the parental contribution to a progeny molecule were confined in a single subunit, two ^{32}P decays would invariably release light stretches of the single-stranded DNA. Knowing the specific activity of the ^{32}P, the extent of this release was calculated for either possibility. We observed that the resulting decays promptly released single-stranded areas of pure, light, parental density. The amount of release fitted well to the theoretical curve of a single parental subunit contribution to a progeny molecule.[5] This indicates that the number of exchanges (recombinations) is very close to those calculated in genetic experiments by Delbruck and Visconti.

This short overview describes how the basic semiconservative mode of DNA replication was extended to T4 bacteriophage, and how recombination was proven to occur by breakage and reunion (molecular exchanges) of DNA, followed by covalent repair (joining) of the participating DNA fragments.

III. INTRACELLULAR EVENTS IN REPLICATION AND RECOMBINATION OF T4 PHAGE DNA

The experiments described above, proving breakage, exchanges, and covalent repair, were performed on the level of the progeny phage where the fate of the parental DNA was followed into the progeny phage DNA. The intracellular events in the processes of replication and recombination of T4 phage DNA will now be described. Here, the sequentiality of temporal events of the replicative and recombinational processes will be briefly summarized. I will start by commenting on the rate of synthesis of the phage DNA. The first noticeable uptake of radioactive precursors occurs at about 5 min after infection. Initially, the uptake is slow, reaching at 8 to 10 min the maximum rate of 5 to 10 phage equivalents per min per infected cell. (For a given medium and temperature, the rate is rather repeatable). This rate continues until the lysis of bacteria. The steady rate is frequently interpreted as a result of withdrawal of progeny DNA from the vegetative pool due to encapsidation (maturation), which supposedly occurs so that the size of the replicative pool remains constant. This is certainly not so as the addition of chloramphenicol (CM), which prevents encapsidation, results in a rate similar to that observed in the suspension without the drug. Even more importantly, mutants defective in encapsidation produce progeny DNA at this same rate in permissive and nonpermissive hosts.[54] Rate limiting factor(s) remain to be identified.

Let us consider the fate of parental and progeny DNA from the moment of infection. Until approximately 5 min after infection, there are abundant transcriptional activities, but there is no replication of the parental molecule. During this time period, parental DNA remains intact (i.e., no breaks are inflicted) and there is no parent-to-parent DNA recombination. If, however, parental DNA is irradiated with UV or acquired ^{32}P decay scissions, resulting discontinuities are repaired, presumably by the host polymerase I as repair is hampered in *E. coli* defective in this enzyme.[6]

The first signs of synthesis of progeny DNA occur 5 min after infection. Newly synthesized DNA is not covalently joined to the parental template. It has, compared with parental

DNA, identical density in CsCl and Cs_2SO_4 gradients — thus it is glycosylated to the same extent.[7] Electron microscopy of parental molecules at this time reveals, within the length of the molecule, several replicative loops (see Figure 2).[7] The fine details of these loops include two 3'-hydroxyl-ending, single-stranded whiskers located in the *trans* position at the two branches of the loop. There are also observable reinitiations (secondary loops) located in the geometric middle of the primary loop.[7-9] Hybridization of the progeny label to cloned segments shows specificity of initiation to multiple sites along the genetic map.[10]

If one adds CM at 5 to 6 min after infection and continues incubation, one observed accumulation of short partial replicas, specific (as shown by hybridization) to areas of initiation. We called the phenomenon "gene amplification".[11]

At about 7 to 8 min past infection, there are abundant parent-to-progeny recombinations, but such recombinants are not covalently joined.[12] Addition of CM at this time allows fast replication and recombination, but even upon extended periods of incubation, no covalent joining of recombinant structures is observed.[12] It is puzzling that despite the vigorous rate of synthesis (approximately 70% of the maximum rate, which is some five phage equivalents per min, recombinants remain unjoined *as if* an enzyme, different from those involved in the rapid synthesis, were missing. This might deserve further study.

After 9 to 10 min past infection, the rate of synthesis becomes maximal and recombinants, upon denaturation, remain covalently joined.[12] At this time, there is also the appearance of concatenated single-stranded DNA.[13,14] Such concatemers, from which progeny phage mature, are assembled by recombination.[15]

The following sections will explore the still controversial nature of nascent fragments residing in the replicative fork and their joining (chase) into larger progeny strands. They will examine where DNA replication initiates, whether it occurs at single or multiple sites, and whether it proceeds in one direction (unidirectionally) or in two directions (bidirectionally) from a given origin. Discussions include the initiation and conclusion of the process of recombination and its role in the formation of concatenated DNA, maturation from which results in the phenomenon of circular permutation.

A. REPLICATIVE FORK: NASCENT FRAGMENTS

In 1968, Okazaki[16] showed that a short pulse of radioactive label is incorporated *exclusively* into very short, single-stranded fragments of progeny DNA. Since replication was assumed to involve both parental strands of DNA,[17] Okazaki at first concluded: "the fact that virtually all the radioactive label was found in the short chains after an extremely short pulse favors a two strand discontinuous mechanism."[16] However, his view of the fork evolved to assume that one progeny strand is "leading" and the other is "lagging" (a semidiscontinuous replication). This lagging strand was then postulated to be the sole location of the nascent fragments.

We had strengthened the leading and lagging strand hypothesis by visualization of replicative DNA.[7] Electron micrographs of replicative DNA revealed two whiskers in the *trans* position at both ends of a replicative loop and of approximately 0.5 μm in length. These whiskers are single stranded as partial melting did not resolve them into tiny "melted" loops (Figure 2). Treatment with exonuclease I (a 3'-hydroxyl-end-specific exonuclease) removed the whiskers. We interpreted the appearance of the whiskers to be a result of the displacement of Okazaki fragments due to branch migration (see Figure 3A).[7] Similar exonuclease-I-sensitive whiskers were observed later in *Drosophila* by Kriegstein and Hogness.[18]

Recently, we repeated and confirmed some of Okazaki's original data: a short pulse of radioactive label was, indeed, found *solely* in short fragments of DNA. According to the leading-lagging strand hypothesis, one would, of course, expect label to be equally incorporated into large fragments (leading strand) and short nascent fragments (lagging strand). In fact, careful scrutiny of numerous publications revealed that in many other organisms, including eukaryotes, only small fragments are synthesized upon a short radioactive pulse.[19,20]

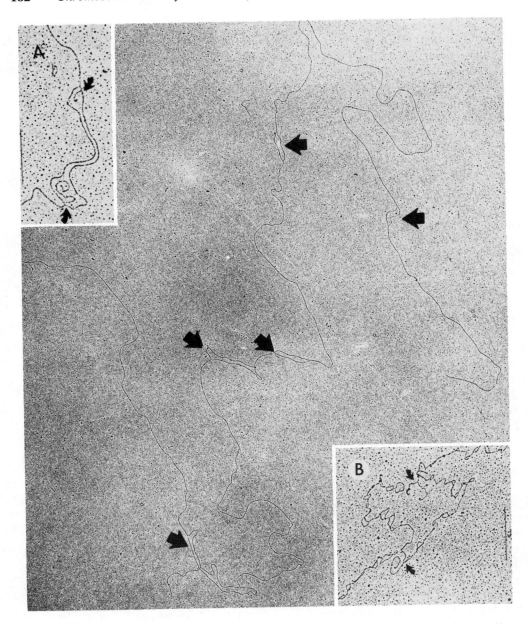

FIGURE 2.   Electron micrographs of partially replicated parental molecules. Insert A represents a replicative loop with two single-stranded whiskers (average length of 0.5μ) in *trans* position. Insert B represents a partially denatured replicative loop. Note the small "melted puffs". Except for the single-stranded whiskers in the branches of the forks, partial melting did not reveal discontinuities in either side of the replicative loop (such breaks would result in multiple single-stranded whiskers within the length of the loop). Large structure represents a partially replicated molecule of genomic length displaying multiple loops. Note the absence of single-stranded whiskers as this moiety was digested with exonuclease I (a 3'-hydroxyl-end-specific enzyme). (From Delius, H., Howe, C., and Kozinski, A. W., *Proc. Natl. Acad. Sci. U.S.A.*, 68, 3049, 1971. With permission.)

A   B   C

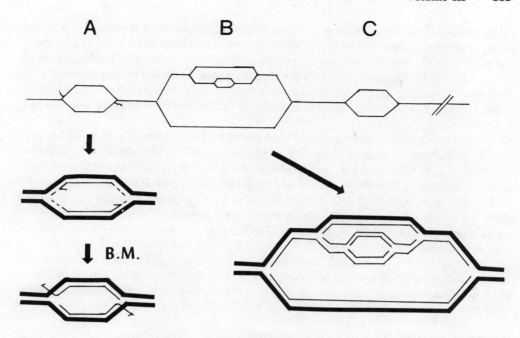

FIGURE 3.   Schematization of electron microscopic pictures. (A) Branch migration displaces Okazaki fragments. (Note that we assume the existence of short fragments on both sides of fork, but more on the lagging 5′ end.) (B) Note the central location of multiple reinitiation loops, providing evidence for bidirectional replication.

The best interpretation of these results is a model where short fragments are synthesized on both arms of the fork, with one notable difference: on the leading end where the 3′ end projects toward the fork, there are fewer fragments than on the lagging end (see Figure 3). We are in no position to assume that short fragments are being synthesized *a priori* as discontinuous segments. It is possible that on the leading end there is continuous synthesis followed immediately by corrective editorial nicking.

The label which is incorporated over a very short period of time eventually becomes integrated into large strands of progeny DNA. As is frequently mentioned in monographs, the obvious candidates for joining the fragments to larger strands are polymerase and ligase — but there are also contradictions: phage-containing amber mutations in gene 30 (ligase) produce no progeny in nonpermissive hosts and exhibit markedly reduced DNA synthesis, and the progeny DNA is extremely short. There is also extensive endonucleotic degradation of the parental phage DNA. We have been fascinated by the fact that one can obtain progeny DNA of a large size and even viable phage in a ligase-negative (gene 30) T4 mutant provided there is, in addition, an rII mutation.[21] Likewise, arresting the synthesis of endonucleotic enzymes through the timely addition of CM resulted in the production of large-size progeny DNA and viable phage.[22] It is very unlikely that an rII mutation or the addition of CM restores ligase activity to gene 30 amber mutants or stimulates bacterial ligase function on phage DNA. Hence, our observation of abundant DNA synthesis, the large size of progeny DNA produced, and restoration of viability in the absence of ligase are not in accord with the suggestion of Okazaki that gene 30 polynucleotide ligase functions *in vivo* to join short stretches of newly synthesized DNA. Our results can best be explained by the assumption that a primary role of phage ligase is to repair DNA nicks resulting from endonuclease action on intracellular phage DNA.[23,24] We observed that in the presence of ligase, the parental DNA of a nonreplicative DO mutant (gene 44) remains intact over an extended period of time. Does this mean that no nicks are inflicted? Clearly not, since removal of phage ligase by an additional mutation resulted in extensive fragmentation of the parental DNA.[24] Evidently, even nonreplicating DNA is subject to a constant barrage of nicking. Ligase masks the nicking as deletion of this enzyme results in revealing considerable breakdown of the DNA. (Obviously, the appearance of these nicks cannot be incidental to the process of DNA replication as nicking affects nonreplicative DNA.) Equilibrium between the damaging endonucleases and the repairing ligase is an example of dialectic unity between two opposing enzymes.

At this point, one should also recall Gellert and Bullock's intriguing data, which show that the rate of chasing Okazaki fragments into large strands is essentially the same in ligase-deficient, ligase-overproducing, and wild-type *E. coli*. They concluded: "We thus find normal joining of DNA fragments even with greatly depressed DNA ligase activity."[25] Oddly, this paper of Gellert and Bullock is quoted by Ogawa and Okazaki[26] as providing proof for the role of ligase in the joining of nascent fragments: "Nascent DNA fragments are normally present in only small amounts but accumulate when DNA ligase is inhibited, indicating a critical role . . . "

It should be obvious that the mechanism of production of short nascent fragments, as well as their joining, is not a closed story and deserves further study which might unveil novel qualities of the replicative process. Seemingly unjustified, self-inflicted nicking invites reflection on its meaning.

B. DIRECTIONALITY OF REPLICATION

To understand the evolution of the concepts and importance of directionality of replication, one should recall the puzzling phenomenon of circular permutation of the T4 phage genome. The best, if not the only, explanation for the formation of circularly permuted DNA is the "headful hypothesis" postulated by Streisinger et al.[27] According to this hypothesis, a capsid of T4 phage packages, from a concatenated structure, segments of DNA which slightly exceed the length of a complete genome. This would result in the formation of terminally redundant and circularly permuted molecules. Frankel[13] and then Thomas[28] were the first to postulate a mode of replication of T4 DNA in which the parental molecule does not divide after duplication, but continues to elongate into a concatenated supermolecule. This hypothesis was renamed and presented as the "rolling circle mode" by Gilbert and Dressler[29] at the 1968 Cold Spring Harbor Symposium. This initiated a flurry of papers, seemingly confirming the rolling circle mode of DNA replication.

Mosig implied that DNA replication starts at gene 43 and proceeds unidirectionally along the circular template.[30,31] Petite phage contain incomplete genomes and circularization of their DNA is not possible. Mosig claimed to have observed only partial and nonrepetitive replication in petite phage DNA, thus strengthening the hypothesis of unidirectional replication. However, we found this conclusion to be incorrect by showing that the DNA molecules of petite phage do replicate completely and repeatedly.[32]

The evidence against the rolling circle model is overwhelming. Even before DNA was proven to be the carrier of genetic information, Luria[33] designed an elegant experiment: he tested whether genetic material (whatever it was) replicated in an exponential or linear manner. In the exponential manner, one would replicate into two, two would replicate into four, etc. The linear mode would require the parental template to act as a rubber stamp, with progeny never participating in consecutive replication. In a single-burst experiment, Luria analyzed the clonal distribution of mutants and found it to be consistent with the exponential mode of replication. Prior to the formulation of the rolling circle model, we commented[12] that Luria's experiments were incompatible with the hypothesis of a giant DNA molecule "growing at one end and maturing at the other end."

There are numerous other arguments against the possibility of continuous unidirectional growth. Our electron microscopic studies (Figures 2 and 3), demonstrating two *trans*-positioned whiskers and centrally located reinitiations, strongly supported bidirectional replication.[7]

There is no covalent attachment of newly synthesized progeny DNA to the parental strands.[7,23] This is incompatible with the rolling circle model where a nicked parental strand would act as a primer and therefore should be joined to progeny DNA.

There is repeated replication of the parental DNA. This was proven in a double density experiment showing that the parental DNA continuously changes progeny partner strands during the replicative phase.[14] According to the rolling circle model, only 50% of the parental DNA will undergo repeated replication (i.e., the closed circular template), while the other parental strand should replicate only once.

Concatemers appear only at late times after infection, coincidentally with the onset of recombinational events.[14] Concatemers were not observed at early times after infection when recombinational events are not fully expressed. According to the rolling circle model, concatemers should be observed shortly after the onset of replication. Rolling circle proponents, however, promptly proposed that the delay in the formation of concatemers was due to the fact that replication was at first bidirectional, switching at later times to a rolling circle mode.[34] This was disproven by an experiment of Kosturko and Kozinski,[35] where a potent mutagen (5-BrdUrd) was introduced at late times after infection when some 100 phage equivalents of progeny DNA had already accumulated. If the rolling circle mode becomes operational by then, the addition of the mutagen should not result in a clonal distribution of mutants. In contrast, the observed clonal distribution compared very well with the theoretical graph expected for exponential replication.

We observed[14] that both parental (^{32}P) and progeny (^{3}H) label mature into phage particles at the same rate. The rolling circle mode would predict that 50% of the parental label which resided in the circular template should be exempt from prompt maturation.

The most direct proof that large concatenates were assembled by recombination was provided by the study of DNA in giant phage.[15] Consider bacteria simultaneously infected with two genetically distinguishable parents. If recombinations were responsible for generating the concatemer (see Figure 4), it would be likely that consecutive genomes within the length of the concatemer might bear markers from one or the other coinfecting mutants as well as double mutants and wild-type recombinants. This possibility was experimentally approached by utilizing a mutation in gene 23 which leads to the formation of "giant" phage. These phage exhibit elongated heads within which several genomic lengths of DNA are packaged in concatenated form.[36] Giants offered a unique system for examining the mechanism of concatemer formation. In the experiment,[15] E. coli B was coinfected with two different amber mutants, amB17 (gene 23) and amE355 (gene 24), bearing in common the ptg191 mutation in gene 23 which determines the formation of giant phage, each at a multiplicity of three. Progeny phage were purified and fractionated in sucrose gradients. Fractions corresponding to complete phage and giant phage were isolated. The genotypes of the purified complete and giant phage were analyzed with respect to genes 23 and 24.

The data can be summarized by saying that in the sample of giant phage, the majority (34 out of 50) was heterozygous, whereas in the sample of complete phage, only one out of 50 was heterozygous. Most of the giant phage, being heterozygous, must therefore have matured using concatemers assembled by the process of recombination.[15]

C. SITES OF EARLY INITIATION OF T4 DNA REPLICATION

This section will first attempt to describe past experiments and dramatically opposite points of view which resulted in what appeared to be a perennial controversy. Then turning to more recent results, it will demonstrate how most of the controversy eventually vanished, bringing opponents to a common view of early replication as being bidirectional and initiated at multiple, yet specific, locations.

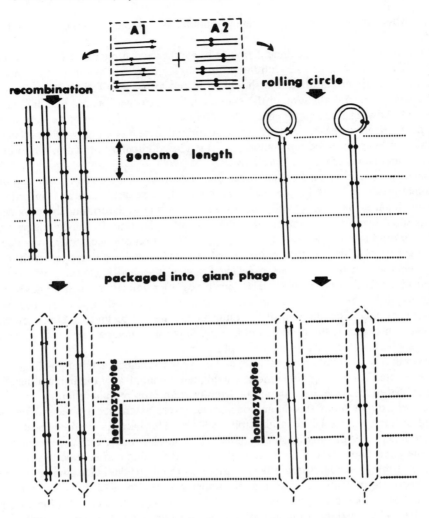

FIGURE 4. Two means of forming concatemers from the replicative pool of phage DNA in bacteria simultaneously infected with two genetically distinct parents, each containing an amber mutation in one of two genes, A1 or A2. Linear DNA molecules in the pool of replicating DNA can circularize and begin to replicate as rolling circles, generating concatemeric tails (right side), or the pool of replicating DNA can recombine to form concatemers (left side). If concatemers produced by both models are used for maturation of phage, the two models predict different results for the genetic character of the giant phage produced. Concatemers produced by the rolling circle should be homozygous for both genes. If concatemers are produced by recombination, the giant phage are likely to be heterozygous for genes A1 and/or A2, and this is what was observed. (From Kozinski, A. W. and Kosturko, L. D., *J. Virol.*, 17, 801, 1976. With permission.)

In the early 1970s, there was a basic controversy concerning both the directionality of replication and the number and location of origins used in the replication of DNA. In previous papers from this laboratory, physicochemical and electron microscopic evidence was presented which showed that the initiation of T4 DNA replication occurred at multiple, but specific, sites.[7-9] Physicochemical approaches which indicated the existence of multiple foci of initiation were threefold.

Gradual shearing — Partially replicated parental molecules (PRM) were isolated from an experiment where parental phage was light and radioactive, while progeny contribution was heavy. It was calculated from the observed displacement from LL reference DNA that the progeny area extended over some one third of the phage equivalent, or approximately 18 μm. If this mass was present in a single unit, shearing of the PRM to 7 μm should have resulted in a substantial portion of the progeny DNA label banding at the hybrid location. The observed lack of a hybrid peak ruled out a structure containing a single replicated region. On the basis of this gradual shearing experiment, the number of replicative regions was calculated to be between three and six.[8]

Hybridization and protection from exonuclease I digestion — Upon hybridization of small single-stranded "donor" fragments of DNA to larger single-stranded "recipient" DNA (extracted from T4 phage), double-stranded regions are created on the recipient. Such areas protect the recipient from hydrolysis by *E. coli* exonuclease I. Depending on the nature of the donor fragments, saturating concentrations of donor DNA will result in complete or less than complete protection of the recipient; if the donor fragments are random, they will effectively cover all parts of the recipient molecules upon hybridization, thus rendering these recipient molecules fully resistant to exonuclease I. On the other hand, if the donor fragments represent a unique portion of the T4 genome, they will hybridize only to a specific region of the recipient fragments, whereas most of the recipient will remain single stranded and unprotected, and thereby sensitive to exonuclease. As a "donor", early progeny DNA was used. A comparison of the observed recipient protection with that theoretically predicted for equally spaced, unique replication sites lead to an estimate of eight to ten specific sites per molecule.[8]

Electron microscopic study[7] — Multiple replicative loops expanding bidirectionally, with the 3' end leading the 5' end by approximately 0.25 to 0.5 μm (see Figures 2 and 3) were demonstrated.

In summary, research from our laboratory revealed bidirectional replication initiated at multiple and specific sites. Multiple initiation along molecules of DNA were not observed previously in prokaryotes or viruses.

In contrast to our results, genetic evidence presented by others[30,31] suggested that replication initiates from a single origin located between genes 42 and 43 and proceeds unidirectionally clockwise along the circular DNA template. Since, more recently, Mosig tacitly withdrew both claims, we will not elaborate further on this apparent contradiction.

With the development of cloning techniques, it has become possible to explore where these origins are. I will discuss (1) the site specificity of initiation of parental DNA and (2) gene amplification — an accumulation of partial replicas specific for the areas of initiation.

1. The Localization of Origins of Replication

This discussion centers on the genetic specificity of the early progeny label which remains hydrogen bonded to the partially replicated parental molecule. The genetic composition of such progeny was determined by hybridization to cloned genetic segments.[10]

To isolate such early replicative progeny DNA, it is necessary to be able to distinguish and isolate parental molecules which have replicated only slightly. If cells are infected with parental phage whose DNA contains a heavy-density label, and then the light, newly synthesized progeny DNA molecules are labeled with [³H]-thymidine, it is possible to observe and purify these partially replicated molecules by their position in CsCl density gradients (see outline of the procedure in Figure 5). This DNA was then used for hybridization to cloned genetic segments immobilized separately on small nitrocellulose filters. As an internal

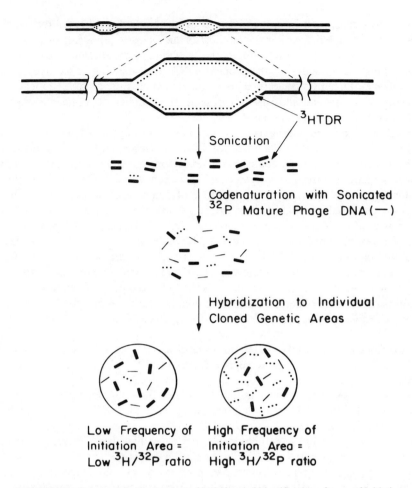

FIGURE 5.   Schematicized procedure of isolation and identification of areas of initiation of DNA replication. (From Kozinski, A. W., *T4 Bacteriophage*, Mathews, C., Ed., American Society for Microbiology, Washington D.C., 1983. With permission.)

reference, ^{32}P-labeled mature phage DNA was cohybridized. This allowed one to represent data as a ratio of experimental DNA (^{3}H-labeled progeny) to ^{32}P-labeled reference DNA.[10]*

The results obtained (Figure 6) allowed us to conclude that, among the genes tested, the region of genes 50 through 5 shows a strong initiation site for DNA replication.[10] The region of genes w through 29 shows a weaker site, whereas the area of genes 40 through 43 (suggested by Mosig) shows no initiation; in fact, we observed the lowest ^{3}H to ^{32}P ratios in this area.[10] Similar results were obtained in experiments using a low multiplicity of infection (MOI).[11] The results are not affected by the use of 5-bromodeoxyuridine density labeling; similar results were seen when the densities were reversed, i.e., the host bacteria

---

*   Kozinski et al.[38] provide a detailed description of the hybridization conditions used in our laboratory, with the range of errors, controls, rationale, and mathematical derivations. One important point should be made here which likely applies to all [^{3}H]-thymidine uptake experiments. The possibility that the uptake patterns reflect a high versus low frequency of adenine-thymine base pairs in the different cloned fragments is eliminated by the observation that in the control cohybridization experiments with [^{3}H]-thymidine and ^{32}P-labeled DNA random labeling, ^{3}H to ^{32}P ratios were identical for all tested genetic areas. In other words, all of the tested areas have, on the average, the same proportional amount of thymidine to phosphorus.[10,38] Second, sonication of [^{3}H]-adenine and ^{32}P-labeled heavy DNA revealed no significant skew of the specific activities of the two labels after CsCl gradient analysis.[8]

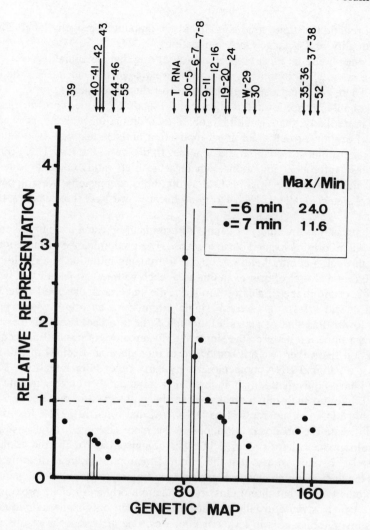

FIGURE 6. Typical pattern of hybridization of early progeny DNA to cloned genetic segments (Kozinski and Ling[38]). The results of hybridization are expressed as relative representations (RR) of genetic segment x in the progeny DNA (Kozinski et al.[11]). The ^{3}H/^{32}P ratio observed for a filter charged with a given genomic area, x, was divided by the ratio of the sums of ^{3}H and ^{32}P hybridized to all of the nitrocellulose filters in a set.

and media were heavy and the parental phage were light or when no density label was used.[37,38]

From the hybridization patterns observed (see Figure 6), one can draw more subtle inferences. For example, the group of genes 40 to 41, 42, and 43 represents the pattern of high, lower, and lowest ratios, respectively. Such a pattern occurred in 15 of 18 hybridizations performed in independent experiments. Likewise, the right-most group of genes 35 to 36, 37 to 38, and 52 represents the pattern of high, low, and high ratios, respectively, which occurred in 13 of 18 hybridizations. One should realize the extreme improbability that these repeatable patterns resulted from random errors. Therefore, the tendency of increasing values on the left side of gene 40 and on the left side of gene 35 has been interpreted[10] as indicative of the presence of other origins located to the left of those two areas. The other important aspect of the observed patterns of hybridization is that areas on both sides of the origin display, as a function of time, gradual gains in the amount of progeny label. This confirms the bidirectionality of T4 phage DNA replication.[10]

We feel one should be most emphatic in observing that, in actuality, every filter tested did hybridize some [³H]-thymidine, with the lowest counts found in the area of genes 40 through 43. There are two possible explanations for this phenomenon. First, all of the DNA molecules probably do not initiate replication at exactly the same moment. Second, which we consider more likely, there is a high probability of initiation at certain very active origins (or clusters of origins), from which most replication initiates, as well as low probability of initiation at other points throughout the genome. In this case, the low ³H to ³²P ratios could be due to these infrequent initiations, whereas the high ratios could be due to frequent initiations at the most active origins. In drawing these conclusions, we emphasize that we may not have detected all of the T4 origins because we have not tested a full genomic complement.*

We also stress that this analysis cannot discriminate between a single potent origin and several less potent origins located close together. The predominance of initiation in the area of genes 50 through 5 could, therefore, be due to frequent initiation at a single potent origin or to initiations at a cluster of origins in this area, each with less probability of being initiated than the single origin in the first case. Similarly, the small peak observed at genes 2 through 29 could be due to a less "attractive" single origin, i.e., an origin whose probability of initiation is lower than that of genes 50 through 5, or to a decreased number of origins in this area. Discrimination between the single vs. clustered origin possibilities requires further experimentation. However, we are compelled to reproduce an electron microscopic picture (Figure 7) of a parental DNA molecule at very early times after infection.[9] The molecule shown is of phage-equivalent length and reveals a large number of very minute loops of 0.1 to 0.3 μm. (Note: no such miniloops have ever been observed in our electron micrographs of mature phage DNA or nonreplicative DNA.) At that time, in 1971,[7] it was not obvious to us how to relate the observed miniloops to the more advanced, large bistranded loops. Similar miniloops were later observed in other organisms and their role as initiation sites was discussed.[39] If one assumes that these miniloops represent areas of initiation, one can deduce that the area in which cluster B resides (see Figure 7) should be frequent in initiation since it contains two (if not three) clusters of putative origins in close proximity. There is also area A, which is more distant. In between, there are only some individual miniloops. If one measures the distances in terms of kilobases, one arrives at the results shown in the adjacent diagram. Perhaps the area in which cluster B resides corresponds to predominant areas of initiation stretching from genes 5—50 to w—29.

Hybridization tests of early progeny DNA as described above cannot, however, pinpoint the exact location of origins within the length of the restriction fragment. I will now describe a completely different, genetic approach to the more specific localization of origins within the cloned segment.

The principle of this genetic approach[40] was the determination of the frequency of recombination (marker rescue) between genes residing in the cloned segment and infecting phage carrying amber mutations for those genes. We will limit our description to two exemplary cloned fragments of similar length — one located in the predominant area of initiation and spanning genes 50, 65, and 5 (plasmid 668) and the other, which represents the area low in the frequency of initiation, in the valley between two initiative peaks and spanning genes 13 to 18, including gene 14 (plasmid TFH 3004).

E. coli carrying either of these plasmids was infected with T4 amber mutants in the corresponding genes. Soon after infection, suspensions were plated to determine the titer of

---

*   Upon publication of our hybridization results indicating multiple origins, there appeared a number of other papers devoted to identification of additional origins. I will not discuss those here as a rather lengthy evaluation and criticism will be needed (see also Reference 41 and 42).

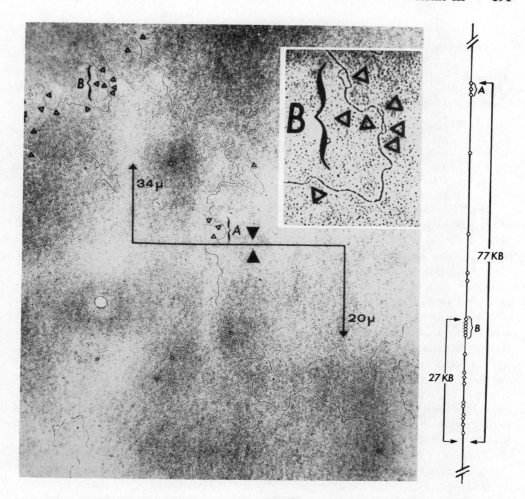

FIGURE 7.   Electron micrograph of DNA at an early initiation stage. Note the abundance of miniloops with noticeable clusters; in upper corner, an enlargement of cluster B. Diagram on the right represents the dimensions of the molecule shown. (From Howe, C., Ph.D. thesis, University of Pennsylvania, Philadelphia, 1972. With permission.)

the infective centers. Plating was performed on two hosts: permissive *E. coli* CR63 and nonpermissive *E. coli* B. The infective center (bacterium), which yields at least one wild recombinant phage, will produce a plaque on *E. coli* B. After lysis, the resulting lysate (final yield) was plated on both detectors and the fraction (%) of wild recombinant phage was determined. The results are summarized in Table 1.

Scrutiny of the data reveals, for gene 5, a surprisingly high frequency of successful recombination between the plasmid-carried gene and the infecting phage: almost 20% of infective centers plated on *E. coli* B and almost 2% of the total final yield was composed of wild phage. On the other hand, genes which were located more distally from gene 5, i.e., genes 65 and 50, recombined progressively less (compare columns showing percent of recombinants per final yield and percent of recombinants per infective center). Note also that plating of infective centers resulted in less pronounced differences among the genes tested, indicating that while a large proportion of infected bacteria did produce some recombinants, the burst size (clone) of recombinants was progressively smaller the further away from gene 5. In contrast, recombinants for gene 14, which is located at the valley between the replicative peaks, revealed a much lower frequency of recombination.

TABLE 1
Recombination Frequencies between T4 Amber Mutants and Cloned Genes

T4 amber mutant gene	Plasmid	T4 genes inserted	Length of insert (kilobases)	Recombinant infective centers (%)	Recombinants in final yield (%)	Recombinants per recombinant infective center (%)
5	668	50—5	4.8	19.48	1.85	9.5
65	668	50—5	4.8	13.50	0.86	6.3
50	668	50—5	4.8	13.70	0.34	1.2
50, 5	668	50—5	4.8	2.68	0.19	7.2
14	TFH 3004	13—18	6.0	3.7	0.23	6.2

We have also tested a double marker rescue for two genes located at the extreme ends of the cloned segment (genes 5 and 50) by infecting with double amber mutant in those genes. It should be noted that while the frequency of bacteria yielding wild phage is smaller, those infective centers which do produce any double recombinants release a rather large burst of recombinant phage (see percent recombinants/infective center).

In the case of infection with the double amber mutant, we have also analyzed their individual genetic makeup. Several thousand individual plaques were scored by plaque picking and testing for their genetic representation with respect to genes 5 and 50. It was found that the average frequency for gene 5 recombinants was 0.035, while the average for recombinants in gene 50 and also for double recombinants was 0.006 — thus, much less.

It is obvious, therefore, that most of the recombinations do not occur by insertion of the entire cloned segment into the infecting genome. Rather, there is a strong preference for gene 5. One could argue that this possibly could be due to an intrinsically higher recombinational frequency for this gene. We reject this possibility as we had tested it by performing a cross between coinfecting wild and double amber mutant phage, and found that the frequency of wild allele recombinants for either gene 5 or 50 was identical.

We now explain our results as a conclusion of the following chain of intracellular events. Upon infection, T4 bacteriophage codes and produces endonucleases which promptly attack nonmodified DNA, while modified phage DNA (i.e., containing glycosylated 5-hydroxy-methylcytosine [5HMC]) remains resistant to these enzymes. This results in inflicting a number of cuts to the plasmids carrying the cloned segment. Those short fragments, however, if endowed with a T4 origin, can be recognized by T4 replicative machinery and replicated further using glycosylated 5HMC. This increases the copy number of the wild-type allele which, in turn, becomes available for recombination with the infecting phage DNA. Plasmids or areas of cloned segments devoid of origins remain nonreplicated, vulnerable to phage-coded nucleases, and will be mostly destroyed.

Summarizing this group of genetic experiments, there is good evidence that the predominant area of initiation can be narrowed to the immediate vicinity of gene 5.[40]

2. Gene Amplification in T4: Appearance of Bistranded Progeny Segments Before Completion of Replication of the Parental Molecule

A simple model for DNA replication in T4 could begin with the following sequence of events. First, initiation at the origins would produce parental molecules containing small, replicative loops. Second, extension of the loops along the full length of the parental molecule would produce hybrid molecules consisting of half parental DNA and half progeny DNA. Third, an additional round of replication would produce molecules consisting of two strands of progeny DNA in addition to hybrid molecules. Consequently, if the parental DNA were labeled with 5-BrdUrd (heavy, nonradioactive) and the newly synthesized progeny DNA were labeled with [³H]-thymidine (light, radioactive), the model predicts that immediately after initiation of replication, the incorporated [³H]-thymidine would band very close to the heavy (HH) density in CsCl gradients. Then, as replication proceeds through the first round, the [³H]-thymidine would move toward the hybrid (HL) density, while the amount of label at the heavy density would decline. Finally, light (LL) density molecules would be observed only after a second round of replication is completed.

Our experimental data[11] did not, however, conform to these simplified predictions. At very early times after infection, the progeny label was found to band not only at the expected heavy density, but also at the light density. At this time, there was still no detectable peak of the progeny label at the hybrid density, which indicates that less than one complete round of replication at the parental molecules had occurred.

When this light moiety of progeny DNA was isolated and assayed for its genetic makeup, it was shown to be biased in its genetic representation. This was especially observable at low MOI.[11] It was shown that, by the timely addition of CM, one can freeze replication in this apparent amplified mode while there is sizable accumulation of partial replicas of areas corresponding to initiation sites. Up to 50 copies of partial replicas of genes 5 through 50, for instance, can be accumulated by 40 min postinfection when CM is added at 5 min.[11] Addition of CM at 10 min, on the other hand, results in the production of progeny DNA representing all tested areas equally. Since the addition of CM at 5 min prevents molecular recombination, whereas addition of CM at 10 min allows full expression of molecular recombination, we postulate[11,37] (see Section III.D below) that without recombination, progeny DNA cannot "extend all the way", and completion of the replication of the parental molecule is prohibited.

We have shown that amplified DNA can replicate autonomously. The strategy of the experiment was similar to that used for "petite" phage.[32] We allowed accumulation of ³H-labeled light (LL)-amplified DNA and then removed the radioisotope and added unlabeled 5-BrdUrd. We observed that upon further incubation, most of the originally light ³H-labeled DNA assumed HL density. Therefore, amplification cannot be due to repetitious "firing" of origins on the parental molecule. If this were the case, amplified DNA would represent a replicative dead end and would not be able to replicate autonomously.[43] Whether amplified DNA is detached from the parental molecule *in vivo* or remains associated by short terminal complementarity, thus resulting in a loop resembling a longitudinal cross-section of an onion, is of no consequence with respect to this conclusion. It is possible that such a multilayered structure is fragile and the weak terminal complementarity is broken during extraction. Multilayered loops have been observed, albeit infrequently, in partially replicated molecules[7] (see also, model in Figure 3B).

Another form of gene amplification is observed upon infection of bacteria with UV-inactivated heavy phage, where a considerable amount of phage DNA synthesis occurs in cells not productively infected. This DNA is biased in its genetic representation, and the accumulated regions correspond to the areas of major origins of T4 DNA replication.[44] These data are consistent with the assumption that UV irradiation, acting upon heavy DNA, imposes termination (most likely by cuts occurring within the host cell). This termination, in turn, leads to the replication and amplification of those segments which are endowed with origins. The sequestration by UV of autonomously replicating subunits complements data on specificity of origins. It also offers a molecular explanation of observed patterns in marker rescue experiments,[45] in which certain genetic areas are rescued better than others.

The process of amplification could have regulatory functions. For instance, it might be involved in late transcription. Most of the amplified areas contain genes that code for late-function proteins, and it is known that DNA replication is required for the expression of late-function genes;[46] therefore, it is possible that amplification is needed to provide proper transcriptional templates for these genes. Preferential and partial replication of specific areas of the genome and the putative role of amplified DNA in transcription could then resemble chromosomal events in the salivary gland of *Drosophila*.

D. LATE EVENTS: INTERRELATION OF REPLICATION AND RECOMBINATION

In the previous sections, we have shown that the amplifying mode of replication can be "fixed" by the early addition of CM, which allows a large number of copies of partial replicas to accumulate over a long period of incubation. Yet the addition of CM at times when molecular recombination is fully expressed results in the accumulation of progeny DNA which does not display a genetic bias. The genetic representation of the progeny DNA accumulated at later times after infection without CM is also unbiased.

What is the mechanism of this equalization? I will now present the argument that the process of recombination provides for equalization of genetic representation among progeny DNA. I will first recall relevant observations.

When bacteria which have accumulated a large pool of HH progeny DNA were superinfected with radioactive phage of light (LL) density, the superinfecting DNA recombined promptly with the pool of heavy progeny DNA. Such recombinants banded, in CsCl gradients, at a "progeny-like" (heavy) density. Sonication of the recombinants revealed, however, that the parental subunits remained nonreplicated, i.e., bistranded LL.* Denaturation of the recombinant moieties released light strands of superinfecting phage DNA quantitatively; thus, such recombinants were not covalently joined to heavy progeny.

At later times after superinfection, the superinfecting genome became replicated (i.e., after sonication, parental label assumed an HL location). At this time, an equivalent volume of parental label invariably became covalently joined to the progeny strands.[24] Taken together, these observations lead to the hypothesis that recombination might provide for initiation at the 3' end of a recombinational intersection. The advancing progeny strand might convert a recombinational intersection into a replicative fork.[24] Such initiation should be random or drastically different from the initiative origins observed early after infection.

It is important here to recall that, as demonstrated by Kozinski et al.,[12] the timely addition of CM can serve as a fine dissecting tool, separating various replicative events. When CM is added at 5 to 6 min postinfection, there is no parent-to-progeny recombination and the progeny DNA displays a genetic bias, being fixed in the amplifying mode.[11] Addition of CM at 9 to 13 min allows for recombination and covalent joining of the parental segments to the progeny DNA.[12]

Therefore, if the working hypothesis of random initiation at recombinant intersections is correct (as contrasted to specific initiation observed before the onset of recombination), two alternatives could be predicted. (1) In the absence of recombination, a short pulse of the radioactive label applied at late times will be divided between label which was involved in the process of reinitiation of origins occurring during the time span of the pulse (which should be of pure density and likely site specific) and label added to the ends of already initiated, elongating progeny strands which would band in CsCl at an intermediate density.

(2) When recombination is expressed and if recombinational 3' ends act as randomly located primers, the short pulse label should find its way either to elongating structures of intermediate density or to progeny strands of pure density in those events when strand displacement and its (presumably) discontinuous replication occur during the span of the labeling period. In this case, the light moiety should not display site specificity. The isolation of the light (L) strands and the hybridization of these to cloned genes should permit discrimination between reinitiation of specific sites and random initiation.

---

*   Since most of the parental contribution resided, upon sonication, in the LL density, the major mechanism of observed recombination calls for terminal joining of bistranded parental subunits to progeny DNA, in contrast to the "invasion" of progeny DNA by a single parental strand. Of course, parental label incorporated as a single strand should, after sonication, assume the form of a hybrid (HL). This also applies to recombination followed by extensive branch migration.

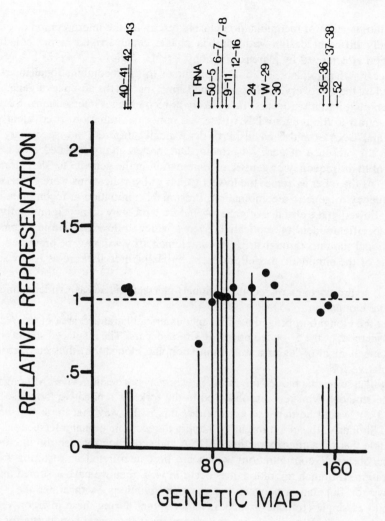

FIGURE 8.   Late reinitiation in the absence or presence of recombination. Hybridization of the light moieties to cloned genes. The thin lines represent the patterns of
hybridization of the light moiety from the CM 6-min branch, and the black circles
represent the results from the CM 13-min branch. The broken line corresponds to the
RR values that would be expected if all of the genetic areas were equally represented.
(From Halpern, M., Mattson, T., and Kozinski, A. W., *J. Virol.*, 42, 422, 1982.
With permission.)

The results of the experiments of Halpern et al.[37] point to the second of these alternatives.
In the branch where recombination was inhibited (by CM added at 6 min), reinitiation
occurred at areas reasonably similar to those involved in the initiation of the parental molecule. In the branch where CM was not added until 13 min postinfection and abundant
molecular recombination was observed, the uptake of [³H]-thymidine to light strands revealed
no apparent site specificity (Figure 8). On the basis of these observations, mechanisms were
proposed for the establishment of covalently joined recombinants. If was postulated that
repair involves mostly replication and displacement of the strand, which is followed, in turn,
by its replication.[37] This proposed model of the conclusion of recombination differs from
the popular (and unproven) view whereupon, after joining (annealing), there is filling of the
gaps and/or excision of uncomplemented areas and eventually ligation. Yet our interpretation

of the concluding steps of recombination closely resembles the interpretation of experiments of completely different design with lambda phage. Stahl arrived at the conclusion that recombination is achieved by "break and copy".[47-49]

Evidence for the covalent addition of precursor to the recombinational intersections is strengthened by the discovery by Kozinski and Ling[38] that, in the absence of gene 44 protein, a late and aborted initiation occurs, most likely at recombinant intersections. Such progeny DNA, in contrast to the progeny DNA observed in normal initiation, is covalently attached to the parental DNA strands and displays dramatically different site specificity from that observed in the presence of gene 44 protein. One preferentially initiated area corresponds to the area of high frequency of genetic recombination in the gene 35 to 36 region described by Mosig.[50] At the other extreme, the lowest values of hybridizations were observed in areas of low frequency in genetic recombination. Presumably, initiation at recombinational intersections is allowed in the absence of gene 44, but such progeny do not significantly elongate.

More specific models of replicative events which follow the joining (annealing) of recombinational subunits through areas of complementarity will now be presented. The basic assumptions of the simplistic model presented by Halpern et al.[37] are as follows:

1.   The 3' hydroxyl ends of recombinational intersections act as primers, initiating replicative elongation.
2.   When the elongating progeny strand encounters a bistranded area of the recombinant, displacement of the 5'-ended opposing strand occurs. The displaced strand is replicated by a mechanism of lagging strand initiation and elongation, thus converting it into a replicative fork.
3.   We will assume that branch migration may or may not occur *in vivo*. Although displaced single-stranded whiskers are observed in the forks of replicative loops (see Figures 2 and 3), it would be naive to assume that this is the way that these whiskers exist *in vivo*. Similarly, this point might well apply to recombinational intersections. It is quite possible that only after extraction of DNA and spreading for electron microscopy, one might observe[51] configurations which are best interpreted as resulting from branch migration. If branch migration does occur *in vivo*, then we have assumed that it simply allows for fluctuation between various configurations, with either the 3' or the 5' termini as single-stranded branches. At any point during these interconversions, replicative enzymes could stabilize the recombinant in a form conducive for replication (e.g., with the 3' end of the strand hydrogen bonded, as shown in Figure 9) and then replicate it. In Figure 9, replicative events leading to the segregation of subunits without single-stranded interruptions are presented. We end the diagram at the formation of heterozygotes since an additional round of replication should result in the segregation of homozygous recombinants. At the top of each panel, a recombinant between subunits BC and ab is drawn. The schemes presented are limited to representations of unmixed patterns of joining, i.e., only gaps or whiskers of the same polarity in each recombinant. If this were not done, then the abundance of combinatorial possibilities would prohibit us from attempting to graph the schematic events. We would like to emphasize that upon the conclusion of replication, the subunits obtained might be complete recombinants (as in AII, BII, and CI) or they could be truncated recombinants containing potentially complementary termini (as in AI, BI, and CII). These complementary termini might provide for preferential recombination either by exonucleolytic denuding of the ends or by a recombinational event such as that proposed in Figure 10A. In addition, it should be noted that in the absence of branch migration, the recombinant containing 3' whiskers represented in CI and CII of Figure 9 cannot replicate until a second recombinational event adds a single-stranded fragment to serve as a primer.

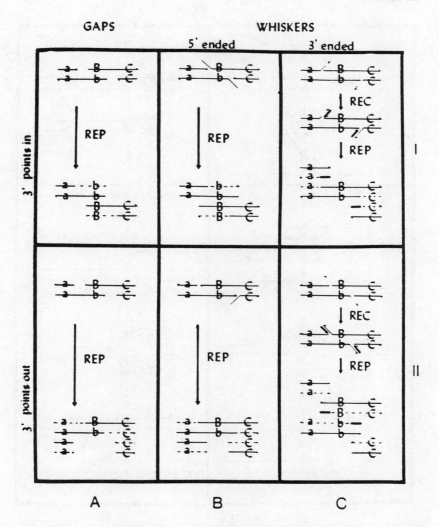

FIGURE 9.   Schematic representation of initiation and elongation of replication in recombinant molecules. We start with a recombinant between subunits BC and ab. Replication is then initiated at the 3′ termini of the recombinational intersections. The results of this replication are depicted at the bottom of each panel. In (C), an additional round of recombination is required to provide a primer for the initiation of replication. (The dotted line represents newly synthesized DNA.) (From Halpern, M., Mattson, T., and Kozinski, A. W., *J. Virol.*, 42, 422, 1982. With permission.)

Figure 10 presents two possible mechanisms which lead to the formation of recombinants containing two terminal genetic markers of one parental subunit, while the centrally located marker originates from a different parent. Such recombinant forms have been attributed to double-switch or double-preferential recombinational events (high negative interference). The panels on side A of the figure do, indeed, involve multiple recombinational events, the first of which occurs between subunits originating from the same parent. The probability of these multiple recombinational events is enhanced by a 3′-ending, single-stranded branch which cannot initiate replication without prior recombination. Then, as shown in the middle of the panel, this probability is further enhanced by the terminal complementarity of the segregated subunits. This model employs terminal joining of complementary areas in such a way as to result in a "replicative loop". In contrast to a true replicative loop, the 3′ ends within the forks are not in the usual *trans* position, but rather, in a *cis* position. However,

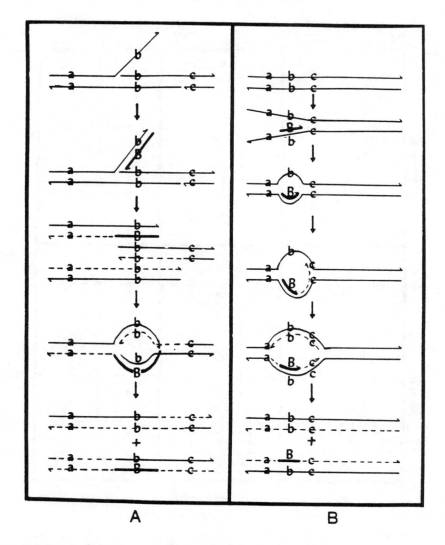

A   B

FIGURE 10.   A schematic representation of possible models resulting in a double crossover event (the phenomenon of high negative interference). The panels present two possible ways of forming recombinants which contain terminal markers of one parent and the central marker from a different parent. (A) involves multiple recombinational events, whereas (B) involves the invasion of a double-stranded molecule by a single-stranded fragment. (From Halpern, M., Mattson, T., and Kozinski, A. W., *J. Virol.*, 42, 422, 1982. With permission.)

there does not seem to be any reason for this type of a loop not to successfully replicate. The schematic events presented in panel B of Figure 10 involve the invasion of a double-stranded molecule by a single-stranded fragment. We assume that this type of recombination might preferentially occur at the termini of molecules due to the "breathing" effect. However, internal invasion should yield identical results.

In summary, we propose mechanisms for the establishment of covalently joined recombinants in which repair involves mostly replication rather than the filling in of gaps, ligation, and/or the excision of single-stranded branches. Indeed, it has been observed that during infection with ligase-deficient mutant phage,[52] there is a great increase in the frequency of recombination, which is followed by an amazingly efficient covalent repair of such recombinants. Furthermore, although both the replicative parental and the progeny DNA are of a very small size, they replicate autonomously.[53] At first glance, one would be at a loss to explain the covalent repair and replication of recombinant molecules in the absence of the ligase. However, according to the model presented here, the repair and replication of the recombinants could proceed quite efficiently in the absence of ligase simply by the proposed mechanism involving initiation at the 3' termini of recombinational intersections, as presented in Figures 9 and 10.

We have reflected[37] that the proposed union of replicative and recombinational events carries the flavor of the theory of the ''copy choice'' mechanism of recombination which we, in the distant past,[1,2] were first to so bluntly contradict while demonstrating the breakage and rejoining of DNA molecules.

REFERENCES

1. **Kozinski, A. W.,** Fragmentary transfer of ^{32}P-labeled parental DNA to progeny phage, *Virology,* 13, 124, 1961.
2. **Kozinski, A. W. and Kozinski, P. B.,** Fragmentary transfer of ^{32}P-labeled parental DNA to progeny phage. II. Average size of the transferred parental fragment, *Virology,* 20, 213, 1963.
3. **Kozinski, A. W. and Uchida, H.,** Phage DNA subunits in the phage precursor pool, *J. Mol. Biol.,* 3, 267, 1961.
4. **Meselson, M. and Stahl, F. W.,** The replication of DNA in *Escherichia coli, Proc. Natl. Acad. Sci. U.S.A.,* 40, 783, 1958.
5. **Shahn, E. and Kozinski, A. W.,** Fragmentary transfer of ^{32}P labeled parental DNA to progeny phage. III. Incorporation of a single parental fragment to the progeny molecule, *Virology,* 30, 455, 1966.
6. **Carlson, K., Lorkiewicz, A., and Kozinski, A. W.,** Host-mediated repair of discontinuities in DNA from T4 bacteriophage, *J. Virol.,* 12, 310, 1973.
7. **Delius, H., Howe, C., and Kozinski, A. W.,** Structure of the replicating DNA from bacteriophage T4, *Proc. Natl. Acad. Sci. U.S.A.,* 68, 3049, 1971.
8. **Howe, C. C., Buckley, P. J., Carlson, K. M., and Kozinski, A. W.,** Multiple and specific initiation of T4 DNA replication, *J. Virol.,* 12, 130, 1973.
9. **Howe, C.,** Early intracellular events in the replication of bacteriophage T4 DNA: structure of replicating T4 DNA, Ph.D. thesis, University of Pennsylvania, Philadelphia, 1972.
10. **Halpern, M. E., Mattson, T., and Kozinski, A. W.,** Origins of phage T4 DNA replication as revealed by hybridization to cloned genes, *Proc. Natl. Acad. Sci. U.S.A.,* 76, 6137, 1979.
11. **Kozinski, A. W., Ling, S.-K., Hutchinson, N., Halpern, M. E., and Mattson, T.,** Differential amplification of specific areas of phage T4 genome as revealed by hybridization to cloned genetic segments, *Proc. Natl. Acad. Sci. U.S.A.,* 77, 5064, 1980.
12. **Kozinski, A. W., Kozinski, P. B., and James, R.,** Molecular recombination in T4 bacteriophage deoxyribonucleic acid. I. Tertiary structure of early replicative and recombining deoxyribonucleic acid, *J. Virol.,* 1, 758, 1967.
13. **Frankel, F. R.,** Studies on the nature of replicating DNA in T_4 infected *E. coli, J. Mol. Biol.,* 18, 127, 1966.
14. **Miller, R. C., Jr., Kozinski, A. W., and Litwin, S.,** Molecular recombination in T4 bacteriophage deoxyribonucleic acid. III. Formation of long single strands, *J. Virol.,* 5, 368, 1970.
15. **Kozinski, A. W. and Kosturko, L. D.,** Late events in T4 bacteriophage production. II. Giant bacteriophages contain concatemers generated by recombination, *J. Virol.,* 17, 801, 1976.
16. **Okazaki, R.,** *In vivo* mechanism of DNA chain growth, *Cold Spring Harbor Symp. Quant. Biol.,* 33, 129, 1968.
17. **Kozinski, A. W.,** Unbiased participation of T4 phage DNA strands in replication, *Biochem. Biophys. Res. Commun.,* 35, 294, 1969.
18. **Kriegstein, H. J. and Hogness, D. L.,** Mechanism of DNA-replication in *Drosophila:* chromosome structure of replication forks and evidence for bidirectionality, *Proc. Natl. Acad. Sci. U.S.A.,* 71, 135, 1974.
19. **Matthes, M. and Denhardt, D. T.,** The mechanism of replication of ϕX174 DNA. XVI. Evidence that the ϕX174 viral strand is synthesized discontinuously, *J. Mol. Biol.,* 136, 45, 1980.
20. **Narkhammar-Meuth, M., Kowalski, J., and Denhardt, D. T.,** Both strands of polyoma DNA are replicated discontinuously with ribonucleotide primers *in vivo, J. Virol.,* 39, 21, 1981.

21. **Berger, H. and Kozinski, A. W.,** Suppression of T4 ligase mutations by rIIA and rIIB mutations, *Proc. Natl. Acad. Sci. U.S.A.,* 64, 897, 1969.

22. **Kozinski, A. W. and Mitchell, M.,** Restoration by chloramphenicol of bacteriophage production in *E. coli* B infected with ligase-deficient amber mutant, *J. Virol.,* 4, 823, 1969.

23. **Kozinski, A. W.,** Molecular recombination in the ligase negative T4 amber mutant, *Cold Spring Harbor Symp. Quant. Biol.,* 33, 375, 1968.

24. **Hutchinson, N., Kazic, T., Lee, S. J., Reyssiguier, C., Emanuel, B., and Kozinski, A. W.,** Late replication and recombination in the vegetative pool of T4, *Cold Spring Harbor Symp. Quant. Biol.,* 48, 517, 1978.

25. **Gellert, M. and Bullock, M. L.,** DNA ligase mutants of *Escherichia coli, Proc. Natl. Acad. Sci. U.S.A.,* 67, 1580, 1970.

26. **Ogawa, T. and Okazaki, T.,** Discontinuous DNA replication, *Annu. Rev. Biochem.,* 42, 421, 1980.

27. **Streisinger, G., Emrich, J., and Stahl, M. M.,** Chromosome structure in phage T4. III. Terminal redundancy and length determination, *Proc. Natl. Acad. Sci. U.S.A.,* 57, 292, 1967.

28. **Thomas, C. A.,** Recombination of DNA molecules, *Prog. Nucleic Acid Res.,* 5, 315, 1966.

29. **Gilbert, W. and Dressler, D.,** DNA replication, the rolling circle model, *Cold Spring Harbor Symp. Quant. Biol.,* 33, 473, 1968.

30. **Mosig, G.,** Preferred origin and direction of bacteriophage T4 DNA replication. I. A gradient of allele frequencies in crosses between normal and small T4 particles, *J. Mol. Biol.,* 53, 503, 1970.

31. **Marsh, R. C., Breschkin, A. M., and Mosig, C.,** Origin and direction of bacteriophage T4 DNA replication. II. A gradient of marker frequencies in partially replicated T4 DNA as assayed by transformation, *J. Mol. Biol.,* 60, 213, 1971.

32. **Kozinski, A. W. and Doermann, A. H.,** Repetitive DNA replication of the incomplete genomes of phage T4 petite particles, *Proc. Natl. Acad. Sci. U.S.A.,* 72, 1734, 1975.

33. **Luria, S. E.,** The frequency distribution of spontaneous bacteriophage mutants as evidence for exponential rate of phage reproduction, *Cold Spring Harbor Symp. Quant. Biol.,* 16, 463, 1951.

34. **Bernstein, H. and Bernstein, C.,** Circular and branched circular concatenates as possible intermediates in bacteriophage T4 DNA replication, *J. Mol. Biol.,* 77, 355, 1973.

35. **Kosturko, L. D. and Kozinski, A. W.,** Late events in T4 bacteriophage production. I. Late DNA replication is primarily exponential, *J. Virol.,* 17, 794, 1976.

36. **Doermann, A. H., Eiserling, F. A., and Boehner, L.,** Genetic control of capsid length in bacteriophage T4, *J. Virol.,* 12, 374, 1973.

37. **Halpern, M., Mattson, T., and Kozinski, A. W.,** Late events in T4 bacteriophage DNA replication. III. Specificity of DNA reinitiation as revealed by hybridization to cloned genetic fragments, *J. Virol.,* 42, 422, 1982.

38. **Kozinski, A. W. and Ling, S.-K.,** Genetic specificity of DNA synthesized in the absence of T4 phage gene 44 protein, *J. Virol.,* 44, 256, 1982.

39. **Baldari, C. T., Amaldi, F., and Buongiorno-Nardelli, M.,** Electron-microscope analysis of replicating DNA of sea urchin embryos, *Cell,* 15, 1095, 1978.

40. **Rosenbaum, J.,** Recombination between plasmid cloned T4 inserts and T4 bacteriophage, Ph.D. thesis, University of Pennsylvania, Philadelphia, 1983.

41. **Kozinski, A. W.,** Origins of T4 DNA replication, in *T4 Bacteriophage,* Mathews, C., Ed., American Society for Microbiology, Washington D.C., 1983, III.

42. **Kozinski, A. W.,** In vivo replication and recombination of bacteriophage T4 DNA, *Acta Microbiol. Pol.,* 37, 111, 1986.

43. **Hutchinson, N. I.,** Gene amplification in bacteriophage T4, Ph.D. thesis, University of Pennsylvania, Philadelphia, 1983.

44. **Ling, S.-K., Vogelbacker, H. H., Restifo, L. L., Mattson, T., and Kozinski, A. W.,** Partial replication of ultraviolet light irradiated T4 phage DNA results in amplification of specific genetic areas, *J. Virol.,* 40, 403, 1981.

45. **Womack, F.,** Cross-reactivation differences in bacteriophage T4D, *Virology,* 26, 758, 1965.

46. **Riva, S., Cascino, A., and Geiduschek, E. P.,** Coupling of late transcription to viral replication in bacteriophage T4 development, *J. Mol. Biol.,* 54, 85, 1970.

47. **Stahl, F. W., McMilin, K. D., Stahl, M. M., Crasemann, J. M., and Lam, S.,** The distribution of crossovers along unreplicated lambda bacteriophage chromosomes, *Genetics,* 77, 395, 1973.

48. **Stahl, F. W. and Stahl, M. M.,** Red-mediated recombination in bacteriophage lambda, in *Mechanisms in Recombination,* Grell, R. F., Ed., Plenum Press, New York, 1974, 407.

49. **Stahl, F. W.,** *Genetic Recombination, Thinking About It in Phage and Fungi,* W. H. Freeman, San Francisco, 1979.

50. **Mosig, G.,** A map of distances along the DNA molecule of phage T4, *Genetics,* 59, 137, 1968.

51. **Broker, T. R. and Lehman, I. R.,** Branched DNA molecules, intermediates in T4 recombination, *J. Mol. Biol.,* 60, 131, 1971.

52. **Kozinski, A. W. and Kozinski, P. B.,** Covalent repair of molecular recombinants in the ligase-negative amber mutant of T4 bacteriophage, *J. Virol.,* 3, 85, 1969.

53. **Kozinski, A. W. and Kozinski, P. B.,** Autonomous replication of short DNA fragments in the ligase negative T4 AM H39X, *Biochem. Biophys. Res. Commun.,* 33, 670, 1968.

54. **Kozinski, A. W.,** unpublished data.

Chapter 9

DOUBLE-STRANDED DNA PACKAGED IN BACTERIOPHAGES: CONFORMATION, ENERGETICS, AND PACKAGING PATHWAY

Philip Serwer

TABLE OF CONTENTS

I. INTRODUCTION

Viruses with a genome of double-stranded DNA have been isolated from a variety of prokaryotic hosts. The DNA of these viruses is contained within an outer shell of protein. For some (reviewed in References 1 to 4), but not most, of these viruses, the outer shell is surrounded by a layer of the membrane of the host. In addition to the outer shell, proteins are found both internal and external to the outer shell; the entire protein of the virus is referred to as the capsid. The membrane-free, double-stranded DNA viruses with bacterial hosts (bacteriophages) have the following structural features in common: (1) an outer shell that is either an icosahedral lattice of subunits (spherical) or two icosahedral hemispheric lattices separated by a cylindrical array of subunits (elongated), (2) an internal ring (connector) that has sixfold rotational symmetry, but contacts the outer shell at an axis of the fivefold rotational symmetry of the outer shell, (3) an external projection (tail) with fibers; most of the tail has sixfold symmetry and the tail joins the connector to form a grommet-like junction with the outer shell, (4) internal proteins that for the related, spherical bacteriophages, T7 and T3, form a cylinder coaxial with the tail connector (reviewed in References 4 to 10; details for *Escherichia coli* bacteriophages T7 and T3 are illustrated in Figure 1b). The tail with fibers attaches the bacteriophage to its host for subsequent injection of DNA and initiation of the reproduction of progeny (reviewed in References 1, 3 to 5, and 11; see Reference 12 for a more recent and detailed study of T4). By use of three-dimensional reconstruction from electron micrographs, the connector of the elongated *Bacillus subtilis* bacteriophage, φ29, is a grommet with one end missing. The end of the grommet present in the connector has 12-fold symmetry and the narrower neck has sixfold symmetry.[13] Although not studied in comparable detail, the data suggest, with one exception, the same structure for the connector of the unrelated *E. coli* bacteriophages λ and T3.[14] The exception is the presence of mass in the axial hole of the T3 connector.[14]

In 0.3 to 0.6 M sodium iothalamate, the hydration of DNA packaged in bacteriophages λ and T7 is lower than the hydration of DNA expelled from the bacteriophage capsid (free DNA).[15] The observation that DNA deleted from λ and T7 is replaced by bound water indicates that the exclusion of water is the result of space limitations in the capsid. Based on this conclusion, the volume available for the packaged T7 DNA was found to be 2.2 × the volume of DNA, assumed to be a flexible cylinder with a radius of 1.0 nm.[15] By use of low angle X-ray scattering, this conclusion has been reached for P22[16] and T7.[17,18] The average distance separating nearest neighboring segments of packaged DNA, determined from an X-ray ring at higher angle, is in agreement with the data from low angle scattering for P22 and T7. The latter data indicate a mean DNA-DNA spacing of 2.64 nm for T4 and 2.73 to 2.75 nm for φ29, λ, T2, and T7 (reviewed in References 17 and 19).

To compare the condensation of bacteriophage DNA with the condensation of eukaryotic chromatin, it is noted that the diameter of bacteriophage outer shells is approximately 1/250 × the length of the DNA packaged.[20] The ratio of the average dimension of a eukaryotic nucleus to the length of the 30-nm fiber of the chromosome (30-nm fibers are described in References 21 and 22) is approximately the same as for bacteriophage.[21] Thus, comparisons between the condensation of eukaryotic and bacteriophage DNA will be most productive if the 30-nm chromosome fiber, rather than the chromosomal DNA, is compared to the bacteriophage DNA.

The discussion below will review evidence that indicates packaging of bacteriophage DNA by the following pathway: (1) assembly of a DNA-free capsid (procapsid) and then (2) DNA binding by and entry into the capsid. The packaging of DNA in extracts of infected cells (*in vitro*) has been accomplished for φ29, λ, P2, P22, T1, T3, T4, and T7 (reviewed for all but T1 in references 19, 23, and 24; data for T1 are in Reference 25). With conclusions (1) and (2) assumed, the review presented here will outline research designed to understand details of the process of DNA packaging by double-stranded DNA bacteriophages. Attempts have been made to build on previous reviews.[8,19,23] The questions to be discussed are the following. What is the conformation of packaged DNA? What is the amount and source of energy needed to package DNA? What is the pathway of packaging? That is, what precursors of the mature bacteriophage are present during packaging? When isolated, precursors may be altered; altered and unaltered precursors are referred to as DNA packaging intermediates.

II. THE CONFORMATION OF PACKAGED DNA

A. TYPES OF NUCLEOTIDE SEQUENCE

The nucleotide sequence of the mature DNA packaged in the bacteriophages to be discussed here is in one of the following categories: (1) linear with a unique nucleotide sequence repeated at the ends (linear, nonpermuted, terminally repetitious: λ, P2, P4, T3, T7), (2) linear with a nonunique nucleotide sequence repeated at the ends (linear, permuted, terminally repetitious: P22, T1, T2, T4), and (3) linear with a protein covalently attached at the 5' end of both polynucleotide chains (ϕ29). Variations of these types include a single-stranded terminal repetition (λ, P2, P4) and different degrees of permutation. T3 is related to T7; T2 is related to T4. These observations are reviewed in References 19, 23, and 26.

B. SITES OF PERTURBED DUPLEX

The secondary structure of double-stranded DNA must be modified during the bending that accompanies the packaging of DNA. Based on the assumption of a discontinuous bend (kink), models have been built with unstacking of some base pairs,[27] sometimes to the point of breaking hydrogen bonds.[28] In the case of coliphage S_d, similar in outer shell and tail morphology to T7,[1] approximately 20% of the packaged DNA reacts with several compounds that do not react with free double-stranded DNA, but do react with single-stranded DNA. These compounds include O-methylhydroxylamine, bisulfite, glyoxal, and formaldehyde (reviewed in Reference 20). The reactive regions are presumed to be regions of altered duplex, but delayed reactivity to nitrous acid suggests the absence of free amino groups.[20] In the case of S_d reacted with formaldehyde, the regions of unreactive DNA have both an increased temperature of denaturation (while still packaged) and a decreased cooperativity of denaturation (i.e., the temperature range of denaturation increased). The interpretation of the latter observation was interspersal of stabilized duplex segments and regions of perturbed duplex.[29] The absence of free amino groups and reactivity with bisulfite and O-methylhydroxylamine suggested binding of protein both at regions of perturbed S_d duplex and in regions (not identified) of packaged λ, P22, and T7 DNA.[20,30] The predominant amino acid bound in S_d was lysine; in the case of P22, lysine and arginine were bound. The proteins bound have apparently been identified only in the case of P22 and are p1, p4, p20, and p26 (proteins will be referred to by p, followed by the gene number of the protein),[31] all proteins that are part of the P22 tail-connector complex.[32] However, room for 20% of the nucleotides of P22 to simultaneously attach to this complex does not exist (see drawings in Reference 32). Perhaps other regions of the capsid bind DNA or binding sites on the DNA are exchanged.

Sites of perturbed duplex in packaged DNA have also been found by measurement of the equilibrium binding of the intercalator, ethidium. In the case of T7,[33] P22,[34] ϕ29, T3, and T4,[161] Scatchard plots of the binding have revealed high affinity sites not present on free DNA. These sites in T7 have a negative $\Delta H°$ of binding $1.5 \times$ that of the binding to free DNA.[33] Because ethidium kinks DNA during binding,[35] it has been proposed that the high-affinity sites in bacteriophages are kinks.[33] The same proposal has been made for high-affinity methylene blue binding sites found in the nucleosomal constituent of chromatin.[36] In support of this idea, DNA known to be kinked has a high-affinity binding site for ethidium.[37]

Thus far, no studies have been made of the distribution of high-affinity ethidium binding sites along packaged DNA. The relationship between these high affinity sites and the sites of altered reactivity to single-stranded, DNA-specific reagents (above) is not known. However, the working hypothesis that these two sites overlap is made here and assumed in the discussion below.

C. ORIENTATION OF PACKAGED DNA SEGMENTS

The width of the ring produced by X-ray scattering from the packaged DNA of λ, P22, and T7 (all spherical) indicates that neighboring segments of packaged DNA are primarily in parallel bundles that cover a domain that averages 12 nm in diameter.[17,38] For the elongated bacteriophage, T4, this domain is 14.5 nm and increases to 21.5 nm for genetically produced T4 variants longer than the wild-type T4 (giants).[38] By use of both X-ray scattering from oriented specimens and electron microscopy of particles not completely full of DNA, the DNA of giants was found to be primarily parallel to the long axis (and tail) of T4.[38] This conclusion is also reached by electron microscopy of vitrified giants[39] and linear dichroism of oriented wild-type T4.[40,41] When observed by electron microscopy of negatively stained specimens, DNA being expelled from ruptured outer shells appears in either concentric circles (λ, P2, T4, T5, T7; Reference 42) or a toroid, sometimes pear-shaped in outline, that also sometimes branches to form thinner fibers (T2, a T4 relative; Reference 43). The concentric circles of Reference 42 were interpreted by the assumption of unidirectional winding of the DNA. If so, then the toroid of Reference 43 is unidirectionally wound (to be referred to as a wound toroid) and is produced by postexpulsion rearrangement of the DNA. However, a pear-shaped toroid has been observed for DNA condensed in the absence of a capsid,[44] and this toroid is, for the following reasons, believed not to be a wound toroid (to be referred to as a folded toroid). (1) The toroids in Reference 44, produced by use of a cationic condensing agent (either spermidine or polylysine) at ammonium acetate concentrations above 150 mM, were accompanied by, and presumed to be interconvertible with, rods of equal thickness. This conversion to a rod is not possible for a wound toroid. (2) In addition, closed circular DNA formed rods and toroids indistinguishable from those formed by linear DNA and (3) the thickness of the rod, but not its length, increased as the length of the DNA increased. The latter two observations were interpreted to indicate 180° folding of DNA at either end of the rod (i.e., without any net winding of the DNA duplex) to produce an array such as that found[45] in a crystal of multiple DNA molecules. Thus, considering only the concentric circles observed by electron microscopy and the results of X-ray diffraction, a folded toroid is equivalent to a wound toroid for explaining the data. However, the branching of the toroid of DNA released from T2 is better explained by a folded toroid than it is by a wound toroid. The hypothesis of folded DNA in a bacteriophage capsid explains the sites of perturbed duplex (above) as kinks within 180° folds. The distribution of perturbed duplex sites throughout packaged S$_d$ DNA (above) supports this hypothesis. From the results of model building described in Reference 27, a 180° fold would have to consist of at least two kinks over a space of not less than five nucleotide pairs.

D. FURTHER COMPARISON OF MODELS FOR THE CONFORMATION OF PACKAGED DNA

Although a definitive discrimination between folding and unidirectional winding of DNA in bacteriophage capsids has not yet been achieved, observations that will have to be explained by the correct model have been made.

1. In the case of T4[46] and λ,[47] the first end packaged is preferentially protected from ion etching, and, therefore, is presumably enriched at the center of the DNA-containing region of the capsid. If winding were unidirectional, this would require that the initial DNA packaged concentrate in turns at the center of the capsid, in disagreement with the results

of experiments designed to understand the forces present during packaging (below). In Reference 46, the following was proposed: folding of the initial DNA packaged at the center of the capsid, followed by outwardly progressive folding of DNA as it entered the capsid. The result is a "unitary basic arrangement". This proposal has the same limitation as unidirectional winding at the center of the capsid. However, the previous proposal[48] that the DNA between folds disperses throughout the capsid at all stages in packaging improves the energetics. According to the model in Reference 48, the DNA between folds would form loops independent of each other and these loops would form the basic elements of the condensate (loop-fold hypothesis). If, upon entry, loops were shunted to the inner edge of the outer shell of the capsid (this is, perhaps, a function of the T7 internal cylinder), then the DNA already packaged would be progressively forced to the center of the capsid, thereby explaining the finding that the first end packaged is found preferentially at the center of the DNA condensate.

To complete the discussion of this section, the author notes the observation that the innermost end[47] of packaged λ DNA is preferentially cross-linked by the intercalator, 8-methoxypsoralen.[49] Based on the assumption that intercalators penetrate uniformly throughout the outer shell of λ, the preferential outermost location of the cross-linked end was proposed in Reference 49. However, the data from ion etching suggest that the intercalator is penetrating at the tail connector and binding preferentially to the innermost DNA in this location, possibly because kinks are preferentially here (see also, the data above for P22).

2. Measurements of dichroism of either packaged DNA or dyes bound to packaged DNA reveal an average DNA orientation incompatible with any simple toroidal or folded arrangement (P22, T7;[50] λ, T4[51]). The best-fit model for a wound toroid of DNA in T7 has an axis tilted at 46° relative to the axis of the internal cylinder of T7.[50] This is, for reasons of packing energetics and economy, an unlikely arrangement. In agreement with the dichroism, the condensation of loops proposed in the loop-fold hypothesis could not produce a simple structure. For example, a folded toroid such as that reported in Reference 44 has a shape different from that of a bacteriophage capsid. Thus, several (connected) folded toroids would have to form to fill an icosahedral capsid, each folded toroid of different size and orientation. The branches observed on the DNA toroid released from T2[43] can be explained as a residual of a second toroid that has not completely merged with the larger toroid as DNA was expelled from the T2 capsid. It is tempting to speculate that the domain size determined by the width of the X-ray diffraction ring of packaged DNA is the diameter of the average folded toroid in a bacteriophage capsid. The increase in the average size of this domain as the T4 capsid becomes longer (above) can be explained by a progressive merging of the folded toroids as the distance from a capsid end increases and, therefore, the steric constraint for separate toroids is released. These ideas are further supported by the absence of axis-parallel domains at the two ends of vitrified T4 giants.[39]

3. The DNA segment-DNA segment, intercalator-induced cross-linking of packaged λ DNA, followed by the analysis of restriction endonuclease digests of the cross-linked DNA by either electron microscopy[52] or agarose gel electrophoresis,[53,54] has revealed a pattern of cross-linking indistinguishable from a random pattern, with the exception of one cross-link. In agreement, the cross-linking of packaged DNA to the λ capsid by use of bromodeoxyuracil-labeled DNA and radiation produces a random pattern of restriction endonuclease fragments cross-linked to the capsid.[55] If the assumption is made that binding of the intercalating cross-linker is random on the DNA, any form of wound toroid with progressive winding is in disagreement with these data. However, a wound toroid that is wound both inside and outside of the previously packaged DNA has been proposed[56] to avoid disagreement with the data in References 52 to 55. The finding of two kinetically and thermodynamically defined binding sites for the intercalator, ethidium (λ,[51] P22,[34] T7[33,57]), suggests that the assumption of random binding used above should be subjected to additional testing.

The loop-fold hypothesis is designed in part to explain the above results obtained by cross-linking. However, by use of these data alone, the loop-fold hypothesis cannot be distinguished from a hypothesis that proposes inside-outside unidirectional winding, such as that in Reference 56. All sequentially ordered unique structures, including the folded structure of Reference 46, are incompatible with the data obtained by cross-linking with λ. Given these data, either the conformation of packaged DNA is not unique or it is not sequentially ordered (or both). The finding, by ion etching, of some sequential ordering of packaged λ DNA[47] is not necessarily in disagreement with the results obtained by cross-linking. Partial sequential ordering, perhaps by the mechanism described above, could explain both results. A more quantitative comparison of these data is needed.

In summary, in the author's opinion, the data favor a multiple, folded, toroid-type conformation of packaged DNA. A wound-toroid-type conformation was favored in a previous review.[19] However, conclusive experiments have not yet been done to determine the conformation of DNA packaged in the double-stranded DNA bacteriophages. The existence of strand crossing in electron micrographs of DNA released from capsids, cited as evidence for a wound toroid,[19] can also be explained by a folded toroid.

Whatever its conformation, the two ends of packaged DNA in P2 capsids that have not added tails are close enough to join. The result is a knotted circle after the extraction of DNA from its capsid.[58] Evidence for the joining of the first end of the DNA injected to either the connector or tail has been obtained by DNA-protein cross-linking, followed by electron microscopy of P2 and λ[59] (see also, the review in Reference 19). For λ, absence of this capsid-DNA binding is associated with loss of infectivity, suggesting that this binding is necessary for injection and is the mechanism for specifying the direction of injection for the DNA.[60]

E. ANALOGIES WITH THE CONDENSATION OF CHROMATIN

Studies of the conformation of the 30-nm eukaryotic chromosome fiber have suggested the presence of radial loops. Such loops were observed by electron microscopy of (1) the DNA of histone-depleted chromosomes — the DNA was attached to a scaffold of nonhistone proteins,[61] (2) thin sections of swollen chromosomes,[62,63] and (3) critical point-dried, metal-coated chromosomes (using scanning electron microscopy).[62,63] These loops are conformationally analogous to the loops proposed (above) for packaged bacteriophage DNA. A possible reason for the evolutionary selection of a conformation with loops in the case of both packaged bacteriophage DNA and 30-nm chromosome fibers is that packaging in loops can eliminate toroidal winding in both cases. Toroidal winding requires axial rotation of the strand wound in relation to its surroundings. In the case of bacteriophage DNA, the DNA either inside or outside of the capsid would become supercoiled or would have to rotate around its axis. The latter possibility would require either topoisomerase activity or the expenditure of additional energy (in unknown quantity). In the case of the 30-nm fiber, the equivalent of topoisomerase activity appears impossible. Thus, the author's working hypothesis is that minimization of unidirectional winding has been selected during evolution for both packaged bacteriophage DNA and the 30-nm eukaryotic chromosome fiber.

III. ENERGETICS

No direct experimental determination has been made of the energy required to condense intracellular DNA in a bacteriophage capsid. In the presence of some multivalent cations (and in the absence of a capsid), including spermine, polylysine (reviewed in Reference 19), and cobalt hexammine,[64] extracellular condensed DNA is more stable than a random coil. In these experiments, no energy is required to condense the DNA to interhelix distances as small as 2.92 nm for spermine, 2.89 nm for polylysine,[19] and 2.8 nm for cobalt hex-

sation of DNA, the DNA would either condense or (more likely; see Reference 65) precipitate outside of the capsid, thereby making packaging in a preformed procapsid impossible. By this reasoning, DNA packaging would have to require energy. The evolution of a procapsid could have occurred because of the absence of a means, such as the intrapolynucleotide chain base-pairing of single-stranded RNA, to even partially condense intracellular DNA without precipitation. Single-stranded RNA bacteriophages assemble by polymerization of the capsid around the RNA (reviewed in Reference 66). Following this thought further, how do eukaryotic chromosomes condense without the precipitation? Presumably, this process also requires energy purposefully channeled.

The energy required to condense DNA presumably varies with both the concentration of condensing agents present and the conformation of the condensed DNA. For example, the presence of kinks in the condensed DNA would alter the energy of condensation to an unknown extent. In the absence of cationic condensing agents, DNA can be condensed in solutions of neutral polymers to an inter-DNA segment spacing that decreases with an increasing concentration of polymer. The osmotic removal of water from the condensate to the polymer phase is the source of the pressure that promotes condensation.[67] Osmotic pressure has also been used to force together membrane fragments and cylindrical rods such as tobacco mosaic virus. In all of these cases, the energy required for condensation becomes independent of ionic strength as spacings decrease.[67,68] Thus, charge-charge repulsion, initially thought to be the major enthalpic contribution to the free energy of packaging,[69] is found experimentally to make a smaller contribution than water-water interactions. This conclusion is in agreement with the data from buoyant density centrifugation that indicate dehydration of packaged λ and T7 DNA.[15,70]

In solutions of polymers, the pressure-volume work done to condense DNA to the 2.75 nm spacing of bacteriophage capsids is 0.1 to 0.4 kcal/mol of nucleotide pair (depending on the ions present; differences were found by the use of Li^+, Na^+, Cs^+, $Mg,^{2+}$ $Ca,^{2+}$ and putrescine^{2+}),[67] or 4 to 16 $\times$ 10^3 kcal/mol of T7 DNA. The calculated entropy-derived change in free energy during DNA condensation is 35 to 140$\times$ smaller.[69] Because the effect of cationic, and possibly other, condensing agents is not known in the case of bacteriophage DNA packaging, this value is an upper limit to the free energy of DNA packaging, possibly omitting energy required for kinking DNA. It is not known whether neutral polymer-induced condensation kinks DNA. At least in the case of T3, cationic condensing agents are not necessary for, but do stimulate (spermidine was used), in vitro DNA packaging.[71]

Because the procapsid of all studied bacteriophages expands and becomes more stable during conversion to a mature capsid (reviewed for λ, P22, T3, T4, and T7 in References 5, 8, 10, and 19), the author[15,72] and others[73] initially proposed that (1) at least some of the energy for packaging was stored in the procapsid and (2) this energy was used by creating a pressure gradient during expansion of a water-impermeable capsid. The isolation of an intermediate form of the T7 capsid impermeable to metrizamide, but permeable to water, suggested a revision of this hypothesis to include an osmotic pressure gradient generated by expansion and maintained by the impermeability of the capsid to intracellular solutes.[74] The difference between the volume of the mature capsid and the procapsid is not great enough to package all DNA this way[19] unless the capsid hyperexpanded and then contracted[17] during packaging.

Although the data initially indicated that ATP is required for at least one event of packaging (reviewed for φ29, λ, P2, P22, T3, and T4 in Reference 19; data for T7 are in Reference 75), these data did not indicate the events of packaging that require ATP. Metabolic inhibitors appear to inhibit events of packaging that occur before T3 DNA enters a capsid in vivo.[76] However, more recently a rigorous demonstration that T3 DNA entry into the capsid requires ATP in vitro has been made.[77] During in vitro φ29 packaging, one ATP per two nucleotide pairs packaged (9×10^3 per DNA) is cleaved by use of conditions such that less than one half of the ATP was cleaved (Figure 1 in Reference 78). Thus, the free energy obtained from this cleavage is at least 3.7 kcal/mole of packaged nucleotide pair ($\Delta G° = -7.3$ kcal/mol[79]), at least an order of magnitude more than the free energy needed to condense the DNA (above). The experiments described in Reference 77 and 78 were done with T3 and ɸ29 *in vitro* packaging performed in mixtures of purified components (more details are below).

The major outer shell protein of at least one form of the capsid binds ATP in the case of T3 (two other T3 proteins also bind ATP; see below)[80] and T4,[81] suggesting that the outer shell of the capsid is part of the energy transducing apparatus. A supporting, much older observation is the finding of ATPase activity associated with two bacteriophages with non-contractile tails, T1[1] and T5.[82] Unlike T2 and T4,[1,83] contraction of the tail cannot account for the ATPase activity, leaving the outer shell as a candidate. ATP-induced contraction of the outer shell can explain an observed 5 to 10 mM ATP-induced expulsion of DNA from a variety of bacteriophages, including T2 (the tails were not contracted),[1] Sd,[1] T4, and T7.[161] GTP does not work for T4 and T7. That contraction of the T2 capsid, induced by a molecule (probably ATP) released from a host cell, is part of the mechanism for injecting DNA has been hypothesized and supporting data presented.[1,84] Contraction of the outer shell of T7 after elevated temperature-induced expulsion of DNA has been observed by use of low angle X-ray scattering.[17]

A. OSMOTIC PUMP HYPOTHESIS

Some of the above data were integrated by the author to form the hypothesis, basically a revision of the above hypothesis of osmotic pressure-driven DNA packaging, that bacteriophage capsids provide energy for DNA packaging by acting as an ATP-driven osmotic pump. This pump has two phases, repeated cyclically (osmotic pump hypothesis):[85] (1) expansion in a state of impermeability, thereby producing an osmotic pressure gradient across the capsid which drives DNA into the capsid, and (2) subsequent ATP-driven contraction in a state of increased permeability at all locations but that of the connector; thus, non-DNA molecules are squeezed out of the capsid and DNA is trapped in the connector and cannot exit. The mass observed in the center of the T3 connector (above) is a candidate for the DNA-trapping agent. The ATP-driven contraction would provide the energy for the entry of DNA during packaging. An expansion to sizes greater than that of the mature capsid would facilitate osmotic pumping, but isolation of a hyperexpanded capsid has not been reported. Procedures of gel electrophoresis for recognition of such a capsid have been developed and used to characterize a P22 capsid with a radius intermediate to that of the P22 procapsid and mature bacteriophage P22.[86]

If the role of ATP in DNA packaging and injection is correctly represented above, ATP-driven contraction of the outer shell of the capsid provides energy for both the packaging and injection of DNA. The direction (during packaging and injection) of DNA motion is controlled by gating at the connector. This gating would answer the question posed in the review of Reference 23: how does DNA find it energetically favorable to both enter and leave the capsid? A second question posed in a previous review[19] is also answered by the osmotic pump hypothesis: how are bacterial proteins, molecules that occupy one half the cytoplasmic volume,[87] excluded from the mature bacteriophage? The exclusion by internal procapsid proteins was proposed in Reference 19. However, the subsequent finding, by electron microscopy of thin sections, that bacterial cytoplasm does penetrate the T4 procapsid[88] suggests that the latter explanation is not correct.

The osmotic pump hypothesis predicts that lowering of water activity by molecules larger, but not smaller, than a critical size (the critical size is determined by the leakage rate into the capsid during pumping) will stimulate packaging. This prediction has been confirmed *in vitro* for T7,[24] P22,[89] T4,[90] and T3.[71] For at least T3, the requirement occurs during entry of DNA into the capsid.[77] The added molecule (usually a neutral polymer such as a dextran or a polyethyleneglycol) would provide the osmotic pressure needed. In contrast, neutral polymers are not needed during *in vitro* φ29 DNA packaging with purified components. Most of the osmotic pressure in the mixtures used would have to be provided by φ29 p16, present at a concentration of 14 μg/ml (100 molecules per molecule of DNA).[91] By comparison with the osmotic pressure of the comparably sized ovalbumin,[92] the expected osmotic pressure in these extracts is approximately 10^{-5} atmospheres, 5 to 6 orders of magnitude lower than the pressure needed to condense the DNA in the absence of a condensing agent. Perhaps the inner surface of the φ29 outer shell, 35 to 40% of which can contact the packaged DNA,[19] acts as a condensing agent and reduces the osmotic pressure needed. Because φ29 is the smallest of the bacteriophages studied, its outer shell would contact the greatest percentage of the packaged DNA.[19] Alternatively, the osmotic pump hypothesis may not apply to φ29.

B. ADDITIONAL HYPOTHESES

Two additional hypotheses describe mechanisms for the ATP cleavage-driven entry of DNA into bacteriophage capsids: (1) condensation either before[93] or after[94] packaging caused by topoisomerase-driven supercoiling of the DNA and (2) threading of the DNA into the capsid by ATP-driven rotation of the connector;[95] the macroscopic analogy, for DNA not unidirectionally coiled in the capsid, is a worm gear. The author adds (3) a reach and push (or pull) mechanism such as that proposed for the motion of sliding filaments during muscle contraction.[96] If several fibrous proteins alternately reached and pushed (or pulled), the DNA would be pushed into the capsid; the result would be analogous to an insect on a treadmill. Mechanism (1) is incompatible with the formation of a folded toroid of DNA in the capsid and also contradicts the observation that nicked DNA is packaged *in vivo* for T4 and T7 (reviewed in Reference 85) and *in vitro* for T3.[97] Because of the latter observations, hypothesis 1 is not considered viable. No direct test of either hypothesis 2 or 3 has been made. However, the *in vitro* packaging of a T3[97] DNA duplex distorted[98] by cross-linking with a psoralen and the *in vivo* packaging of DNA with single-stranded segments by the *Erwinia herbicola* bacteriophage, Erhl,[99] are difficult, though not impossible, to explain by use of either hypothesis 2 or hypothesis 3. Even more difficult to explain by hypothesis 2 and 3 is the observation that λ can efficiently package a heteroduplex with a 19-base-pair, single-stranded loop both *in vitro*[100] and *in vivo*;[101] T4 does the same *in vivo*.[102]

C. DISTRIBUTION OF SEGMENTS OF PACKAGED DNA DURING ENTRY

Excluding interactions with internal capsid proteins, the predominant force guiding the distribution of DNA segments during entry into a capsid is the repulsive "hydration" force described above. This force increases exponentially as the distance between two DNA segments decreases.[67] Because internal proteins are not distributed throughout the interior of capsids (see Figure 1 for T3 and T7), the prediction is that, during entry, packaged DNA segments will maximize their average separation and fluctuations caused by Brownian motion will be superimposed. This prediction is confirmed for mature bacteriophages by the observation that genetically deleting nonpermuted DNA from its capsid results in a quantitative increase in the average DNA-DNA spacing. This was observed by either low angle X-ray scattering[103] for λ or buoyant density sedimentation[15] for λ and T7. Comparable studies have not yet been done for DNA packaging intermediates with a subgenome length of packaged DNA (referred to as spDNA capsids). However, DNA in spDNA capsids isolated from P22-infected cells does not undergo the restriction of ethidium binding found[34] for DNA in mature P22 capsids, suggesting that the DNA in these spDNA capsids is not concentrated in one region of the capsid.[104] The observations of this paragraph support models for DNA packaging in which DNA is not concentrated in one region of the capsid during entry into the capsid.

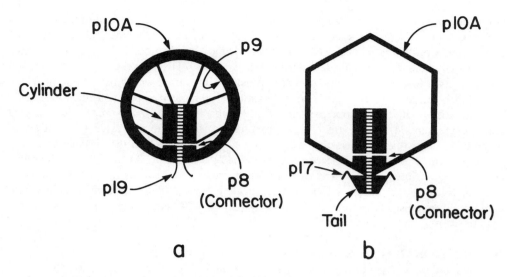

FIGURE 1.   Bacteriophage T7 (and T3) and procapsid. (a) Procapsid; (b) bacteriophage.

IV. DNA PACKAGING PATHWAY

A. PROCAPSIDS

To determine the mechanisms of DNA packaging, attempts are made to isolate and characterize the bacteriophage precursors in the DNA packaging pathway (such precursors may be altered during isolation; either altered or unaltered precursors are referred to as intermediates). Initially, attempts were made to detect DNA packaging intermediates by use of electron microscopy. A capsid-like particle, referred to as a doughnut, was observed in lysates of T2-infected cells.[105,106] To determine whether doughnuts were intermediates, their appearance kinetics were determined. The concentration of doughnuts in an infected cell initially increases and then decreases as the mature bacteriophage appears.[106] Assuming that the doughnuts are a capsid, these kinetics identify the doughnuts as a precursor of some other capsid, presumed to be the capsid of the mature bacteriophage because the mature bacteriophage is the only other capsid-like particle observed. Doughnuts are presumed to be either one or more of the capsids subsequently found (see below) in the DNA packaging pathway.

Among the capsids in the DNA packaging pathway of ϕ29, λ, P2, P22, T1, T2, T3, T4, and T7 is a capsid that is smaller, rounder, and less stable than the mature capsid (reviewed in References 5 to 9 and 19). The negative solid-support-free electrophoretic mobility (μ_o) is also higher in magnitude for all sufficiently studied isolates of this capsid than the negative μo of the mature bacteriophage outer shell (ϕ29;[163] reviewed in Reference 11 for P22, T3, and T7). Because it appears in a DNA-free form before any other capsid, this physically different capsid is referred to as a procapsid. Evidence that DNA is subsequently packaged in the procapsid has been previously reviewed.[5,19,23] These data include appearance kinetics *in vivo* (for T7 and T3, see Reference 11) and the packaging of DNA *in vitro* by a procapsid with efficiency independent of the dilution of the procapsid (for T7, see Reference 107). Variability of the μ_o and composition of the T3 procapsid has been detected.[76]

The detection and quantitation of procapsids by agarose gel electrophoresis has simplified determination of their appearance kinetics and also the detection of DNA-capsid complexes.[11,77,86,104] During conversion of all of the above procapsids to a mature bacteriophage-like capsid, either one or more proteins associated with the outer shell is lost by either cleavage or dissociation. The details are reviewed in References 5, 8, 19, and 23.

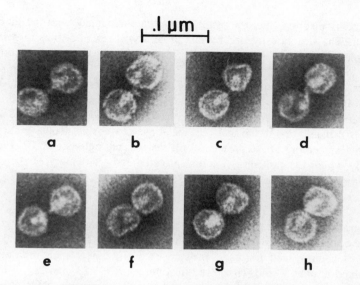

FIGURE 2.   Evidence for an external, fibrous projection from the connector of the T7 procapsid. Dimers of the T7 procapsid were isolated and prepared for electron microscopy, as previously described.[108]

The evolution of the comparatively high magnitude of the negative μ_0 of the capsid can be explained by electrostatic repulsion-caused prevention of the nonproductive binding of procapsid to DNA. If so, then (by analogy to the tail fibers of P22, T4, and T7 [see Reference 11]) a projection, probably an externally positively charged protein, from either the procapsid or the DNA must overcome this repulsion to promote productive binding to DNA. In the case of T7, evidence for such a fibrous projection was obtained by observation of previously isolated[108] dimers of the procapsid. Of 151 dimers observed, 94 (62%) had a visible internal cylinder in both constituent procapsids and 96% of the latter procapsids were dimerized at the cylinder-connector joining region of the outer shell (Figure 2, all particles). At the point of dimerization, some of the latter dimers had fibrous material connecting the two cylinder connectors (Figure 2, dimers a to e). The fibrous projection is presumably also in other dimers (Figure 2, dimers f to h), but is not visible. Although the protein(s) that forms the fiber has not been rigorously identified, the following observations suggest that p19 is one, possibly the only, constituent of the fiber: (1) p19 is a minor constituent of the T7 procapsid, and unlike known internal proteins of this procapsid, is susceptible to digestion with trypsin[109] (dimers are converted to monomers by trypsin),[164] (2) T3 p19, partially homologous to T7 p19,[110-113] aggregates the T3 procapsid,[114] and (3) when p19 is removed from an infected cell by the use of an amber mutant, DNA packaging is blocked even though DNA is synthesized and the procapsid is assembled (T7,[115] T3[76,110]). Removal of a second protein, p18, also has this result. Neither p18 nor p19 is present in mature T7.[109,115] Six copies of p19 are found on the T3 procapsid.[114]

Although the external projection of the T7 procapsid has not been observed in the procapsids of other bacteriophages, the connector has been observed in the procapsid, and two proteins, whose removal has the effect described above for p18 and p19 of T7 and T3, have been observed for all sufficiently studied bacteriophages (reviewed in References 19 and 23). In all cases, these two proteins differ in mass by at least a factor of 1.4. The larger protein (p19 for T7 and T3) will be referred to as the large accessory protein (LAP) and the smaller protein will be referred to as the small accessory protein (SAP). Further studies of the role of LAP and SAP in the DNA packaging pathway are described below.

Procapsids contain at least one nonaccessory protein (i.e., a protein needed for procapsid assembly) that, by electron microscopy, is in continuous contact with the inner surface of the outer shell. The structure formed by this protein(s) is referred to as a scaffold. By the time of completion of DNA packaging, the scaffold has been removed from the capsid (reviewed for φ29, λ, P2, P22, T3, T4, and T7 in References 5 to 10 and 116). The scaffold of T7 consists of only one known protein, p9 (reviewed in Reference 10 and illustrated in Figure 1a); the scaffold of T4 consists of at least six proteins.[117] The structure and fate of the scaffold varies among the different bacteriophages discussed here (reviewed in References 5 to 10 and 116). In addition to the scaffold, the above procapsids also have internal proteins attached to the connector. These proteins form the internal cylinder of T7 (Figure 1). The mechanism by which the procapsid internal proteins assist assembly of the outer shell is under investigation, but will not be discussed here.

B. DNA

The substrate of the procapsid during packaging is a linear end-to-end polymer (concatemer) of the monomeric DNA in the case of λ, P1, P22, T1, T3, T4, and T7. For P2, the substrate is a circle and for φ29, the substrate is a linear monomer with a protein covalently joined at both 5' ends (reviewed in References 19, 23, and 26). Either during or after packaging, circles and concatemers are cut to a linear mature DNA. Why are concatemers and circles formed *in vivo* and then cut to mature, linear DNA? Concatemerization is not necessary for *in vitro* DNA packaging in the case of λ (both linear and circular monomeric DNA is packaged[118]), P22,[119] and T3.[71] The intracellular loss of the ends of DNA in P1[120]- and T7[125]- infected cells suggests that concatemers evolved to increase the chance that some genomes would survive to be packaged with both ends intact. By this (unproven) hypothesis, the covalently bound protein of φ29, known to protect φ29 DNA 5' ends (but not its 3' ends) from DNA-metabolizing enzymes,[122] removes the evolutionary pressure for the formation of concatemers. That is, concatemerization and φ29 DNA terminal protein accomplish the same objective, an objective apparently accomplished by the telomeres of eukaryotic chromosomes.[123]

Concatemerization can occur by either (1) ligation of the double-stranded ends of monomeric DNA (blunt-end ligation), (2) circularization of monomeric DNA, followed by priming of DNA synthesis at a nick in the circle and multiple rounds of replication (rolling circle), or (3) joining of complementary base pairs at the terminally repetitious ends of linear DNA. Concatemers of the nonpermuted DNA of λ are formed *in vivo* by a rolling circle (reviewed in Reference 124). Although evidence has been obtained for a comparatively small amount of T7 concatemerization *in vivo* by blunt-end ligation,[125] more T7 concatemers are formed *in vivo* by mechanism 3.[125,126] No evidence for the *in vivo* formation of circles has been obtained for the permuted DNA of T4 and the form of replicating DNA for T4 resembles that for T7.[127,128] Therefore, T4 concatemers should be formed by mechanism 3. However, this mechanism is not possible for the permuted T4 genome unless the two progeny of DNA replication join to form a concatemer before mixing with other intracellular DNA.

C. INITIATION OF PACKAGING

The events of DNA packaging will be divided into three categories: (1) *initiation*, those events that occur before DNA begins to enter a capsid, (2) *entry* of DNA into a capsid, and (3) *cutting* of DNA to a mature length, the latter for all of the above bacteriophages except φ29. In the present section, initiation is discussed; in the last two sections, entry and cutting are discussed.

The first event of initiation must be binding of the procapsid to DNA. The data discussed in Section III.A indicate that LAP and SAP participate in binding DNA. This conclusion is also indicated by the following observations. (1) When packaging DNA from two different, but related, bacteriophages *in vitro*, the specificity of DNA packaging is determined by either LAP or SAP (T3;[110] reviewed for λ in References 129 and 130). (2) Procapsid-DNA complexes (DNA outside of the capsid) that are intermediates detected during DNA packaging *in vitro* do not form unless LAP and SAP are present (φ29,[131] λ,[129,132] T3[114]). (3) Mutation of the P22 SAP[133] changes the specificity of the capsid for DNA.[134,135] By constructing hybrid LAP and SAP proteins consisting of amino acids from λ and a λ relative, and by observing the specificity of these hybrids for procapsid and DNA, the following sequence of interactions has been deduced for λ: DNA→NH$_2$-SAP-COOH→NH$_2$-LAP-COOH→procapsid.[130] The binding of T7 and T3 LAP to the procapsid was discussed above.

The binding of SAP to DNA has also been demonstrated biochemically. After expression under control of a foreign promoter, LAP and SAP have been purified for T3[136] and T7;[137,138] SAP has been purified for λ.[139] SAP is the protein covalently bound to the 5' ends of φ29 DNA; φ29 LAP has also been purified[91] and, like the LAP of λ, T3, and T7, binds to the φ29 procapsid.[131] Use of a foreign promoter was needed to increase the comparatively small amount of LAP and SAP in infected cells (for λ, see Reference 140). SAP was isolated bound to DNA in the case of λ[139] and T3.[114] In the case of λ, nucleotide sequences protected by SAP from DNase I digestion are at a known specific recognition site (cos B) described below.[139] The above similarities in the properties of LAP and SAP among unrelated bacteriophages are presumed to be the result of evolutionary pressure. However, the source of this pressure is not known. In addition to LAP and SAP, φ29 requires a procapsid-associated RNA for initiation.[141]

The following observations suggest the hypothesis, previously mentioned in a review of λ DNA packaging,[129] that after procapsid binding of DNA, ATP is cleaved during initiation to accelerate a ''search'' for a specific binding site. (1) In the case of T3, both LAP and SAP cross-link to a photoactivatable ATP-analog.[80] (2) A change in the T3 exterior, apparently needed for packaging, is prevented by metabolic inhibitors *in vivo*.[76] (3) The amount of ATP cleaved during *in vitro* φ29 packaging is greater than that needed to condense DNA (above). (4) ATP is required for the binding of procapsid to DNA in the case of φ29,[131] λ,[129,132] and T3.[114] (ATP can be replaced by a noncleavable analog in the case of T3,[114] but not in the case of φ29.[131]) (5) Most of the ATPase activity of T3 *in vitro* packaging extracts is stimulated by DNA not homologous to T3 DNA as well as by T3 DNA.[114] However, no direct test of this hypothesis has been performed.

In the case of nonpermuted and partially permuted bacteriophages (i.e., λ, P1, P22, T1, T3, T7) the procapsid has to recognize at least one nucleotide sequence before packaging. That at least one sequence recognized is not the eventual terminal sequence for T7 is suggested by the isolation of T7 capsid-DNA complexes, some containing concatemeric DNA, in which the capsid is near, but not usually at (i.e., within 7% of the mature genome length) the genetic right end of T7 DNA.[142] That a nucleotide sequence in this region is recognized is confirmed by observation of the specificity of the *in vitro* packaging of T7/T3 hybrids.[112] By observing the packagability of λ DNA after either removing or altering nucleotide sequences, two sequences were found necessary for packaging: (1) approximately 46 base pairs that include the junction between two genomes (referred to as cos N; reviewed in References 130 and 139) and (2) nucleotides approximately 50 to 130 base pairs from the genetic left end (referred to as cos B[130,139]), a site presumably analogous to the capsid-binding site observed for T7. Concatemeric P22 DNA is also recognized at a site distal from the eventual end of its mature genome; however, after recognition, the cutting does not have a unique specificity.[143,144] The finding of P22 cuts within a limited region determined by recognition at a distal site is an observation also made with type I restriction endonucleases.[145]

Although no initiation events have yet been determined by isolation and characterization of intermediates, the observation of circles in lysates of cells infected by wild-type T7, but not in lysates of cells infected with T7 that doesn't produce T7 LAP, suggests the formation of a loop during packaging.[125] Recombination of the two ends of a genome in the loop would produce the circles as an abortive by product. This loop could form during either initiation or entry. The formation of the loop involves procapsid binding to two sequential genomes in a concatemer, a process whose evolution can be explained by assuming evolutionary pressure for the capsid recognizing an intact genome before attempting packaging. Binding of two sequential genomes also explains several additional observations: (1) the failure of λ to package a circular monomer *in vivo*;[146] however, the requirement for a concatemer can be bypassed *in vitro* for λ, P22, and T3 (above), (2) homologous DNA selectivity of a T3 extract greater for concatemeric than for monomeric DNA,[71] (3) the requirement *in vitro* for an amount of LAP and SAP that is two times greater for concatemeric DNA than it is for monomeric DNA (reviewed in Reference 129), and (4) processive packaging of genomes along a concatemer. Processive packaging is the explanation for the limited permutation of the ends of mature P22 and P1 DNA.[19,147] Processive packaging has also been found for λ (reviewed in References 129 and 130) and T1.[148]

D. ENTRY OF DNA INTO A CAPSID

The signal for beginning the entry of DNA into the capsid has not yet been discovered. Because capsids produced by abortive attempts at packaging do not accumulate in amounts comparable to the amount of the mature bacteriophage (for example, see data for T3 in Reference 11 and for T2 in Reference 106), the working assumption is made here that nucleotide sequence recognition for nonpermuted and partially permuted genomes occurs before entry begins. If, as suggested by the data of Section III.C, these genomes are bound at a site near, but not at, the site of eventual cutting, most DNA would be packaged in one direction (right to left for T7; left to right for λ). However, the DNA between the binding site and the nearest end would have to be packaged in the opposite direction. In support of right-to-left packaging for most T7 DNA: (1) during coupled *in vitro* recombination-packaging of endogenous T7 DNA (i.e., DNA from T7 used to prepare the extract) with exogenous DNA, recombinants with the right end of the endogenous DNA are favored, as though something bound to the right of gene 19 were needed for packaging that proceeded right to left, and (2) this asymmetry is absent when recombination and packaging are uncoupled.[149] In support of the left-to-right packaging of T7 DNA to the right of the specific capsid-binding site, an entire restriction endonuclease fragment that contains the left-right end joint of a T7 concatemer is converted to a DNase-resistant state and, therefore, is packaged *in vitro*.[150] Evidence for left-to-right packaging of λ DNA *in vivo* has previously been described[151] and reviewed.[129] The left end of λ DNA appears to be equivalent to the right end of T7 and T3 DNA. By analysis of the restriction endonuclease fragments of λ DNA packaged *in vitro*, initiation of entry at the left end of λ DNA has been confirmed.[152] This type of experiment has also revealed initiation of entry at the left end of φ29 DNA.[153]

The force that drives the packaging of DNA has been discussed above. By analysis of the appearance kinetics of DNA packaging intermediates *in vivo*, the upper limit for the time required for T7 DNA entry into a capsid at 30°C was estimated at 1.5 min (26.6 × 10^3 base pairs per min).[142] Unfortunately, the spDNA capsids involved had all extruded their DNA before isolation. Although P22 spDNA capsids formed *in vivo* are stable enough to isolate and quantitate,[104] P22 DNA packaging *in vitro* appeared to be too rapid to isolate spDNA capsids.[89] That is, all detectable DNA made DNase resistant by packaging at 35°C had a length equal to that of mature P22 DNA. More recently, this approach has yielded subgenomic DNA during *in vitro* T3 DNA packaging of mature DNA (no concatemers were present) at a temperature lowered to 20°C to slow packaging. Entry had been synchronized by formation of a procapsid-DNA complex before allowing entry. The rate of entry at 20°C was 5.7 × 10^3 base pairs per minute and, assuming entry to be rate limiting for production of infectious particles, was 22 × 10^3 base pairs per minute at 30°C,[77] not significantly different from the upper limit of the rate estimated *in vivo* for T7. Subgenomic, packaged φ29 DNA was found in crude DNA packaging extracts,[153] but was not observed during packaging in a mixture of purified components.[131] One explanation of this greater apparent packaging speed when extract components are removed is the reduced requirement for pumping non-DNA substances out of the capsid during packaging (see the osmotic pump hypothesis above).

To test hypotheses that describe mechanisms for driving DNA into a bacteriophage capsid, spDNA capsids are isolated and characterized. Interpreting the results obtained will involve determining whether the capsid, as isolated, is either reversibly or irreversibly altered from its state(s) during packaging. Among the spDNA capsids thus far isolated are spDNA procapsids (φ29,[153] λ,[152] T3,[77] and P22[104]). Thus, for these bacteriophages, the procapsid does not irreversibly convert to a bacteriophage-like capsid until at least some (20 to 30%) of the DNA has entered the capsid. Data indicating conversion of procapsids before packaging *in vivo* have been presented (T4,[154] T7[74]), but the pre-entry converted capsid observed was not rigorously shown to be in its native state; that is, the capsid may have converted during its preparation for observation. However, in the case of T4, a converted capsid packages DNA *in vitro*.[90] This apparent difference among the different bacteriophages has not yet been explained and has at least the following possible explanations: (1) DNA enters some bacteriophage capsids before and some after capsid conversion, (2) DNA can enter either before or after conversion, depending on conditions, and (3) the aspects of the conversion needed for packaging (expansion, for example; see Reference 85) are reversible.

E. CUTTING OF DNA FROM CONCATEMERS

In the case of all of the concatemer-producing bacteriophages described above, mutationally removing only capsids from an infected cell prevents the cutting of DNA from concatemers (reviewed in References 19 and 23). For the nonpermuted genome of T4, this observation is explained by the assumption that DNA cutting requires filling of the T4 capsid with DNA. Cutting signaled by filling was originally proposed to explain the increase of the T4 terminal repetition that occurred for T4 deletion mutants.[155] Even though cutting signaled only by filling is incompatible with the nucleotide sequence specificity of the ends of a nonpermuted DNA, the following observations suggest that filling is one of the signals for cutting for nonpermuted DNAs. (1) The efficiency of cutting λ DNA increases with the length of the DNA[156] and (2) some T7 concatemers are isolated attached to capsids; the data suggest that the concatemer-bound capsids had packaged some of the DNA attached before isolation.[142]

Although at least partial filling of a capsid appears necessary for cutting during DNA packaging, removal of at least one λ gene results in capsid-independent, LAP-dependent cutting of intracellular DNA at cos N.[157] A similar observation has been made for P1.[120] Thus, for at least these bacteriophages, the capsid appears to be needed to remove an inhibitor of the cutting enzyme from the DNA. The evolutionary need for both such an inhibitor and the filling before cutting can be explained by the need to prevent damage to the ends of a DNA to be packaged. By preserving the concatemer until packaging is almost complete, the chance of exonucleolytic damage to the mature DNA ends is minimized.

In the case of λ, the purified LAP is an endonuclease with the nucleotide sequence specificity of cos N.[158] Thus, this protein is necessary for both specific binding and cutting. The cuts produced are two staggered nicks that result in separation of a genome with self-complementary, single-stranded ends. However, in the case of T3 and T7, the unique ends are double stranded, requiring either duplication or loss of one terminally repeated sequence during packaging. That the T7 terminally repeated sequence is duplicated during DNA packaging has been proposed.[159,160] After the packaging-cutting *in vitro* of a restriction endonuclease fragment that contains the left-right joint of a T7 concatemer, the left and right ends were found.[150] Thus, the terminal repetition of T7 is duplicated during cutting from a concatemer. The cutting observed required a T7 $5' \rightarrow 3'$ exonuclease for production of a mature left, but not right, end.[150] Apparently, details of enzymology of cutting blunt ends for other bacteriophages have not been investigated.

V. CONCLUSION

From the perspective of biophysics and molecular physiology, the problems addressed during the study of bacteriophage DNA packaging overlap problems in several other areas. Key words that describe some of the research above might include contractile proteins, biological control of pore radius, ATP-requiring directed motion, folding of biological threads, nucleotide sequence-specific DNA-protein binding, osmotic effects, nucleic acid enzymology, and structural dynamics of supramolecular complexes. In addition, bacteriophages and their assembly intermediates are used to characterize agarose gels which, in turn, are models for connective tissue and other fibrous networks. The condensation of DNA in bacteriophages serves as a test problem for understanding the condensation of the fibers of eukaryotic chromosomes. Thus, pursuit of a more complete understanding of bacteriophage morphogenesis is basically a pursuit of biophysics and molecular physiology in general. The author suggests the following analogy: the study of bacteriophage genetics is to modern molecular genetics as the study of bacteriophage assembly is to future molecular physiology.

ACKNOWLEDGMENTS

The author thanks Paul M. Horowitz for critically reading and Linda C Winchester for typing this manuscript. Research in the author's laboratory was supported by the National Institutes of Health (grant GM-24365) and the Robert A. Welch Foundation (grant AQ-764).

REFERENCES

1. **Tikhonenko, A. S.,** *Ultrastructure of Bacterial Viruses,* Plenum Press, New York, 1970.
2. **Mindich, L.,** Bacteriophages that contain lipid, in *Comprehensive Virology,* Vol. 12, Fraenkel-Conrat, H. and Wagner, R. R., Eds., Plenum Press, New York, 1978, 271.
3. **Fraenkel-Conrat, H. and Kimball, P. C.,** *Virology,* Prentice-Hall, Englewood Cliffs, NJ, 1982, chap. 8.
4. **Ritchie, D. A.,** Bacteriophages, in *Topley and Wilson's Principles of Bacteriology, Virology and Immunity,* Vol. 1, Wilson, G. and Dick, H. M., Eds., Williams & Wilkins, Baltimore, 1983, chap. 7.
5. **Wood, W. B. and King, J.,** Genetic control of complex bacteriophage assembly, in *Comprehensive Virology,* Fraenkel-Conrat, H. and Wagner, R. R., Eds., Plenum Press, New York, 1979, 581.
6. **Eiserling, F. A.,** T4 structure and initiation of infection, in *Bacteriophage T4,* Mathews, C. K., Kutter, E. M., Mosig, G., and Berget, P. B., Eds., American Society for Microbiology, Washington, D.C., 1983, 11.
7. **Hendrix, R. W.,** Shape determination in virus assembly: the bacteriophage example, in *Virus Structure and Assembly,* Casjens, S., Ed., Jones and Bartlett, Boston, 1985, 170.
8. **Bazinet, C. and King, J.,** The DNA translocating vertex of dsDNA bacteriophage, *Annu. Rev. Microbiol.,* 39, 109, 1985.
9. **Carrascosa, J. L.,** Bacteriophage morphogenesis, in *Electron Microscopy of Proteins,* Vol. 5, Harris, J. R. and Horne, R. W., Eds., Academic Press, London, 1986, 37.
10. **Steven, A. C. and Trus, B. L.,** The structure of bacteriophage T7, in *Electron Microscopy of Proteins,* Vol. 5, Harris, J. R. and Horne, R. W., Eds., Academic Press, London, 1986, 1.
11. **Serwer, P., Watson, R. H., Hayes, S. J., and Allen, J. L.,** Comparison of the physical properties and assembly pathways of the related bacteriophages, T7, T3 and φII, *J. Mol. Biol.,* 170, 447, 1983.
12. **Riede, I., Drexler, K., Schwarz, H., and Henning, U.,** T-even-type bacteriophages use an adhesin for recognition of cellular receptors, *J. Mol. Biol.,* 194, 23, 1987.

13. **Carazo, J. M., Donate, L. E., Herranz, L., Secilla, J. P., and Carrascosa, J. L.,** Three-dimensional reconstruction of the connector of bacteriophage φ29 at 1.8 nm resolution, *J. Mol. Biol.,* 192, 853, 1986.

14. **Carazo, J. M., Fujisawa, H., Nakasu, S., and Carrascosa, J. L.,** Bacteriophage T3 gene 8 product oligomer structure, *J. Ultrastruct. Mol. Struct. Res.,* 94, 105, 1986.

15. **Serwer, P.,** Buoyant density sedimentation of macromolecules in sodium iothalamate density gradients, *J. Mol. Biol.,* 92, 433, 1975.

16. **Earnshaw, W. C., Casjens, S., and Harrison, S. C.,** Assembly of the head of bacteriophage P22: x-ray diffraction from heads, proheads and related structures, *J. Mol. Biol.,* 104, 387, 1976.

17. **Stroud, R. M., Serwer, P., and Ross, M. J.,** Assembly of bacteriophage T7: dimensions of the bacteriophage and its capsids, *Biophys. J.,* 36, 743, 1981.

18. **Ronto, G., Agamalyan, M. M., Drabkin, G. M., Feigin, L. A., and Lvov, Y. M.,** Structure of bacteriophage T7: small-angle x-ray and neutron scattering study, *Biophys. J.,* 43, 309, 1983.

19. **Earnshaw, W. C. and Casjens, S. R.,** DNA packaging by the double-stranded DNA bacteriophages, *Cell,* 21, 319, 1980.

20. **Tikchonenko, T. I.,** Structure of viral nucleic acids *in situ,* in *Comprehensive Virology,* Vol. 5, Fraenkel-Conrat, H. and Wagner, R. R., Eds., Plenum Press, New York, 1975, 1.

21. **Alberts, B., Bray, D., Lewis, J., Raff, M., Roberts, K., and Watson, J. D.,** *Molecular Biology of the Cell,* Garland Publishing, New York, 1983, chap. 8.

22. **Williams, S. P., Athey, B. D., Muglia, L. J., Schappe, R. S., Gough, A. H., and Langmore, J. P.,** Chromatin fibers are left-handed double helices with diameter and mass per unit length that depend on linker length, *Biophys. J.,* 49, 233, 1986.

23. **Casjens, S.,** Nucleic acid packaging by viruses, in *Virus Structure and Assembly,* Casjens, S., Ed., Jones and Bartlett, Boston, 1985, 76.

24. **Serwer, P., Masker, W. E., and Allen, J. L.,** Stability and *in vitro* DNA packaging of bacteriophages: effects of dextrans, sugars and polyols, *J. Virol.,* 45, 665, 1983.

25. **Liebeschuetz, J., Davison, P. J., and Ritchie, D. A.,** A coupled *in vitro* system for the formation and packaging of concatemeric phage T1 DNA, *Mol. Gen. Genet.,* 200, 451, 1985.

26. **Thomas, C. A., Jr., Kelly, T. J., Jr., and Rhoades, M.,** The intracellular forms of T7 and P22 DNA molecules, *Cold Spring Harbor Symp. Quant. Biol.,* 33, 417, 1968.

27. **Crick, F. H. C. and Klug, A.,** Kinky helix, *Nature (London),* 255, 530, 1975.

28. **Gourévitch, M., Puigdoménech, P., Cavé, A., Etienne, G., Méry, J., and Parello, J.,** Model studies in relation to the molecular structure of chromatin, *Biochimie,* 56, 967, 1974.

29. **Tikchonenko, T. I. and Dobrov, E. N.,** Peculiarities of the secondary structure of bacteriophage DNA *in situ.* II. Reaction with formaldehyde, *J. Mol. Biol.,* 42, 119, 1969.

30. **Sklyadneva, V. B., Chekanovskaya, L. A., Nikolaeva, I. A., and Tikchonenko, T. I.,** The secondary structure of bacteriophage DNA *in situ.* VIII. The reaction of sodium bisulfite with intraphage cytosine as a probe for studying the DNA-protein interaction, *Biochim. Biophys. Acta,* 565, 51, 1979.

31. **Chekanovskaya, L. A., Sklyadneva, V. B., and Tikchonenko, T. I.,** Identification of proteins and amino acids reacting with DNA in the composition of bacteriophage P22, *Vopr. Virusol.,* 26, 51, 1981.

32. **Hartweig, E., Bazinet, C., and King, J.,** DNA injection apparatus of phage P22, *Biophys. J.,* 49, 24, 1986.

33. **Griess, G. A., Serwer, P., and Horowitz, P. M.,** Binding of ethidium to bacteriophage T7 and T7 deletion mutants, *Biopolymers,* 24, 1635, 1985.

34. **Griess, G. A., Serwer, P., and Horowitz, P. M.,** Binding of ethidium to bacteriophages T7 and P22, *Biophys. J.,* 49, 19, 1986.

35. **Sobell, H. M., Tsai, C.-C., Jain, S. C., and Gilbert, S. G.,** Visualization of drug-nucleic acid interactions at atomic resolution. III. Unifying structural concepts in understanding drug-DNA interactions and their broader implications in understanding protein-DNA interactions, *J. Mol. Biol.,* 114, 333, 1977.

36. **Hogan, M. E., Rooney, T. F., and Austin, R. H.,** Evidence for kinks in DNA folding in the nucleosome, *Nature (London),* 328, 554, 1987.

37. **Hatfull, G. F., Noble, S. M., and Grindley, D. F.,** The γδ resolvase induces an unusual DNA structure at the recombinational crossover point, *Cell,* 49, 103, 1987.

38. **Earnshaw, W. C., King, J., Harrison, S. C., and Eiserling, F. A.,** The structural organization of DNA packaged within the heads of T4 wild-type, isometric and giant bacteriophages, *Cell,* 14, 559, 1978.

39. **Lepault, J., Debochet, J., Baschong, W., and Kellenberger, E.,** Organization of double-stranded DNA in bacteriophages: a study by cryo-electron microscopy of vitrified samples, *EMBO J.,* 6, 1507, 1987.

40. **Basu, S.,** Molecular arrangement of DNA in bacteriophage T4, *Biopolymers,* 16, 2299, 1977.

41. **Hall, S. B. and Schellman, J. A.,** Flow dichroism of capsid DNA phages. I. Fast and slow T4B, *Biopolymers,* 21, 1991, 1982.

42. **Richards, K. E., Williams, R. C., and Calendar, R.,** Mode of DNA packaging within bacteriophage heads, *J. Mol. Biol.,* 78, 255, 1973.

43. **Klimenko, S. M., Tikchonenko, T. I., and Andreev, V. M.,** Packing of DNA in the head of bacteriophage T2, *J. Mol. Biol.,* 23, 523, 1967.

44. **Eickbush, T. H. and Moudrianakis, E. N.,** The compaction of DNA helices into either continuous supercoils or folded-fiber rods, *Cell,* 13, 295, 1978.

45. **Lerman, L. S., Wilkerson, L. S., Venable, J. H., and Robinson, B. H.,** DNA packing in single crystals inferred from freeze-fracture-etch replicas, *J. Mol. Biol.,* 108, 271, 1976.

46. **Black, L. W., Newcomb, W. W., Boring, J. W., and Brown, J. C.,** Ion etching of bacteriophage T4: support for a spiral-fold model of packaged DNA, *Proc. Natl. Acad. Sci. U.S.A.,* 82, 7960, 1985.

47. **Brown, J. C. and Newcomb, W. W.,** Ion etching of bacteriophage λ: evidence that the right end of the DNA is located on the outside of the phage DNA mass, *J. Virol.,* 50, 564, 1986.

48. **Serwer, P.,** Arrangement of double-stranded DNA packaged in bacteriophage capsids: an alternative model, *J. Mol. Biol.,* 190, 509, 1986.

49. **Shurdov, M. A. and Popova, T. G.,** Mapping of 8-methoxypsoralen binding sites in DNA within phage λ particles, *FEBS Lett.,* 147, 89, 1982.

50. **Kosturko, L. D., Hogan, M., and Dattagupta, N.,** Structure of DNA within three isometric bacteriophages, *Cell,* 16, 515, 1979.

51. **Hall, S. B. and Schellman, J. A.,** Flow dichroism of capsid DNA phages. II. Effect of DNA deletions and intercalating dyes, *Biopolymers,* 21, 2011, 1982.

52. **Haas, R., Murphy, R. F., and Cantor, C. R.,** Testing models of the arrangement of DNA inside bacteriophage λ by crosslinking the packaged DNA, *J. Mol. Biol.,* 159, 71, 1982.

53. **Welsh, J. and Cantor, C. R.,** Studies on the arrangement of DNA inside viruses using a breakable bis-psoralen crosslinker, *J. Mol. Biol.,* 198, 63, 1987.

54. **Welsh, J. and Cantor, C. R.,** The packaging of DNA in bacteriophage lambda, in *Bacterial Chromatin,* Gualerzi, C. O. and Pon, C. L., Eds., Springer-Verlag, Berlin, 1987, 30.

55. **Widom, J. and Baldwin, R. L.,** Tests of spool models for DNA packaging in phage lambda, *J. Mol. Biol.,* 171, 419, 1983.

56. **Harrison, S. C.,** Packaging of DNA into bacteriophage heads: a model, *J. Mol. Biol.,* 171, 577, 1983.

57. **Griess, G. A., Serwer, P., Kaushal, V., and Horowitz, P. M.,** Kinetics of ethidium's intercalation in packaged bacteriophage T7 DNA: effects of DNA packaging density, *Biopolymers,* 25, 1345, 1986.

58. **Liu, L. F., Perkocha, L., Calendar, R., and Wang, J. C.,** Knotted DNA from bacteriophage capsids, *Proc. Natl. Acad. Sci. U.S.A.,* 78, 5498, 1981.

59. **Chattoraj, D. K. and Inman, R. B.,** Location of DNA ends in P2, 186, P4 and lambda bacteriophage heads, *J. Mol. Biol.,* 87, 11, 1974.

60. **Thomas, J. O., Sternberg, N., and Weisberg, R.,** Altered arrangement of the DNA in injection-defective lambda bacteriophage, *J. Mol. Biol.,* 123, 149, 1978.

61. **Paulson, J. R. and Laemmli, U. K.,** The structure of histone-depleted metaphase chromosomes, *Cell,* 12, 817, 1977.

62. **Marsden, M. P. F. and Laemmli, U. K.,** Metaphase chromosome structure: evidence for a radial loop model, *Cell,* 17, 849, 1979.

63. **Adolph, K. W., Kreisman, L. R., and Kuehn, R. L.,** Assembly of chromatin fibers into metaphase chromosomes analyzed by transmission electron microscopy and scanning electron microscopy, *Biophys., J.,* 49, 221, 1986.

64. **Widom, J. and Baldwin, R. L.,** Monomolecular condensation of λ-DNA induced by cobalt hexammine, *Biopolymers,* 22, 1595, 1983.

65. **Post, C. B. and Zimm, B. H.,** Theory of DNA condensation: collapse versus aggregation, *Biopolymers,* 21, 2123, 1982.

66. **Hohn, T.,** Packaging of genomes in bacteriophages: a comparison of ss RNA bacteriophages and ds DNA bacteriophages, *Philos. Trans. R. Soc. London Ser. B,* 276, 143, 1976.

67. **Rau, D. C., Lee, B., and Parsegian, V. A.,** Measurement of the repulsive force between polyelectrolyte molecules in ionic solution: hydration forces between parallel DNA double helices, *Proc. Natl. Acad. Sci. U.S.A.,* 81, 2621, 1984.

68. **Parsegian, V. A., Rand, R. P., Fuller, N. K., and Rau, D. C.,** Osmotic stress for the direct measurement of intermolecular forces, *Methods Enzymol.,* 127, 400, 1986.

69. **Riemer, S. C. and Bloomfield, V. A.,** Packaging of DNA in bacteriophage heads: some considerations on energetics, *Biopolymers,* 17, 785, 1978.

70. **Costello, R. C. and Baldwin, R. L.,** The net hydration of phage lambda, *Biopolymers,* 11, 2147, 1972.

71. **Hamada, K., Fujisawa, H., and Minagawa, T.,** A defined *in vitro* system for packaging of bacteriophage T3 DNA, *Virology,* 151, 119, 1986.

72. **Serwer, P.,** Analysis of empty phage T7 capsids and capsid-DNA complexes, Abstr., ICN Assembly Meet., Squaw Valley, CA, 1974.

73. **Hohn, B., Wurtz, M., Klein, B., Lustig, A., and Hohn, T.,** Phage lambda DNA packaging *in vitro, J. Supramol. Struct.,* 2, 302, 1974.

74. **Serwer, P.,** A metrizamide-impermeable capsid in the DNA packaging pathway of bacteriophage T7, *J. Mol. Biol.,* 138, 95, 1980.
75. **Masker, W. E.,** *In vitro* packaging of bacteriophage T7 DNA requires ATP, *J. Virol.,* 43, 365, 1982.
76. **Serwer, P., Watson, R. H., and Hayes, S. J.,** Heterogeneity of the procapsid of bacteriophage T3, *J. Virol.,* 55, 232, 1985.
77. **Shibata, H., Fujisawa, H., and Minagawa, T.,** Characterization of the bacteriophage T3 DNA packaging reaction *in vitro* in a defined system, *J. Mol. Biol.,* 196, 845, 1987.
78. **Guo, P., Peterson, C., and Anderson, D.,** Prohead and DNA-gp3-dependent ATPase activity of the DNA packaging protein gp16 of bacteriophage φ29, *J. Mol. Biol.,* 197, 229, 1987.
79. **Stryer, L.,** *Biochemistry,* W. H. Freeman, San Francisco, 1981, chap. 11.
80. **Hamada, K., Fujisawa, H., and Minagawa, T.,** Characterization of ATPase activity of a defined *in vitro* system for packaging bacteriophage T3 DNA, *Virology,* 159, 244, 1987.
81. **Rao, V. B. and Black, L. W.,** Evidence that a phage T4 DNA packaging enzyme is a processed form of the major capsid gene product, *Cell,* 42, 967, 1985.
82. **Dukes, P. P. and Kozloff, L. M.,** Phosphatases in bacteriophages T2, T4 and T5, *J. Biol. Chem.,* 234, 534, 1959.
83. **Kozloff, L. M. and Lute, M.,** A contractile protein in the tail of bacteriophage T2, *J. Biol. Chem.,* 234, 539, 1959.
84. **Cummings, D. J. and Kozloff, L. M.,** Various properties of the head protein of T2 bacteriophage, *J. Mol. Biol.,* 5, 60, 1962.
85. **Serwer, P.,** The source of energy for bacteriophage DNA packaging: an osmotic pump explains the data, *Biopolymers,* 27, 165, 1988.
86. **Serwer, P.,** Agarose gel electrophoresis of bacteriophages and related particles, *J. Chromatogr. Biomed. Applic.,* 418, 345, 1987.
87. **Kennel, D. and Riezman, H.,** Transcription and translation initiation frequencies of the *Escherichia coli* lac operon, *J. Mol. Biol.,* 114, 1, 1977.
88. **Schaerli, C. and Kellenberger, E.,** Head maturation pathway of bacteriophages T4 and T2. V. Maturable ε-particle accumulating in acridine-treated bacteriophage T4-infected cells, *J. Virol.,* 33, 830, 1980.
89. **Gope, R. and Serwer, P.,** Bacteriophage P22 *in vitro* DNA packaging monitored by agarose gel electrophoresis: rate of DNA entry into capsids, *J. Virol.,* 47, 96, 1983.
90. **Rao, V. B. and Black, L. W.,** DNA packaging of bacteriophage T4 proheads *in vitro:* evidence that prohead expansion is not coupled to DNA packaging, *J. Mol. Biol.,* 185, 565, 1985.
91. **Guo, P., Grimes, S., and Anderson, D.,** A defined system for *in vitro* packaging of DNA-gp3 of the *Bacillus subtilis* bacteriophage φ29, *Proc. Natl. Acad. Sci. U.S.A.,* 83, 3505, 1986.
92. **Castellino, F. J. and Barker, R.,** Examination of the dissociation of multichain proteins in guanidine hydrochloride by membrane osmometry, *Biochemistry,* 7, 2207, 1968.
93. **Witkiewicz, H. and Schweiger, M.,** A model of λ DNA arrangement in the viral particle, *J. Theor. Biol.,* 116, 587, 1985.
94. **Black, L. W. and Silverman, D. J.,** Model for DNA packaging into bacteriophage T4 heads, *J. Virol.,* 28, 643, 1978.
95. **Hendrix, R.,** Symmetry mismatch and DNA packaging in large bacteriophages, *Proc. Natl. Acad. Sci. U.S.A.,* 75, 4779, 1978.
96. **Skolnick, J.,** Possible role of helix-coil transitions in the microscopic mechanism of muscle contraction, *Biophys. J.,* 51, 227, 1987.
97. **Fujisawa, H., Hamada, K., Shibata, H., and Minagawa, T.,** On the molecular mechanism of DNA translocation during *in vitro* packaging of bacteriophage T3 DNA, *Virology,* 161, 228, 1987.
98. **Tomic, M. T., Wemmer, D. E., and Kim, S. H.,** Structure of psoralen cross-linked DNA in solution by nuclear magnetic resonance, *Science,* 238, 1722, 1987.
99. **Kozloff, L. M., Chapman, V., and DeLong, S.,** Defective packing of an unusual DNA in a virulent *Erwinia* phage, Ehrl, in *Bacteriophage Assembly,* Dubow, M. S., Ed., Alan R. Liss, New York, 1981, 253.
100. **Pearson, R. K. and Fox, M. S.,** Effects of DNA heterologies on bacteriophage λ packaging, *Genetics,* 118, 5, 1988.
101. **Pearson, R. K. and Fox, M. S.,** Effects of DNA heterologies on bacteriophage λ recombination, *Genetics,* 118, 13, 1988.
102. **Mosig, G. and Powell, D.,** Heteroduplex loops are packaged in gene 49 (endonuclease VII) mutants of bacteriophage T4, Abstr., Annu. American Society for Microbiology Meet., 1985, 209.
103. **Earnshaw, W. C. and Harrison, S. C.,** DNA arrangement in isometric phage heads, *Nature (London),* 268, 598, 1977.
104. **Serwer, P. and Gope, R.,** Bacteriophage P22 capsids with a subgenome length of packaged DNA, *J. Virol.,* 49, 293, 1984.

105. **DeMars, R. I., Luria, S. E., Fisher, M., and Levinthal, C.,** The production of incomplete bacteriophage particles by the action of proflavine and the properties of the incomplete particles, *Ann. Inst. Pasteur Paris,* 84, 113, 1952.

106. **Levinthal, C. and Fisher, H.,** The structural development of a bacterial virus, *Biochim. Biophys. Acta,* 9, 419, 1952.

107. **Masker, W. E. and Serwer, P.,** DNA packaging *in vitro* by an isolated bacteriophage T7 procapsid, *J. Virol.,* 43, 1138, 1982.

108. **Serwer, P.,** A technique for electrophoresis in multiple-concentration agarose gels, *Ann. Biochem.,* 101, 154, 1980.

109. **Serwer, P., Hayes, S. J., and Watson, R. H.,** The structure of a bacteriophage T7 procapsid and its *in vivo* conversion product probed by digestion with trypsin, *Virology,* 122, 392, 1982.

110. **Fujisawa, H. and Yamagishi, M.,** Studies on factors involved in *in vitro* packaging of phage T3 DNA, in *Bacteriophage Assembly,* Dubow, M. S., Ed., Alan R. Liss, New York, 1981, 239.

111. **Dunn, J. J. and Studier, F. W.,** Complete nucleotide sequence of bacteriophage T7 DNA and the locations of T7 genetic elements, *J. Mol. Biol.,* 166, 477, 1983.

112. **Yamagashi, M., Fujisawa, H., and Minagawa, T.,** Isolation and characterization of bacteriophage T3/T7 hybrids and their use in studies on molecular basis of DNA-packaging specificity, *Virology,* 144, 502, 1985.

113. **Yamada, M., Fujisawa, H., Kato, H., Hamada, K., and Minagawa, T.,** Cloning and sequencing of the genetic right end of bacteriophage T3 DNA, *Virology,* 151, 350, 1986.

114. **Shibata, H., Fujisawa, H., and Minagawa, T.,** Early events in DNA packaging in a defined *in vitro* system of bacteriophage T3, *Virology,* 159, 250, 1987.

115. **Roeder, G. S. and Sadowski, P. D.,** Bacteriophage T7 morphogenesis: phage-related particles in cells infected with wild-type and mutant T7 phage, *Virology,* 76, 263, 1977.

116. **Murialdo, H. and Becker, A.,** Head morphogenesis of complex double-stranded deoxyribonucleic acid bacteriophages, *Microbiol. Rev.,* 42, 529, 1978.

117. **Caldenty, J., Lepault, J., and Kellenberger, E.,** Isolation and reassembly of bacteriophage T4 core proteins, *J. Mol. Biol.,* 195, 637, 1987.

118. **Gold, M., Hawkins, D., Murialdo, H., Fife, W. L., and Bradley, B.,** Circular monomers of bacteriophage λ DNA as substrates for *in vitro* packaging, *Virology,* 119, 35, 1982.

119. **Strobel, E., Benisch, W., and Schmieger, H.,** *In vitro* packaging of mature phage DNA by *Salmonella* phage P22, *Virology,* 133, 158, 1984.

120. **Sternberg, N. and Coulby, J.,** Recognition and cleavage of the bacteriophage P1 packaging site (pac). I. Differential processing of the cleaved ends *in vivo, J. Mol. Biol.,* 194, 453, 1987.

121. **DeWyngaert, M. A. and Hinkle, D. C.,** Characterization of the defects in bacteriophage T7 DNA synthesis during growth in the *Escherichia coli* mutant tsnB, *J. Virol.,* 33, 780, 1980.

122. **Ito, J.,** Bacteriophage φ29 terminal protein: its association with the 5′ termini of the φ29 genome, *J. Virol.,* 28, 895, 1978.

123. **Murray, A. W. and Szostak, J. W.,** Artificial chromosomes, *Sci. Am.,* 257, 62, 1987.

124. **Furth, M. E. and Wickner, S. H.,** Lambda DNA replication, in *Lambda II,* Hendrix, R. W., Roberts, J. W., Stahl, F. W., and Weisberg, R. A., Eds., Cold Spring Harbor Laboratory, Cold Spring Harbor, NY, 1983, 145.

125. **Serwer, P., Watson, R. H., and Hayes, S. J.,** Multidimensional analysis of intracellular bacteriophage T7 DNA: effects of amber mutations in genes 3 and 19, *J. Virol.,* 61, 3499, 1987.

126. **Langman, L., Paetkau, V., Scraba, D., Miller, R. C., Roeder, G. S., and Sadowski, P. D.,** The structure and maturation of intermediates in bacteriophage T7 DNA replication, *Can. J. Biochem.,* 56, 508, 1978.

127. **Mosig, G.,** Relationship of T4 DNA replication and recombination, in *Bacteriophage T4,* Mathews, C. K., Kutter, E. M., Mosig, G., and Berget, P. B., Eds., American Society for Microbiology, Washington, D.C., 1983, 120.

128. **Curtis, M. J. and Alberts, B.,** Studies on the structure of intracellular bacteriophage T4 DNA, *J. Mol. Biol.,* 102, 793, 1976.

129. **Feiss, M. and Becker, A.,** DNA packaging and cutting, in *Lambda II,* Hendrix, R. W., Roberts, J. W., Stahl, F. W., and Weisberg, R. A., Cold Spring Harbor Laboratory, Cold Spring Harbor, NY, 1983, 305.

130. **Feiss, M.,** Terminase and the recognition, cutting and packaging of λ chromosomes, *Trends Genet.,* 2, 100, 1986.

131. **Guo, P., Peterson, C., and Anderson, D.,** Initiation events in *in-vitro* packaging of bacteriophage φ29 DNA-gp3, *J. Mol. Biol.,* 197, 219, 1987.

132. **Becker, A., Murialdo, H., and Gold, M.,** Studies on an *in vitro* system for the packaging and maturation of phage λ DNA, *Virology,* 78, 277, 1977.

133. **Poteete, A. R. and Botstein, D.,** Purification and properties of proteins essential to DNA encapsulation by phage P22, *Virology,* 95, 565, 1979.

134. **Jackson, E. N., Laski, F., and Andres, C.,** Bacteriophage P22 mutants that alter the specificity of DNA packaging, *J. Mol. Biol.,* 154, 551, 1982.
135. **Casjens, S., Huang, W. M., Hayden, M., and Parr, R.,** Initiation of bacteriophage P22 DNA packaging series. Analysis of a mutant that alters the target specificity of the packaging apparatus, *J. Mol. Biol.,* 194, 411, 1987.
136. **Hamada, K., Fujisawa, H., and Minagawa, T.,** Overproduction and purification of the products of bacteriophage T3 genes 18 and 19, two genes involved in DNA packaging, *Virology,* 151, 110, 1986.
137. **White, J. H. and Richardson, C. C.,** Gene 18 protein of bacteriophage T7: overproduction, purification and characterization, *J. Biol. Chem.,* 262, 8845, 1987.
138. **White, J. H. and Richardson, C. C.,** Gene 19 of bacteriophage T7: overexpression, purification and characterization of its product, *J. Biol. Chem.,* 263, 2469, 1988.
139. **Shindler, G. and Gold, M.,** The Nul subunit of bacteriophage lambda terminase binds to specific sites in *cos* DNA, *J. Virol.,* 62, 387, 1988.
140. **Murialdo, H., Davidson, A., Chow, S., and Gold, M.,** The control of lambda DNA terminase synthesis, *Nucleic Acids Res.,* 15, 119, 1987.
141. **Guo, P., Erickson, S., and Anderson, D.,** A small viral RNA is required for *in vitro* packaging of bacteriophage φ29 DNA, *Science,* 236, 690, 1987.
142. **Serwer, P. and Watson, R. H.,** Capsid-DNA complexes in the DNA packaging pathway of bacteriophage T7: characterization of the capsids bound to monomeric and concatemeric DNA, *Virology,* 108, 164, 1981.
143. **Casjens, S. and Huang, W. M.,** Initiation of sequential packaging of bacteriophage P22 DNA, *J. Mol. Biol.,* 157, 287, 1982.
144. **Backhaus, H.,** DNA packaging initiation of *Salmonella* bacteriophage P22: determination of cut sites within the DNA sequence coding for gene 3, *J. Virol.,* 55, 458, 1985.
145. **Winnacker, E.-L.,** *From Genes to Clones,* VCH, Weinheim, 1987, chap. 2.
146. **Dawson, P., Skalka, A., and Simon, L.,** Bacteriophage lambda head morphogenesis: studies on the role of DNA, *J. Mol. Biol.,* 93, 167, 1975.
147. **Tye, B.-K. and Botstein, D.,** P22 morphogenesis II: mechanism of DNA encapsulation, *J. Supramol. Struct.,* 2, 225, 1974.
148. **Ramsey, N. and Ritchie, D. A.,** Uncoupling of initiation site cleavage from subsequent headful cleavages in bacteriophage T1 DNA packaging, *Nature (London),* 301, 5897, 1983.
149. **Roberts, L., Sheldon, R., and Sadowski, P. D.,** Genetic recombination of bacteriophage T7 DNA *in vitro.* IV. Asymmetry of recombination frequencies caused by polarity of DNA packaging *in vitro, Virology,* 89, 252, 1978.
150. **White, J. H. and Richardson, C. C.,** Processing of concatemers of bacteriophage T7 DNA *in vitro, J. Biol. Chem.,* 262, 8851, 1987.
151. **Emmons, S. W.,** Bacteriophage lambda derivatives carrying two copies of the cohesive end site, *J. Mol. Biol.,* 83, 511, 1974.
152. **Hohn, B.,** DNA sequences necessary for packaging of bacteriophage λ DNA, *Proc. Natl. Acad. Sci. U.S.A.,* 80, 7456, 1983.
153. **Bjornsti, M.-A., Reilly, B. E., and Anderson, D. L.,** Morphogenesis of bacteriophage φ29 of *Bacillus subtilis:* oriented and quantized *in vitro* packaging of DNA protein gp3, *J. Virol.,* 45, 383, 1983.
154. **Hsiao, C. L. and Black, L. W.,** Head morphogenesis of bacteriophage T4. III. The role of gene 20 in DNA packaging, *Virology,* 91, 26, 1978.
155. **Streisinger, G., Emrich, J., and Stahl, M. M.,** Chromosome structure in phage T4. III. Terminal redundancy and length determination, *Proc. Natl. Acad. Sci. U.S.A.,* 57, 292, 1967.
156. **Feiss, M. and Siegele, D. A.,** Packaging of the bacteriophage lambda chromosome: dependence of cos cleavage on chromosome length, *Virology,* 92, 190, 1979.
157. **Murialdo, H. and Fife, W. L.,** Synthesis of a transacting inhibitor of DNA maturation by prohead mutants of phage λ, *Genetics,* 115, 3, 1987.
158. **Wang, J. C. and Kaiser, A. D.,** Evidence that the cohesive ends of mature λ DNA are generated by the A gene product, *Nature New Biol.,* 241, 16, 1973.
159. **Kelly, T. J. and Thomas, C. A., Jr.,** An intermediate in the replication of bacteriophage T7 DNA molecules, *J. Mol. Biol.,* 44, 459, 1969.
160. **Watson, J. D.,** Origin of concatemeric T7 DNA, *Nature New Biol.,* 239, 197, 1972.
161. **Griess, G. A. and Serwer, P.,** unpublished observations, 1987.
162. **Parsegian, V. A.,** personal communication, 1988.
163. **Anderson, D.,** personal communication, 1988.
164. **Serwer, P. and Watson, R. H.,** unpublished data, 1988.

Chapter 10

BACTERIOPHAGE P1 DNA PACKAGING

Nat Sternberg

TABLE OF CONTENTS

I. INTRODUCTION

Bacteriophage P1 is a temperate bacteriophage that can exist in cells either as a unit-copy extrachromosomal plasmid prophage or as a lytic phage.[1,2] During lytic growth, the virus replicates its DNA to 100 to 200 copies per cell, packages that DNA into phage capsids, and then lyses the cell. It is the specificity and regulation of the packaging process that will be the main topic of this review. In addition to packaging viral DNA, P1 can also package host DNA into capsids and can transfer that DNA to cells by phage infection, a process called generalized transduction.[1-3] In the final section of this review, the packaging of phage and the packaging of bacterial DNA will be compared in an effort to draw conclusions about the latter process.

II. PHAGE PHYSICAL STRUCTURE

A. THE STRUCTURE OF PHAGE PARTICLES

The infectious P1 phage particle consists of an icosahedral head that is about 85 nm wide and 95 nm long.[4-6] Attached to one of the vertices of the head is a head-tail connector and a 215-nm tail that consists of a tail tube and a contractile sheath.[5,6] Attached to the tail, at the end distal to the head, is a base plate and a number of kinked fibers.[6]

During the vegetative cycle, P1 normally produces three classes of particles with the same size tails, but with isometric heads of different widths.[4,5,7] The largest particles (P1B) are infectious, have 85-nm-wide heads, and contain approximately 1.1 genome equivalents of P1 DNA. They constitute approximately 80% of phage particles in wild-type lysates. Two classes of smaller phage particles are also found in lysates. These phage, designated small (P1S) and minute (P1M), have heads that are 65 and 45 nm wide, respectively. The P1S particle contains 0.4 to 0.45 genome equivalents of P1 DNA. Both types of particles inject their DNA, but are noninfectious on single infection. P1S particles represent 6 to 27% of the particle population and P1M particles represent less than 3% of the particle population.

B. VIRAL DNA

The DNA of plaque-forming P1 phage is a linear double-stranded molecule[8] that is also terminally redundant and cyclically permuted.[9] Its molecular weight is about 66 MDa,[10] or about 105 kilobase pairs, and the extent of terminal redundancy has been estimated as either 12%[9] or 7.9%[10] of the viral DNA. The DNA of P1S particles is 25.5 MDa in size; it lacks terminal redundancy, but is cyclically permuted.[11]

When the DNA of plaque forming particles is injected into cells that are proficient for general recombination, cyclization occurs by recombination between the terminally redundant ends,[9] resulting in the loss of the redundancy. The lack of redundancy in the DNA of P1S particles suggests that this DNA is not cyclized after infection and is probably rapidly degraded by cellular nucleases. However, at high multiplicities of phage infection, these particles can generate infectious phage by recombination. The permutation of DNA in the population of P1S particles is the same as that found in P1B particles, suggesting that the two classes of particles are generated by the same packaging mechanism, a processive headful mechanism[11] (see Section IV).

C. VIRAL PROTEINS

The major head protein of P1 has a molecular weight of 44 kDa. Fourteen other head proteins can be detected by SDS-PAGE analysis of P1 particles, four of which are dispensable internal proteins.[12] The major tail proteins have molecular weights of 21 and 72 kDa and are probably the subunits of the tail tube and the tail sheath, respectively.[12]

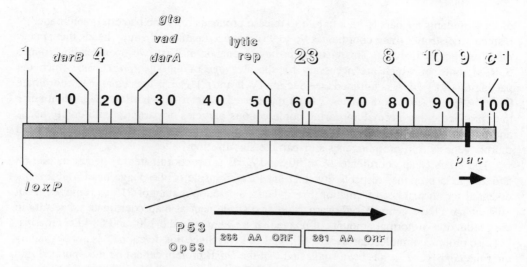

FIGURE 1.   A genetic map of P1 packaging functions. The map is divided into 100 units, with P1 genes, cistrons, and functions located above the stippled bar, and the P1 *lox*P recombination site and packaging site *(pac)* located below the bar. The direction of packaging is designated by the arrow below *pac*. The elements that comprise the lytic replicon of P1 are shown below the bar in expanded form.

D. THE STRUCTURE OF TRANSDUCING PARTICLES

About 0.3 to 0.5% of the total particles in a phage lysate contain exclusively bacterial DNA.[8,13] That DNA is found in both P1B and P1S particles, but, uniquely, a protein appears to be linked to the terminus of the bacterial DNA in the P1B transducing particles.[14] Based on the buoyant density of DNA from the P1B transducing particles, it was estimated that about 500 kDa of protein is associated with each DNA molecule.

III. THE ORGANIZATION AND PROPERTIES OF PHAGE GENES INVOLVED IN VEGETATIVE REPLICATION, HEAD MORPHOGENESIS, AND DNA PACKAGING

A P1 map containing genes (cistrons) involved in DNA packaging and lytic replication is shown in Figure 1. While the map is frequently drawn as a circle,[15] it is, in fact, linear, owing to the presence of a hot spot (*lox*P) for P1-determined, site-specific recombination at map coordinate 0/100.

A. P1 GENES INVOLVED IN HEAD MORPHOGENESIS AND PACKAGING

While the coding capacity of P1 is about 100 genes, only 23 cistrons have been identified from the more than 100 amber mutations isolated by Walker and Walker[16,17] and Scott.[18] Of these cistrons, only four can be unequivocally assigned to head morphogenesis functions and they are dispersed throughout the P1 genetic map. While mutations in each of the cistrons are associated with a unique structural phenotype, the specific role of proteins encoded by the cistrons in the head morphogenesis pathway is frequently not known. Defects in cistrons 8 (map coordinates 85 to 88) and 23 (map coordinates 60 to 62) result in the production of a normal amount of phage tails, but few phage heads, which are unattached to tails and are empty. Since the top of the tail appears to contain a head-tail connector, it has been suggested that the function encoded by these cistrons is involved in stabilizing the DNA-filled head.[6] Defects in cistron 4 (map coordinates 15 to 20) apparently result in the inability to produce a protease necessary for both the processing of the nonessential internal protein encoded by the *dar*A gene (see below) and for the processing of other phage proteins. The phenotype of these mutants appears to be a failure to release proheads from the bacterial membrane.[6,19] Defects in cistron 9 (map coordinates 96 to 98) produce unattached empty heads that appear morphologically mature.[6] This cistron appears to contain at least two open reading frames (ORFs), both of which are necessary for the cleavage of the site (*pac*) from which the packaging of P1 DNA is initiated (see Section IV.B and C). *Pac* is located at map coordinate 96.[20,21] Gene 1 (map coordinates 2 to 4) mutants are defective in particle maturation since they produce either normal appearing noninfectious particles or particles with empty heads and contracted tails. It is not clear whether this gene encodes a protein involved in head morphogenesis, tail morphogenesis or particle stabilization.[6]

The *dar*A (map coordinate 26 to 30) and *dar*B (map coordinate 15) genes encode 68 and 200 kDa proteins, respectively, that are located inside of the phage head, but are not essential for normal head formation.[12,22,23] Thus, a deletion mutant of P1 that removes about 10% of the phage genome rightward from an IS1 element at map coordinate 24 results in the production of normal amounts of phage that lack not only the 200- and 68-kDa proteins, but also proteins of molecular weight 47.5 and 10 kDa. Since this deletion does not physically alter the *dar*B gene, it has been suggested that the *Dar*B protein cannot be incorporated into phage head precursors in the absence of the *Dar*A protein. Following its injection into *Escherichia coli,* P1 DNA is much less sensitive to type I restriction enzymes than it is to type II and type III enzymes. This insensitivity has been attributed to the *Dar*A and *Dar*B proteins.[22,23] Mutants in *dar*B are restricted by *Eco*K and *Eco*B enzymes, but not by the *Eco*A enzyme. Mutants in *dar*A (which are *dar*A$^-$ and *dar*B$^-$) are sensitive to all three restriction systems. The Dar proteins will protect any DNA packaged into a P1 head, but will not protect the DNA injected by another phage; namely, DNA injected by phage λ is not protected even if cells are coinfected with P1. These results suggest that the DarA and DarB proteins are injected into cells along with the phage DNA and remain associated with it long enough to allow it to avoid restriction.

Besides affecting the presence of the two largest internal proteins, the above mentioned deletion mutant also changes the ratio of P1B to P1S particles in a P1 lysate. Normally, P1S particles constitute about one sixth of the yield of a lysate, but with the deletion mutant both classes of particles are produced in equal amounts.[24,25] The gene that affects the proportion of these classes of particles has been designated *vad* (viral architecture determinant), and amber mutations have been isolated in it. Since the protein composition of P1B and P1S particles from the *vad* deletion mutant is similar, it has been suggested that the Vad protein is not found in the finished P1 particles. Using cloned fragments that contain and express *vad,* Walker and Walker[26] showed that Vad can act in *trans* to complement the defect of the P1 deletion mutant. They could also show that P1B and P1S particles are produced in the same infected cell and that the ratio of these two particles does not change with time after infection. The latter result indicates that the increased production of P1S particles is not due to the accumulation of an essential protein needed for P1S production that might result from the delayed-lysis phenotype associated with *vad-dar*A deletion mutant.

B. P1 GENES INVOLVED IN VEGETATIVE REPLICATION

Within the past two years, the location and nature of the primary, if not the only, P1 vegetative (lytic) replicon has been precisely determined. Its position remained elusive until recently because of the failure to identify a single conditional-lethal mutation that rendered it inactive. This failure was probably due to its small size, less than a kilobase of DNA.

The replicon is located between P1 map coordinates 53 and 56. It consists of a promoter at map coordinate 53 (P53) that directs transcription rightward (Figure 1) and two downstream ORFs. The promoter-proximal ORF is 266 amino acids long and the promoter-distal ORF is 281 amino acids long. Much of what we know about this replicon derives from mutagenesis and cloning experiments.[2,27-29] Reviewed below are some of the conclusions that have emerged from these experiments.

1. Regulation of the Replicon Genes

Transcription from P53 is essential for vegetative replication. Cohen cloned the segment of P1 DNA containing the vegetative replicon into a λ vector[30] and could show that the resulting construct would replicate in *E. coli* when the λ replicon was rendered inactive.[28] Replication was blocked if the host contained the repressor of P1 vegetative growth, the *c1* gene product (Figure 1).[31,32] Since a *c1*-binding site (Op53) overlaps the P53 promoter,[29] the repressor presumably works by blocking transcription from P53. To determine whether one or both of the ORFs in this replicon are necessary for replication, each ORF was rendered inactive by mutation and the effects of the mutations on replication were assessed.[29] The conclusion is that the 266-amino-acid ORF is not necessary for vegetative replication, but the 281-amino-acid ORF is.

2. Other Properties of the Vegetative Replicon

P1 DNA starts to be synthesized by the fifth minute after phage infection of *E. coli* at 37°C and by 30 min represents 95% of the total DNA synthesis in the cell.[33] Analysis of replication products by electron microscopy indicates that theta and sigma forms are present in equal numbers during the first 15 to 30 min after infection, but that sigma forms predominate later in infection.[34] Since sigma forms are rarely, if ever, generated by recombination,[35] there must be at least two modes of replication. The shift from theta to sigma replication is blocked in a *rec*A⁻ host,[34,36] and the yield of phage is reduced proportionately. These results are most easily explained if the sigma form of P1 DNA is the substrate for DNA packaging, a conclusion consistent with the demonstration that P1 packages DNA by a processive headful mechanism (see below).

Do the two modes of P1 vegetative replication reflect the existence of two independent phage replicons? While this question cannot be definitely answered yet, it seems more reasonable to propose that the replicon described above is responsible for both modes of P1 replication. Thus, IS50[27] or Tn5[29] insertion mutations located in P1 DNA between P53 and the 281-amino-acid ORF block any detectable vegetative replication following prophage derepression. This result indicates that if P1 contains another vegetative replicon, it must be epistatic to the replicon described here. Since the cloned P1 replicon appears to replicate DNA exclusively by the theta mode,[29] either another P1 gene product or some feature of the P1 DNA itself must be necessary for the conversion from theta to sigma replication.

IV. THE PACKAGING OF P1 DNA

While little is known about the actual mechanics of P1 DNA packaging, progress is being made in understanding the elements necessary for the initiation of that process. Consequently, we will devote most of our efforts here to a discussion of that topic.

A. PROCESSIVE HEADFUL PACKAGING

For the large double-stranded DNA phages such as P1, in which the virion DNA is terminally redundant and cyclically permuted, it has been proposed that packaging takes place by a processive headful mechanism.[37,38] The DNA substrate for packaging is a concatemer of P1 DNA that is generated by the replication system described above and consists of tandemly repeated units of the phage genome arranged in a head-to-tail configuration. Packaging is initiated when the concatemer is cut at a unique site called *pac* (Figure 1). DNA on the one side of the cut is then packaged into a phage prohead and DNA on the other side of the cut is degraded by cellular nucleases.[20] The direction of P1 packaging is from *pac* toward the right end of the P1 genetic map[21] (Figure 1). Once packaging is complete and the head is full of DNA, the packaged DNA is cleaved away from the rest of the concatemer (the headful cut) and a second round of packaging is initiated from the free end generated by that cleavage event. Processivity continues until either one of the free DNA ends generated by a headful cut fails to associate with proteins needed to direct that DNA into a P1 prohead or until the concatemer has been completely packaged. In this packaging scheme, the *pac* site is cut only once per processive series, to initiate the reaction, and all subsequent cleavage events occur by the headful cut. The *pac* cut occurs at a relatively precise location in P1 DNA (see below), but the headful cut is relatively imprecise. How imprecise has not been determined for P1, but for phage P22 the ends generated by the headful nuclease are distributed over about 1.5 kb of DNA or 3.4% of the P22 genome.[39]

The degree of processivity of P1 packaging can be estimated by evaluating the nonstoichiometry of the *pac* end in the population of packaged P1 DNAs. By digesting P1 DNA with a variety of restriction enzymes and comparing the amount of the fragments with one *pac* end to the amount of the fragments with restriction ends, Bächi and Arber[21] were able to localize *pac,* determine the direction of packaging, and predict that packaging is processive for 3 to 4 headfuls. In another determination of processivity, an isolated *pac* site was placed in the chromosome of *E. coli* and the enhanced production of generalized transducing phage after P1 infection was measured.[20] In this experiment, processive packaging was undiminished for about five headfuls and continued for an additional five headfuls with diminishing efficiency. The seemingly contradictory results about the extent of processivity from the above two experiments can be reconciled if the physical length of the P1 concatemer limits the degree of processivity. This possibility is supported by studies showing that the population of P1S particles contains the entire complement of DNA present in P1B particles, including the same-sized *pac* fragments, present in the same amounts relative to other fragments, as in P1B particles.[11]

B. *PAC* PROCESSING GENES

Using an *in vivo* hybridization assay to measure *pac* cleavage that is independent of either DNA packaging or phage particle formation, Sternberg and Coulby[20] have shown that P1 *am* mutants with defects in cistron 9 could not cleave *pac.* In contrast, P1 mutants with defects in tail genes, head genes, or the gene that positively regulates late gene expression (gene 10; Figure 1)[6] are able to cleave *pac* as efficiently as does P1 wild-type. These results indicate that *pac* cleavage can occur normally in the absence of DNA packaging or the production of phage heads. Moreover, since gene 10 mutants fail to synthesize any of the late phage morphogenesis proteins, the synthesis of the *pac* cleavage enzyme(s) would appear not to be regulated in the same way as is the synthesis of other P1 proteins involved in DNA packaging. Additional support for this conclusion comes from the demonstration that *pac* cleavage of a site placed in the host chromosome can be detected as early as 10 to 15 min after P1 infection[20] and is, therefore, probably an early phage function.

The DNA sequence of the region containing the genes involved in *pac* cleavage has recently been determined. Analysis of that sequence indicates that it contains at least two ORFs and possibly a third (Figure 2). Since none of the proteins encoded by these ORFs have yet been identified or isolated, their nature and/or existence must still be considered speculative. The upstream ORF (ORF 1) encodes a protein of 283 amino acids and contains several amber mutations (*am* 9.16, *am* 154) that inactivate *pac* cleavage function. This ORF also contains the *pac* site in its N-terminal half. The downstream ORF (ORF 3) is 271 amino acids long and also contains at least one amber mutation (*am* 43) that inactivates *pac* cleavage function. Between these two ORFs is another ORF (ORF 2) of 133 amino acids whose reading frame is different from that of the ORF 1 and whose termination codon overlaps the initiation codon of ORF 3. ORF 2 could either encode a separate *pac* cleavage protein or it could encode the C-terminal portion of a 413-amino-acid protein whose N-terminal portion is encoded by ORF 1. The 413-amino-acid protein would have to be generated by a translational frame shift (see legend to Figure 2). Two amber mutants (*am*

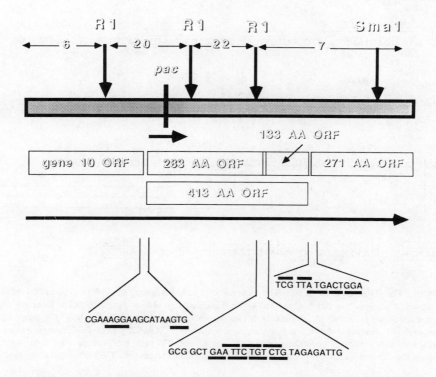

FIGURE 2.   A map of the P1 *pac* cleavage genes. The top of the map shows *Eco*R1 fragments that span the region of the P1 genome containing *pac*-cleavage functions (map coordinates 94 to 97). The positions of the open reading frames and their sizes are indicated below the map. The *pac* site is located in the N-terminal portion of the 283-amino-acid ORF and the direction of packaging is indicated by the arrow below that site. The direction of transcription of the operon containing the ORFs is indicated by the long arrow spanning the entire region. Selected DNA sequences between the ORFs are shown at the bottom of the figure. The sequence on the left shows the GTG start codon of the 283-amino-acid ORF and its upstream ribosomal binding site. The sequence in the middle shows the last four codons of the 283-amino-acid ORF (black bars below the sequence) and the first three codons of the presumptive 133-amino-acid ORF (black bars above the sequence). The 413-amino-acid ORF is generated by a translational frame shift that must occur between the TAG termination codon of the 283-amino-acid ORF and the TGA termination codon just upstream of the 133-amino-acid ORF. The DNA sequence on the right shows the last two codons of the 133-amino-acid ORF (black bars above the sequence) and the first three codons of the 271-amino-acid ORF (black bars below the sequence).

131 and *am* 136) that inactivate *pac* cleavage functions have been mapped to a region of the P1 genome containing the 133-amino-acid ORF,[20] although neither has yet been definitively localized to that ORF. Since all of the amber mutants defective for *pac* cleavage fail to complement each other for phage growth,[20] they are all probably part of the same translational unit. Indeed, only ORF 1 has an identifiable ribosomal binding site. The region immediately upstream of this ribosomal binding site does not contain a detectable transcription promoter, but rather codes for gene 10 of P1. These observations suggest that the gene 10 promoter also transcribes the cistron 9 ORFs and, therefore, that the region contains an operon whose first gene regulates late phage protein synthesis (gene 10) and whose other genes are involved in *pac* cleavage (Figure 2). Gene 10 and the *pac* cleavage genes are probably translated separately since gene 10 mutants complement mutants that are defective in *pac* cleavage.[20]

Δ20 ┐   (8)→   (7)→   (6)→   (5)→
5'-AG|CA|TGATCA|T|TGATCA|CTCTAA|TGATCA|ACATGCAGG|TGATCA|CATTGCGGCTGAAAATAGCGG
        230   220   210   200   190   180

↓↓↓↓↓↓↓↓↓↓↓↓
AAAAAGAAAGAGTTAATGCCGTTGTCAGTGCCGCAGTCGAGAATGCGAAGCGCCAAAATAAG
170   160   CAACAGTCACGG   130   120   110
                    150 ↑ ↑ ↑↑↑↑↑
                            140

        (4)→   (3)→   (2)→
CGC|ATAAA|TGATCG|TTCAGA|TGATCA|TGAC|G|TGATCA|C|CCGCGCCCACCGGACCTTACG
        100   90   80   70   60   50
                                                                    ——Δ3
                                                                    ——Δ5
                                                                    ——Δ6
                                                                    ——Δ7

(1)→   EcoRI
TGATCG|CCTGGAACGCGACACCCTGGATGATGATGGTGAACGCTTT|GAATTC|-3'
        40   30   20   10   1

FIGURE 3.   The DNA sequence of *pac*. The sequence of a 234-bp segment of *Eco*RI fragment 20 containing *pac* is shown. A minimal *pac* site is contained between nucleotide positions 71 and 232. The locations of *pac* cleavage termini are indicated by the arrows between nucleotide positions 137 and 148. The large arrows indicate preferred cleavage points. Seven hexanucleotide elements (TGATC$_G^A$) flank the cleavage termini and have been shown by an analysis of deletion mutants Δ3 to Δ7 and Δ20 to be essential for normal *pac* function. We propose that these elements are binding sites for the *pac* recognition proteins. They are also Dam methylation sites.

C. THE P1 *PAC* SITE

1. Localization of *Pac*

The experiments of Bächi and Arber[21] localized the unique end of the virion DNA, which is generated by cleavage at *pac,* to a position about 5 kb from the right end of the phage genetic map (Figure 1). In order to more precisely localize *pac,* P1 fragments flanking the segment of the P1 chromosome predicted by Bächi and Arber to contain *pac* were cloned either into phage λ DNA or pBR322 plasmid DNA and the resulting constructs placed in bacteria. Following P1 infection of those cells, *pac* cleavage of pBR322- or λ-*pac* substrates could be directly measured by Southern hybridization analysis of restriction enzyme-digested cellular DNA. The results of these experiments place *pac* within the 640-bp *Eco*RI-20 fragment of P1 DNA.[20] Moreover, analyses of a series of deletion mutations that remove DNA from both ends of this fragment define a minimal functional *pac* site of 161-bp located near the right end of *Eco*RI-20 and within the N-terminal portion of ORF 1. The sequence of one strand of the right end of *Eco*RI-20 containing the minimal *pac* site (the region between deletion mutants Δ3 and Δ20) is shown in Figure 3.

2. Position of *Pac* Termini

The 5' and 3' *pac* termini found at the ends of virion DNA are indicated by the arrows above and below the double-stranded region shown in Figure 3 and are located 140 to 150 bp from the *Eco*RI site that separates *Eco*RI-20 from *Eco*RI-22.[40] The termini are distributed over a turn of the DNA helix with every position represented, but with preferred cleavage sites (large arrows) separated by a half-turn of a helix on the strand that generates the 5' end. On the other strand, preferred cleavage sites are separated by 2 bp. Since *Eco*RI digests of P1 virion DNA produce two prominent double-stranded fragments with *pac* ends differing in size by about 5 bp, it is likely that those fragments contain the preferred 3' termini that are located between bases 141 and 142 and 147 and 148, and, therefore, that *pac* cleavage generates 2-bp 3'-protruding ends.

3. Other Features of *Pac*

The 161-bp minimal *pac* sequence contains seven hexanucleotide elements (TGATC$_G^A$), four near one end of the site and three near the other. An eighth element (element 1 in Figure 3) is located outside of the minimal *pac* sequence and is apparently not essential for *pac* cleavage. Each of the hexanucleotide elements contains the internal sequence, GATC, which is recognized and methylated by the *E. coli* DNA adenine methylase (Dam).

Another feature of *pac* is the presence of two regions containing adenine-rich clusters that are about one turn of the DNA helix apart and are separated by a GC-rich sequence. These regions are located between nucleotides 100 to 120 and between nucleotides 160 to 180. Sequences such as these have been implicated in the bending of DNA[41,42] and *pac* DNA may be bent since it migrates anomalously in acrylamide gels.[29]

4. A Mechanistic Model for *Pac* Cleavage — Involvement of the Hexanucleotide Elements, Dam Methylation, and DNA Bending

An analysis of deletion mutations that enter the minimal *pac* site from either side indicates that the hexanucleotide elements are important for *pac* cleavage.[40] A deletion of just one of those elements from either side of the site (e.g., deletion Δ5, Figure 3) reduces cleavage in the *in vivo* assay described above about tenfold. Moreover, the elements appear to be multiplicative in their effect on cleavage: removal of $1^1/_2$ or all of the elements from the right side of *pac* (deletions Δ6 and Δ7) reduces the efficiency of *pac* cleavage 100-fold and greater than 1000-fold, respectively. Exactly how these elements operate to facilitate *pac* cleavage is not known, but the results of *in vitro* experiments, in which proteins in extracts from P1-infected cells were shown to bind specifically to *pac*-containing DNA and to cleave it, shed some light on their role. Using a fragment gel retardation assay as a measurement of binding, it was shown that a *pac* deletion mutant missing nucleotides 189 to 231 (Figure 3), including elements 5 through 8, does not generate bound complexes in gels, but a mutant that is missing elements 2 through 4 (deletion Δ7; Figure 3) does generate those complexes.[29] Neither of these mutants is cleaved *in vivo* or *in vitro*.[29] The simplest interpretation of these results is that the hexanucleotide elements constitute binding sites for one or more *pac* recognition proteins (PRP), with each element facilitating binding to the cluster as a whole. The *in vitro* studies indicate that the cluster containing elements 5 through 8 can bind *pac* proteins in the absence of cluster 2 through 4, but that the reverse is not true. Since *pac* cleavage requires both clusters of elements, we suggest that the two clusters must interact through proteins that are bound to them in order for cleavage to occur.

A reasonable model (Figure 4) is one in which *pac* recognition proteins bind first to the cluster containing elements 5 through 8. This facilitates protein binding to the cluster containing elements 2 through 4 and, in turn, permits the two ends of the *pac* site to interact by protein-protein contacts. The latter interaction is required for *pac* cleavage, perhaps because it bends the DNA in the region where cleavage is destined to occur. Observations on the effects of Dam methylation and proposed *pac* bending sequences lend support to this hypothesis.

a. Adenine Methylation

Since each of the hexanucleotide elements in the minimal *pac* site contains a Dam methylation site, the effects of methylation on *pac* cleavage were assessed. The first experiments were carried out *in vivo* and indicate that the onset of *pac* cleavage in a *dam*⁻ host is delayed about 20 min relative to that observed in a *dam*⁺ host.[2,29] This is not due to a general effect of the *dam*⁻ phenotype on P1 expression since P1 lytic replication is normal in the *dam*⁻ host. At about the time that this experiment was performed, it was shown that P1 encodes its own Dam methylase and that this methylase starts to be produced about midway through the viral life cycle.[43] Considering all of these results, the conclusion

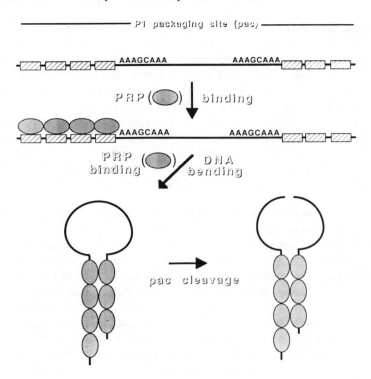

FIGURE 4.   A mechanistic model for *pac* cleavage. The pac site is shown at the top of the figure with its seven hexanucleotide elements (   ) that flank regions of proposed bent DNA (AAAGCAAA). It is postulated that the *pac* recognition proteins (PRP) bind first to the cluster of elements on the left (see text). This, in turn, facilitates binding to the elements on the right. Protein contacts between the two clusters bend the DNA between them and generate a cleavable substrate. Cleavage occurs centrally within the bend.

that emerges is that *pac* cleavage requires methylation of the site, with the source of that methylation being the phage-encoded enzyme at late times in the infection of a *dam⁻* host. To prove the point, an *in vitro pac* cleavage assay was employed using extracts from P1-induced lysogens. In that assay, unmethylated plasmid DNA containing *pac* (DNA prepared in a *dam⁻* host) is not cleaved when incubated with the extract unless it is premethylated by the *E. coli* Dam methylase. Measurements of the amount of methylation required for *pac* cleavage indicate that not all of the 14 GATC sequences (there are two in each of the hexanucleotide elements of the minimal *pac*) have to be methylated for *pac* cleavage. It was also shown that hemimethylated *pac* DNA (one strand fully methylated and the other fully unmethylated) is not cleaved. The cleavage results are, in part, a reflection of the binding properties of *pac*-containing DNA in that methylated DNA binds to proteins in the extract, but unmethylated DNA does not. However, the results of cleavage and binding experiments are not completely complementary since hemimethylated DNA, despite its failure to be cleaved, generates bound complexes with proteins in the extract that are indistinguishable from those found with fully methylated DNA. These results support the contention that the hexanucleotide elements have a role to play in facilitating the binding of *pac*-recognition protein(s) to *pac*, but that binding per se is not sufficient to permit cleavage.

b. Pac Bending Sequences

Two mutants with alterations in regions of *pac* that resemble DNA bending sequences (see Section IV.C.3) have been constructed.[29] In the mutants, the adenine-rich regions between nucleotides 110 and 116 or between nucleotides 166 and 170 were replaced by an *XhoI* linker (CCTCGAGG), generating *pac* sites with a net addition of only a single base to the total number of nucleotides between the two clusters of hexanucleotide elements. The two mutant *pac* sites behave similarly; they form protein-DNA complexes that are indistinguishable from those seen with the wild-type site, but they are not cleaved efficiently. This result supports the contention that DNA bending may be essential for *pac* cleavage, but is not essential for the generation of bound complexes. Presumably, the sequence alterations make it difficult to form the bent structure (Figure 4) needed to create a cleavable substrate. Direct evidence for this point is still lacking.

5. Regulation of *Pac* Cleavage

Several observations lead to the conclusion that *pac* cleavage must be regulated during the phage life cycle if P1 is to produce a normal yield of infectious particles. First, in order for virion DNA to produce a productive infection after its injection into sensitive cells, it must be cyclized. Cyclization is achieved in P1 by homologous recombination between the terminally redundant ends of the virion DNA.[9,44] For P1 to generate phage with terminal redundancy, the packaging system must avoid cutting each *pac* site on the concatemer since that would generate monomer-length DNA. Ideally, the concatemer should be cut once at a *pac* site near the end of the concatemer and packaging should move down the concatemer, bypassing uncleaved *pac* sites and packaging DNA by the headful. Since packaging is unidirectional, if the *pac* site chosen for cleavage were in the middle of the concatemer, the process would be wasteful. Without a means of regulating *pac* cleavage, P1 would have difficulty establishing such a packaging regimen. Most problematical is the fact that *pac* cleavage functions are synthesized early during the phage growth cycle, probably before phage heads are made.[20] If those functions could act at that time, it is difficult to imagine how each *pac* site in the cell could avoid being cut before the DNA is packaged.

The methylation requirements for *pac* binding and cleavage provide the basis for a model (Figure 5) that can account for regulated *pac* cleavage in the cell. According to this model, newly replicated *pac* sites present on P1 rolling circle concatemers early during phage infection are poor substrates for the cleavage reaction because they are hemimethylated (methylation lags behind replication[45]). We postulate that *pac* recognition proteins (PRP) bind to these hemimethylated sites more avidly then does the bacterial Dam methylase and protects them from being activated by methylation. This generates a rolling circle whose tail contains multiple inert *pac* complexes. While this is a desirable state early during the phage growth cycle, it is undesirable later in the cycle when P1 DNA needs to be packaged. One can think of two ways to increase the probability that a *pac* site will become fully methylated late in the phage growth cycle. The first is simply to imagine that some fraction of the newly replicated hemimethylated *pac* sites will be methylated by the host methylase before the *pac*-recognition proteins bind. If, for example, only one in five newly replicated *pac* sites is methylated before *pac* cleavage proteins bind, then on average the P1 concatemers will contain 3 to 4 monomer units before a *pac* site is cleaved to initiate a processive headful packaging process. Note that this also ensures that the *pac* site chosen for cleavage will be at the end of the concatemer. Alternatively, one could imagine that the host methylase is hardly ever able to methylate newly replicated *pac* sites before they are protected and that the ability to activate those sites late in P1 infection is due to the production of the new P1-encoded Dam methylase. That enzyme should be quite abundant late in the infectious cycle since there are many copies of the gene in the cell at that time. Accordingly, it should be able to compete favorably with the *pac*-recognition proteins for hemimethylated *pac* sites, increasing the probability that those sites will be methylated and able to initiate a packaging series. This model makes the prediction that uncleaved *pac* DNA isolated from P1 virions is a poor substrate for *in vitro* cleavage because it is undermethylated. This prediction is in agreement with recent results.

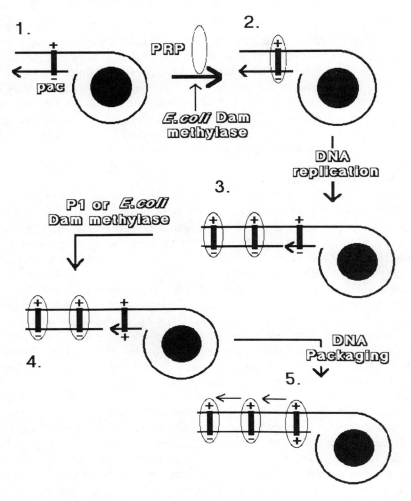

FIGURE 5.   The regulation of *pac* cleavage. We propose that when *pac* is replicated onto the tail of a DNA rolling circle (step 1), the template *pac* strand is fully methylated (+), but the newly replicated strand is unmethylated (−). The resulting hemimethylated site is competed for by the host Dam methylase and the *pac* recognition protein (PRP), and if PRP binds first, it protects the site from further methylation (step 2). The resulting complexes are not cleaved. Continued replication (step 3) increases the length of the tail of the rolling circle and the number of inert *pac*-PRP complexes until a newly replicated site is methylated before PRP binds (step 4). PRP binding to the fully methylated site is followed by *pac* cleavage and the initiation of a series of headful packaging events that move down the tail of the concatemer (step 5).

V. THE PACKAGING OF HOST DNA INTO GENERALIZED TRANSDUCING PHAGE

Since the initial discovery by Lennox[46] of P1-mediated generalized transduction, little progress has been made in understanding how the virion packages host DNA and whether that packaging process differs from the packaging of phage DNA.

A. THE DNA IN TRANSDUCING PARTICLES

Measurements of the source of the DNA in generalized transducing particles by density labeling were interpreted by Ikeda and Tomizawa[8] to mean that transducing particles lack significant amounts of phage DNA (<5%) and contain only fragments of the bacterial chromosome that existed prior to the time of infection. They were also able to show that transducing particles comprise about 0.3% of the total phage particles and that the molecular weight of the DNA in infectious and transducing phage is about the same. A surprising observation was the demonstration that the density of DNA in transducing particles is less than that of bacterial DNA, but could increase to that of bacterial DNA by treatment with proteolytic enzymes.[14] The density of the DNA of infectious particles is unaffected by that treatment. It was concluded that the DNA of transducing particles, but not infective particles, contains an associated protein. Further studies indicate that the protein is covalently linked to the DNA, is not evenly distributed over the DNA molecule, and is not associated with transducing DNA derived from small-headed particles. It was estimated that the total amount of protein bound to DNA is about 500 kDa. Neither the nature of the protein nor the reason for its presence only in transducing particles has ever been clarified. However, it probably plays a role in the cyclization of at least some of the transducing DNA following particle infection since much of that DNA can be found in a protease-sensitive ring in abortively transduced cells.[47,48] This probably accounts for its long-term stability in cells and, therefore, our ability to detect abortive transductants. Since the DNA of plaque-forming P1 particles does not have this protein associated with it, it may be that the packaging of host and phage DNA occurs by different mechanisms.

B. THE INITIATION OF PACKAGING

Of the two possibilities for the initiating of host DNA packaging — nonselective initiation or selective initiation — the former appears to be the more likely for P1. Direct measurements of the relative abundance of bacterial DNA in transducing particles by Southern blot hybridization indicates that, except for DNA from the 2-min region of the *E. coli* chromosome (including the *leu* gene), all sequences are present in approximately equal amounts.[49] Moreover, the nearly 30-fold difference in transduction frequencies exhibited by different chromosomal markers could be accounted for by differences in the stabilization of the transduced DNA once it is injected into the recipient cell.[50,51] These studies show that the packaging of chromosomal DNA into P1 heads is nonselective and, therefore, probably doesn't involve the initiation of packaging at a *pac* or pseudo-*pac* site in the chromosome. However, the results do not rule out the possibility that a much smaller recognition sequence than *pac,* such as a single *pac* hexanucleotide (TGATC$_G^A$) element, directs the packaging of DNA into transducing particles.

C. PROCESSIVITY

To address the questions of processivity in the packaging of host DNA, Harriman[13] carried out a series of elegant studies in which the yield of transducing particles was measured in a single phage burst cycle using prophages as chromosomal markers to increase the efficiency of scoring transduction. The results indicate that regions on the order of 20% of the *E. coli* chromosome (10 P1 headfuls) are packaged into phage by a processive headful mechanism. They also show that cells producing transducers frequently contain several independent processive packaging events, i.e., cells producing transducers are special in that they are prone to initiate processive packaging events. Harriman concluded that his results could be explained if one tenth of the infected cells contain several regions of localized packaging of 20% of the chromosome.

D. MUTANTS THAT AFFECT THE PRODUCTION OF GENERALIZED TRANSDUCING PHAGE

High-frequency transducing mutants of P1 were isolated by Wall and Harriman[52] and were shown to transduce all chromosomal markers 5 to 10 times more efficiently than does the parental P1. Sandri and Berger[47] showed that this was due to an increase in the packaging of the bacterial chromosome into phage in the donor cell. Since one of the mutants (an amber) is suppressible, the increased packaging frequency would appear to be due to the loss of a phage function. Failure to map these mutants has prevented any possible correlation of mutant phenotype with a known P1 packaging function. A second class of mutants that affects transduction maps in the *dar*A operon (Figure 1) and results in an increased transduction frequency. The mutation associated with this phenotype is a deletion of the *dar*A operon and the function responsible has been designated *gta*.

VI. COMPARATIVE BIOLOGY: P22 VS. P1

Bacteriophage P22 also packages DNA by a processive headful mechanism and is a generalized transducer. However, the initiation of P22 packaging appears to differ significantly from that of P1. First, the P22 *pac* recognition sequence appears to be significantly smaller than that of P1, perhaps no more than 12 bp.[53,54] Second, the *pac* termini are distributed over a much larger region of the P22 DNA than are the P1 *pac* termini.[53,54] Third, the cleavage of the P22 *pac* site is probably not regulated *in vivo* like the cleavage of the P1 *pac* site, although a firm conclusion on this point must await the development of a P22 *in vitro* cleavage reaction. Fourth, while the synthesis of *pac*-processing proteins are coordinately regulated with synthesis of phage head proteins during the P22 growth cycle,[55,56] the two classes of protein are not coordinately regulated in the P1 growth cycle. Fifth, unlike P1, the packaging of bacterial DNA by P22 appears to be initiated at several unique chromosomal sites.[57-59] Clearly, the two phage systems have devised very different ways of carrying out the same process.

REFERENCES

1. **Sternberg, N. and Hoess, R.,** The molecular genetics of bacteriophage P1, *Annu. Rev. Genet.,* 17, 123, 1983.
2. **Yarmolinsky, M. B. and Sternberg, N.,** Bacteriophage P1, in *The Bacteriophages,* Plenum Press, New York, 1987, chap. 9.
3. **Masters, M.,** Generalized transduction, in *The Genetics of Bacteria,* Scaife, J. G., Leach, D., and Galizzi, A., Eds., Academic Press, New York, 1985, 197.
4. **Anderson, T. F. and Walker, D. H., Jr.,** Morphological variants of the bacteriophage P1, *Science,* 132, 1488, 1960.
5. **Walker, D. H., Jr. and Anderson, T. F.,** Morphological variants of coliphage P1, *J. Virol.,* 5, 765, 1970.
6. **Walker, J. T. and Walker, D. H., Jr.,** Coliphage P1 morphogenesis. Analysis of mutants by electron microscopy, *J. Virol.,* 45, 1118, 1983.
7. **Ikeda, H. and Tomizawa, J.-I.,** Transducing fragments in generalized transduction by phage P1. III. Studies with small phage particles, *J. Mol. Biol.,* 14, 120, 1965.
8. **Ikeda, H. and Tomizawa, J.-I.,** Transducing fragments in generalized transduction by phage P1. I. Molecular origin of the fragments, *J. Mol. Biol.,* 14, 85, 1965.
9. **Ikeda, H. and Tomizawa, J.-I.,** Prophage P1, an extrachromosomal replication unit, *Cold Spring Harbor Symp. Quant. Biol.,* 33, 791, 1968.
10. **Yun, T. and Vapnek, D.,** Electron microscopic analysis of bacteriophage P1, P1Cm and P7. Determination of genome sizes, sequence homology and location of antibiotic resistance determinants, *Virology,* 77, 376, 1977.
11. **Walker, J. T., Iida, S., and Walker, D. H., Jr.,** Permutation of the DNA in small-headed virions of coliphage P1, *Mol. Gen. Genet.,* 167, 341, 1979.
12. **Walker, J. T. and Walker, D. H., Jr.,** Structural proteins of coliphage P1, in *Progress in Clinical and Biological Research,* Vol. 64, DuBow, M. S., Ed., Alan R. Liss, New York, 1981, 69.

13. **Harriman, P. D.,** A single-burst analysis of the production of P1 infectious and transducing particles, *Virology,* 48, 595, 1972.
14. **Ikeda, H. and Tomizawa, J.-I.,** Transducing fragments in generalized transduction by phage P1. II. Association of DNA and protein in the fragments, *J. Mol. Biol.,* 14, 110, 1965.
15. **Yarmolinsky, M.,** Bacteriophage P1, in *Genetic Maps,* Vol. 4, O'Brien, S. J., Ed., Cold Spring Harbor Laboratory, Cold Spring Harbor, NY, 1987, 38.
16. **Walker, D. H., Jr. and Walker, J. T.,** Genetic studies of coliphage P1. I. Mapping by use of prophage deletions, *Virology,* 16, 525, 1975.
17. **Walker, D. H., Jr. and Walker, J. T.,** Genetic studies of coliphage P1. III. Extended genetic map, *J. Virol.,* 20, 177, 1976.
18. **Scott, J. R.,** Genetic studies on bacteriophage P1, *Virology,* 36, 564, 1968.
19. **Streiff, M. B., Iida, S., and Bickle, T. A.,** Expression and proteolytic processing of the *dar*A anti-restriction gene product of bacteriophage P1, *Virology,* 157, 167, 1987.
20. **Sternberg, N. and Coulby, J.,** Recognition and cleavage of the bacteriophage P1 packaging site *(pac).* I. Differential processing of the cleaved ends *in vivo, J. Mol. Biol.,* 194, 453, 1987.
21. **Bächi, B. and Arber, W.,** Physical map of *Bgl*II, *Bam*HI, *Eco*RI, *Hind*III, and *Pst*I restriction fragments of bacteriophage P1 DNA, *Mol. Gen. Genet.,* 153, 311, 1977.
22. **Krüger, D. H. and Bickle, T. A.,** Bacterial survival: multiple mechanisms for avoiding the deoxyribonucleic acid restriction systems of their hosts, *Microbiol. Rev.,* 47, 345, 1983.
23. **Iida, S., Streiff, M. B., Bickle, T. A., and Arber, W.,** Two DNA anti-restriction systems of bacteriophage P1, *dar*A and *dar*B: characterization of *dar*A⁻ phage, *Virology,* 157, 156, 1987.
24. **Iida, S. and Arber, W.,** Plaque forming specialized transducing phage P1: isolation of P1CmSmSu, a precursor of P1Cm, *Mol. Gen. Genet.,* 153, 259, 1977.
25. **Iida, S. and Arber, W.,** On the role of IS1 in the formation of hybrids between bacteriophage P1 and the R plasmid NR1, *Mol. Gen. Genet.,* 177, 261, 1980.
26. **Walker, D. H. and Walker, J. T.,** personal communication, 1987.
27. **Hanson, E.,** Structure and regulation of the lytic replicon of Phage P1, *J. Mol. Biol.,* 207, 135, 1989.
28. **Cohen, G. and Sternberg, N.,** Genetic analysis of the lytic replicon of bacteriophage P1. I. Isolates and partial characterization, *J. Mol. Biol.,* 207, 99, 1989.
29. **Sternberg, N. and Cohen, G.,** Genetic analysis of the lytic replicon of bacteriophage P1. II. Organization of replicon elements, *J. Mol. Biol.,* 207, 111, 1989.
30. **O'Regan, G. T., Sternberg, N. L., and Cohen, G.,** Construction of an ordered, overlapping library of bacteriophage P1 DNA in λD69, *Gene,* 60, 129, 1987.
31. **Scott, J. R.,** Clear plaque mutants of phage P1, *Virology,* 41, 66, 1970.
32. **Rosner, J. L.,** Formation, induction and curing of bacteriophage P1 lysogens, *Virology,* 48, 679, 1972.
33. **Segev, N., Laub, A., and Cohen, G.,** A circular form of bacteriophage P1 DNA made in lytically infected cells of *Escherichia coli.* I. Characterization and kinetics of formation, *Virology,* 101, 261, 1980.
34. **Cohen, G.,** Electron microscopy study of early lytic replication forms of bacteriophage P1 DNA, *Virology,* 131, 159, 1983.
35. **Bornhoeft, J. W. and Stodolsky, M.,** Lytic cycle replicative forms of bacteriophage P1 and P1*dl* concatemers, *Virology,* 112, 581, 1981.
36. **Segev, N. and Cohen, G.,** Control of circularization of bacteriophage P1 DNA in *Escherichia coli, Virology,* 114, 333, 1981.
37. **Streisinger, G., Emrich, J., and Stahl, M. M.,** Chromosome structure in phage T4. III. Terminal redundancy and length determination, *Proc. Natl. Acad. Sci. U.S.A.,* 57, 292, 1967.
38. **Tye, B. K., Chan, R. K., and Botstein, D.,** Packaging of an oversize transducing genome by *Salmonella* phage P22, *J. Mol. Biol.,* 85, 485, 1974.
39. **Casjens, S. and Hayden, M.,** Analysis *in vivo* of the bacteriophage P22 headful nuclease, *J. Mol. Biol.,* in press.
40. **Sternberg, N. and Coulby, J.,** Recognition and cleavage of the bacteriophage P1 packaging site *(pac).* II. Functional limits of *pac* and location of *pac* cleavage termini, *J. Mol. Biol.,* 194, 469, 1987.
41. **Wu, H.-M. and Crothers, D. M.,** The locus of sequence-directed and protein-induced DNA bending, *Nature (London),* 308, 509, 1984.
42. **Koo, H.-S., Wu, H.-M., and Crothers, D. M.,** DNA bending at adenine thymine tracts, *Nature (London),* 320, 501, 1986.
43. **Coulby, J.,** personal communication, 1987.
44. **Sternberg, N., Hamilton, D., Austin, S., Yarmolinsky, M., and Hoess, R.,** Site-specific recombination and its role in the life cycle of bacteriophage P1, *Cold Spring Harbor Symp. Quant. Biol.,* 45, 297, 1980.
45. **Marinus, M. G.,** DNA methylation in *Escherichia coli, Annu. Rev. Genet.,* 21, 113, 1987.
46. **Lennox, E. S.,** Transduction of linked genetic characters of the hosts by bacteriophage P1, *Virology,* 1, 190, 1955.

47. **Sandri, R. M. and Berger, H.,** Bacteriophage P1-mediated generalized transduction in *Escherichia coli:* fate of transduced DNA in *rec⁺* and *rec⁻* recipients, *Virology,* 106, 14, 1980.

48. **Sandri, R. M. and Berger, H.,** Bacteriophage P1-mediated generalized transduction in *Escherichia coli:* structure of abortively transduced DNA, *Virology,* 106, 30, 1980.

49. **Hanks, M. and Masters, M.,** personal communication, 1987.

50. **Newman, B. J. and Masters, M.,** The variation in frequency with which markers are transduced by phage P1 is primarily a result of discrimination during recombination, *Mol. Gen. Genet.,* 180, 585, 1980.

51. **Masters, M., Newmann, B. J., and Henry, C. M.,** Reduction of marker discrimination in transductional recombination, *Mol. Gen. Genet.,* 196, 85, 1984.

52. **Wall, J. D. and Harriman, P. D.,** Phage P1 mutants with altered transducing abilities for *Escherichia coli, Virology,* 59, 532, 1974.

53. **Casjens, S. and Huang, W. M.,** Initiation of sequential packaging of bacteriophage P22 DNA, *J. Mol. Biol.,* 157, 287, 1982.

54. **Casjens, S., Huang, W., Hayden, M., and Parr, R.,** Initiation of bacteriophage P22 DNA packaging series. Analysis of a mutant that alters the DNA target specificity of the packaging apparatus, *J. Mol. Biol.,* 194, 411, 1987.

55. **Botstein, D., Waddell, C. H., and King, J.,** Mechanisms of head assembly and DNA encapsidation in *Salmonella* phage P22. I. Genes, proteins, structures, and DNA maturation, *J. Mol. Biol.,* 80, 699, 1973.

56. **Susskind, M. M. and Botstein, D.,** Molecular genetics of bacteriophage P22, *Microbiol. Rev.,* 42, 385, 1978.

57. **Chelala, C. A. and Margolin, P.,** Effects of deletions on co-transduction linkage in *Salmonella typhimurium.* Evidence that bacterial chromosome deletions affect the formation of transducing DNA fragments, *Mol. Gen. Genet.,* 131, 97, 1974.

58. **Schmieger, H.,** Packaging signals for phage P22 on the chromosome of *Salmonella typhimurium, Mol. Gen. Genet.,* 187, 516, 1982.

59. **Schmieger, H. and Backhaus, H.,** Altered co-transduction frequencies exhibited by HT-mutants of *Salmonella* phage P22, *Mol. Gen. Genet.,* 143, 307, 1976.

Chapter 11

BACTERIOPHAGE P22 DNA PACKAGING

Sherwood Casjens

TABLE OF CONTENTS

I. INTRODUCTION

Many dsDNA virus particles contain a single, very compact nucleic acid molecule within a protein coat (Table 1). Ever since this structural arrangement was deduced, molecular biologists have been interested in understanding how these particles are built *in vivo,* and the dsDNA tailed bacteriophages have been a particularly useful system in which to study the problem. P22 is a member of this supergroup that was first studied by Zinder and Lederberg in the early 1950s. Its ability to lysogenize its host and its program of gene expression have made it an interesting subject for a variety of molecular biological studies, and its ability to perform generalized transduction has made it a very useful genetic tool in the study of its host, *Salmonella typhimurium.* P22 can be considered to be a member of the lambdoid phage family in that its early genes and program of gene expression are quite analogous to those of phage lambda — that is, two divergent early operons express early genes that control the program of gene expression, the "lytic growth versus establishment of lysogeny" decision, DNA replication, and genetic recombination. The late operon encodes all of the proteins required for assembling the progeny virions and packaging DNA, but it is not homologous with that of lambda. The P22 life cycle has been reviewed by Susskind and Botstein,[5] Susskind and Youderian,[6] and Poteete.[7]

II. THE P22 VIRION

The P22 virion contains nine different protein species and a single linear chromosome (rather simple for a dsDNA tailed bacteriophage). The particle has a hexagonal outline in electron micrographs,[8,9] is about 60 nm in diameter,[9,10] and has a short hexagonal tail structure (Figure 1a and b).[9-12] The coat protein shell is composed of one protein, gp5,* arranged in a T = 7 icosahedral lattice.[14,15] The bulk of the short tail structure is composed of the product of gene 9,[11,16] which is responsible for attachment to susceptible cells before ejection.[12,17] Four less abundant proteins (gp1, gp4, gp10, and gp26; 2 to 20 molecules of each) are located at the point of tail attachment.[18] In addition, three minor proteins whose virion locations are not known, gp7, gp16, and gp20, are required for successful DNA ejection by these particles.[16,19-21] Of these seven minor proteins, only gp1 is required for successful DNA condensation within the coat protein shell.[16]

Although many linear, dsDNA virus chromosomes have unique ends, several virus types, the iridoviruses and some tailed bacteriophages, do not have unique termini (reviewed by Casjens[3]). P22 is a member of the latter group. Its virions contain linear dsDNA chromosomes about 43,400 bp long, which are circularly permuted, about 3.8% terminally redundant,[22-26] and are thought to have blunt ends.[27] Thus, all virions contain dsDNA molecules with the same sequence information, but different particles contain different circular permutations of the information. The DNA (in B-form) is packed very tightly within the P22 phage coat protein shell, at approximately the same density as is found in crystals of short dsDNA molecules — about 1.5 g water per gram DNA (see Table 1).[2,10,28] There appear to be no proteins analogous to histones bound along the length of the packaged DNA molecule. The actual arrangement of the intravirion DNA is not known in detail. Solution X-ray diffraction and transient electric dichroism studies indicate that there is considerable order in the DNA and that helix segments run locally parallel.[10,29] It has been suggested that in dsDNA phage virions, the nucleic acid is arranged as a solenoid,[28-31] in a folded structure,[32,33] or in a partially ordered nematic liquid crystal.[34] When P22 DNA is chemically cross-linked to the shell proteins and then the structure is partially disrupted, supertwisted loops of DNA are released, suggesting that the DNA in the head contains topological

---

* gpX refers to the protein gene product of gene X.[13]

TABLE 1
Approximate Double-Stranded Nucleic Acid Packing Densities in
Virus Particles

Virus	Common hosts	Outside diameter (nm)	~bp/100 nm³
dsDNA			
Papovaviruses[a]	Mammals	40—55	12
Adenoviruses[a]	Mammals, birds	70—75	18
Herpesviruses	Vertebrates	60—80	57
Iridoviruses[b]	Insects, vertebrates	130—160	11
Tailed-bacteriophages	Eubacteria, archaebacteria		
Podoviruses[b] (P22, T7, φ29)		60	56
Myoviruses[b] (T4)		65 × 95	57
Styloviruses[b] (lambda)		60	56
Calimoviruses	Plants	50	12
Hepadnaviruses	Mammals, birds	28	26
Poxviruses	Mammals, birds	200 × 300	2
Baculoviruses	Insects	65 × 280	16
Tectiviruses (phage PRD1)	Eubacteria	65	25
Corticoviruses (phage PM2)	Eubacteria	60	8
dsRNA			
Reoviruses[b]	Mammals, insects, plants	60—70	47
Birnaviruses	Fish, birds, insects	64	5
Cystovirus (phage φ6)	Eubacteria	50	24
dsDNA crystals	—	—	50—60
dsDNA in solution[c]	—	—	0.008

Note: Where diffraction data on nucleic acid density or capsid inside diameter are available they were
used. Where such data are not available, calculations assume all the space within the virion (or
core, where appropriate) is available to the nucleic acid, and so are likely to underestimate the
actual nucleic acid density. See References 2 and 3 for structural references and Reference 4 for
a discussion of virus taxonomic groups.

[a]   Known to have basic proteins bound along the length of the virion DNA.
[b]   X-ray diffraction measurements of DNA density or capsid diameter available.
[c]   The state of condensation of unpackaged DNA *in vivo* is not accurately known. The value given was
calculated for the bacteriophage lambda chromosome (48.5 kbp) within the interior volume of its *E.
coli* host (~6 × 10⁸ nm³). This represents the theoretical maximally decondensed state for the phage
DNA.

supertwists (Figure 2).[35] In P22, it is not known if the DNA ends occupy a special position
in the virion, but in several other dsDNA phages the last end to enter the particle during its
assembly has been shown to be in contact with the tail.[36-40]

III. MECHANISM OF P22 DNA PACKAGING

A. GENERAL PATHWAY OF DNA PACKAGING

DNA packaging in a number of tailed bacteriophage systems has been studied in some
detail and all are known to first assemble a protein particle, called a *prohead,* and then insert
the nucleic acid into this structure (called headful packaging;[41] reviewed in References 2,
3, and 42). The proteins required for phage P22 assembly were identified and the general
pathway of the assembly process was determined by Botstein, King, and co-workers about
a decade ago.[11,16,21,43,44] Since then, many details of individual steps have been elucidated.
P22 DNA packaging requires only five phage-encoded proteins. Three of these, gp*1,* gp*5,*
and gp*8,* are structural components of proheads, and two, gp*2* and gp*3,* are transiently

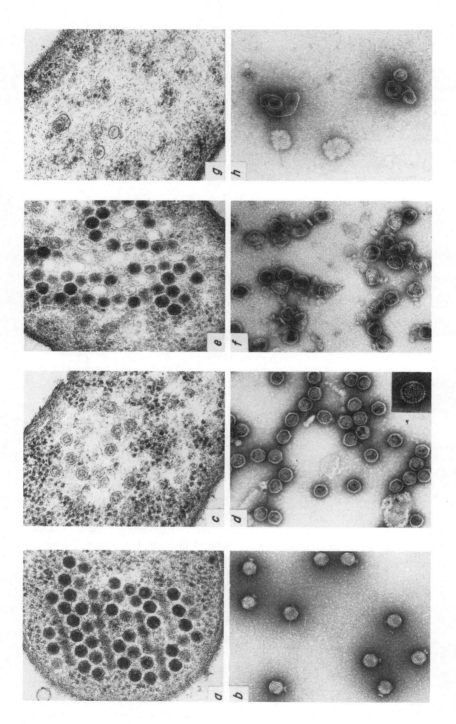

FIGURE 1.   Electron microscopy of bacteriophage P22 and related structures. The upper panels are ultrathin sections of infected cells stained with osmium tetroxide and uranyl acetate, and the lower panels are the isolated particles negatively stained with uranyl acetate. (a,b) Normal virons; (c,d) proheads produced by a P22 gene 2 defective infection; (e,f) emptied heads from a P22 gene 26 defective infection; (g,h) aberrant coat protein structures from a scaffolding protein-defective infection. (From Casjens, S. and King, J., *J. Supramol. Struct.*, 2, 202, 1974. With permission.)

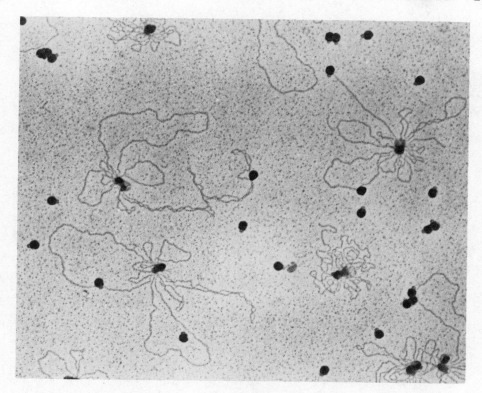

FIGURE 2.   Partially disrupted P22 virions. Phage particles were treated with formaldehyde before Pt-Pd shadowing. About one fifth of the particles were disrupted. Most disrupted particles show at least one supercoiled, extruded loop. About 40 to 60% of the chromosome is extruded in the particles shown.

required during the DNA packaging reaction; none of these proteins are known to undergo any covalent changes during the assembly process.[11,16,43] Figure 3 summarizes the DNA packaging portion of the P22 assembly pathway as it is currently understood. The packaging reaction is quite complex; the bacteriophage DNA must be recognized and caused to enter the prohead, the scaffolding protein leaves the prohead, the coat protein shell expands, and gp2 and gp3 must act. The genes which encode these five proteins lie in a contiguous cluster in the late operon (Figure 4). After DNA is condensed within the capsid, three additional phage-encoded proteins, gp4, gp10, and gp26, stabilize the structure.[11,16] These and the other late operon genes are coordinately expressed at late times after infection from a single promoter.[46,47] Since they are synthesized simultaneously, their order of action in the assembly pathway is dictated by the protein molecules themselves (see References 48 to 50).

B. DNA METABOLISM IN PACKAGING

Newly injected P22 DNA is circularized by homologous recombination between the terminally redundant ends[51] and is eventually replicated into head-to-tail concatemeric structures by rolling circle replication.[11,52] These concatemers serve as the substrate for DNA packaging, and mature chromosomes generated during packaging are linear, circularly permuted, and terminally redundant.[22] Tye et al.[53] first found that, in fact, the circular permutation is not random in that most chromosome ends fall within a 15 to 25% fraction of the genome. These studies and subsequent work[26,54-56] strongly support the model shown in Figure 5, in which multiple DNA packaging events occur in a "processive" series on each DNA concatemer. Such packaging *series* initiate at a specific location, called the *pac* site,

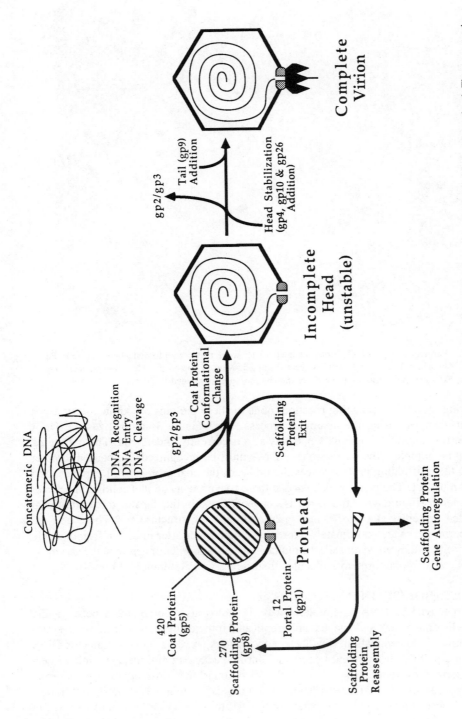

FIGURE 3.   Bacteriophage P22 DNA packaging pathway. The numbers above the prohead protein components give the molecules per prohead. The coat protein shells of all three particles have T = 7 geometry, which implies exactly 420 molecules, but it is likely that the portal structure replaces some of them. Most likely, 5, gp 7, gp 14, gp 16, and gp 20, to 15 coat protein molecules are replaced, so there are probably 415 to 405 molecules of gp 5 per prohead and virion. Four additional phage-encoded proteins (not shown) are required for virion infectivity, but are not required for normal DNA packaging.

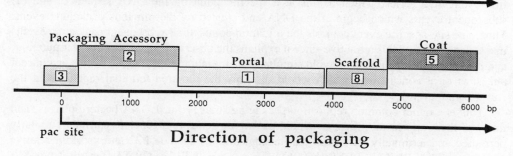

FIGURE 4.   Bacteriophage P22 DNA packaging genes. The gene numbers are indicated in the rectangles that represent each gene.[45]

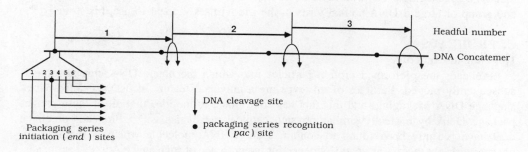

FIGURE 5.   Sequential P22 DNA packaging series. Three head-to-tail genome repeats of concatemeric phage DNA are shown. The first packaging event (1) begins with the recognition of a *pac* site (●), followed by condensation within the prohead rightward from that site. After a headful of DNA (103.8% of the genome) has been packaged, the DNA is cut (↓) from the concatemer by the headful nuclease. The terminal redundancy and headful cleavage regions are magnified here for clarity. Subsequent packaging events proceed rightward from the end created by the previous packaging event. Below, the series initiation region is expanded and shows schematically that individual packaging series begin at one of six *end* regions or sites.

on the P22 DNA sequence. Initiation of a packaging series is complex and requires gp*2* and gp*3*.[56-60]

Before or during initiation of packaging series, the left ends of the first members of packaging series are generated. These ends fall in a 120-bp region within gene *3*.[27,61] After initiation, according to the packaging series model, DNA insertion into the prohead proceeds unidirectionally (rightward on the standard P22 map) from that point along the concatemer until about 103.8% of a genome length of DNA is encapsidated. Then a nuclease, called the "headful nuclease", cleaves the DNA, releasing the packaged DNA from the concatemer. Subsequent, similar packaging events then occur which initiate at the concatemer end generated by the previous packaging event. The concatemer ends thus generated are not detectably "nibbled" before the subsequent event, suggesting that these ends might be protected in some way.[26] Such packaging series normally average 2.5 to 5 events in length, but can under some conditions be up to 12 events long.[23,26,53-55] Thus, chromosome ends are normally found mainly within a one quarter region of the genome sequence since series are not long enough to move the endpoints more than about $5 \times 4\% = 20\%$ from the initiation site. The series model for DNA handling by the packaging apparatus has been very strongly supported by the analysis of mature DNA isolated from wild-type virions and ones which carry deletion mutations[23,26,54] by the fact that average series length increases with time after infection[55] and by the observation that bacterial chromosomal deletions can affect the transduction of distant markers (see below).[62,63]

Note that there are two types of recognition events in such a packaging scheme: (1) the first event of a series, which begins at a specific point on the DNA sequence, and (2) subsequent events which begin at the DNA end created by the previous packaging event. Thus, one site-specific event controls the initiation points of all subsequent packaging events in a series. The model is attractive since it explains the observed *partial* circular permutation, and it also enables the phage to avoid simultaneous initiation of packaging events at adjacent *pac* sites on a concatemer (which would disallow the required terminal repetition in the "upstream" event).

This is a rather common replication/packaging strategy for dsDNA bacteriophages, and many of them appear to utilize sequential packaging series that generate partially circularly permuted and terminally redundant chromosomes; these include P22 and its close relative phage L,[64] T1,[65] P1,[66,67] SPP1,[68] φ5006M,[69] α3,[70] φH,[71] T12,[72] CP-T1,[73] φ149,[74] Mx-8,[75] SF1,[76] φ138,[77] MB78,[78] and φSE6.[79] These phages, which include both lytic and lysogenic phages, infect a number of host bacterial species, both Gram-negative and -positive, and their DNA packaging proteins have no obviously close evolutionary relationship. In addition, one group of large dsDNA animal viruses, the iridoviruses, appear to use this strategy.[80]

C. PROHEADS

1. Prohead Structure

Proheads are preformed protein particles into which the phage DNA chromosome is subsequently moved. Particles of this type are a universal feature of dsDNA tailed-bacteriophage DNA packaging, and this fact has contributed to the idea that all of these phages package DNA by basically similar physicochemical mechanisms.[2,3,42] Proheads of phage φ29 have recently been found to contain a short RNA molecule which is required for packaging,[81] but there is as yet no evidence for the presence of such an RNA in P22 proheads. P22 proheads contain three essential structural proteins, the products of genes *5*, *8*, and *1*. The *coat protein* that forms the icosahedral shell in both proheads and virions is gp5. The prohead outer shell structure, like virions, has T = 7 geometry[14,15] and contains about 420 molecules of gp5. The other major protein component of the prohead is gp8, or *scaffolding protein*. There are 275 ± 25 molecules of gp8 in the interior of each prohead (Figure 6).[10,43,82] The structural arrangement of these gp8 molecules within the prohead is unknown, although it has been argued that they may not be arranged with icosahedral symmetry.[50] King, Botstein, and co-workers[11,16] first found that the gp8 is removed from the prohead during DNA packaging and that proper prohead assembly fails in its absence. It was named scaffolding protein since it is needed to build the proheads, but is removed later. King and Casjens[82] showed that P22 scaffolding protein is, in fact, recycled during assembly, and in lysis-inhibited infections each gp8 molecule participates in an average of 5 to 6 prohead assembly events. The third prohead structural protein is gp1. It is present at one location in the shell (the icosahedral corner to which tails will eventually attach) as a ring-shaped dodecamer (Figure 7).[18,43,83] This protein has been called the *portal protein* because it is thought to form the portal through which DNA enters the prohead during packaging and leaves during ejection.[42] Solution X-ray scattering data have shown that the scaffolding protein is in the interior of the prohead and that in proheads the coat protein shell is thicker and about 11% smaller in diameter than in phage particles (Figure 6).[10]

2. Prohead Assembly

Assembly of the prohead is not well understood. Experiments done *in vivo* with conditional lethal mutant phages have shown that in the absence of coat protein, neither scaffolding protein cores nor portal structures assemble.[43,82-84] In the absence of portal protein, apparently normal proheads assemble that lack the portal protein and are unable to package DNA.[11] When scaffolding protein is missing, coat protein assembles mainly into aberrant

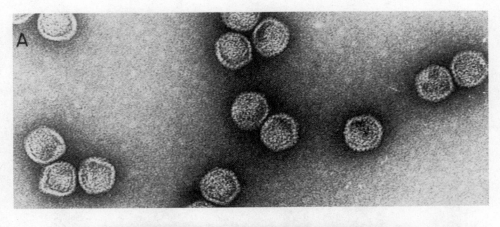

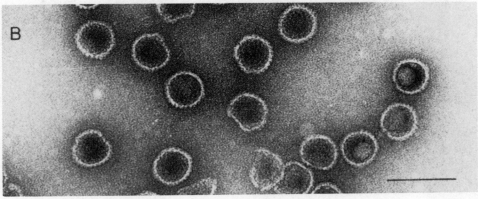

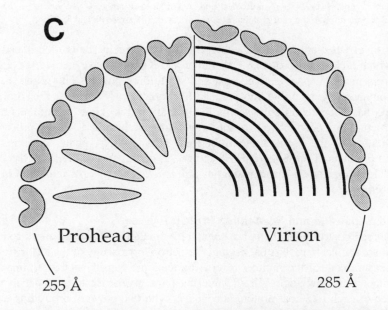

FIGURE 6.   P22 prohead structure. (A) Electron micrographs of negatively stained proheads; (B) proheads treated with 0.8% SDS at room temperature which have lost their scaffolding protein. The bar represents 100 nm. (C) Schematic structures of a P22 virion and a prohead. The numbers of outer coat protein and internal scaffolding protein molecules, and the DNA (curved bold lines) spacing are diagrammatic and not meant to be accurate. (Panels A and B from Casjens, J. and King, J., *J. Supramol. Struct.*, 2, 202, 1974. With permission.)

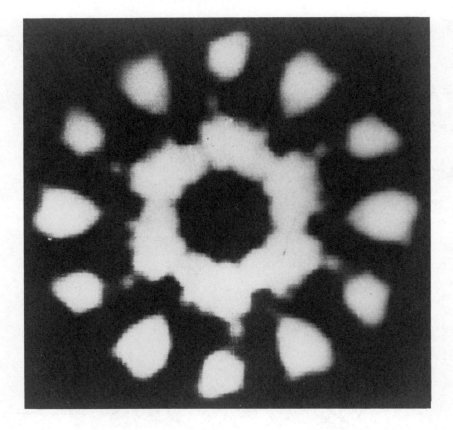

FIGURE 7.   Computer-enhanced P22 *gp1* portal structure. The structure is about 15 nm in diameter, contains twelve *gp1* molecules, and is positioned in proheads and virions so that it forms a pore through the coat protein shell. (Courtesy of J. Carrascosa and J. King.)

shells with incorrect diameters, but a few empty coat protein shells of prohead diameter assemble which lack portal protein.[11,16,85] Fuller and King[86-88] have studied the assembly of proheads from purified, unassembled coat and scaffolding proteins and found that the two proteins spontaneously assemble into prohead-like structures. No small, mixed oligomer of gp8 and gp5 has been found so it seems likely that the two kinds of subunits add to the growing structure independently. It is not known how the natural assembly process ensures the incorporation of the portal protein. In some other dsDNA phage systems, two host proteins, the products of the *E. coli groES* and *groEL* genes, are required for correct prohead assembly.[89] It is not known if the analogous *Salmonella* proteins participate in *in vivo* P22 prohead assembly.

3. Prohead Expansion and Scaffolding Protein Release

The prohead coat protein shells are about 11% smaller in diameter, more rounded, and thicker than are shells after DNA packaging.[10] Electron microscopy of the P22 virion surface, physical measurement of the number of coat proteins per capsid, and determination of the capsid symmetry by low-angle X-ray scattering have given rise to a model in which the outer surface of each P22 coat protein is concave, with two outward protruding ends of the molecule participating in the formation of different capsomeres (surface bumps) — trimers at one end of the protein and hexamers at the other.[14,15] This shape of coat protein lends itself to a shell expansion model in which the angle between the two ends increases to cause expansion, but this is currently speculative. The negative charge density on the outside surface of the prohead decreases upon expansion.[90] Treatment of proheads with low concentrations of protein-denaturing agents causes the quantitative release of scaffolding protein[43] and the expansion of the coat protein shell to mature phage diameter.[10] Earnshaw and King[85] found that in scaffolding protein-defective infections, a small number of coat protein shells the diameter of proheads are formed, and these shells expand exactly like proheads in response to mild denaturing agent treatment. This shows that expansion is a property of coat protein alone since these shells also lack the portal protein. The natural signal for expansion during DNA packaging is not known and could result from DNA entry. It seems unlikely from the above experiments that expansion is caused by scaffolding protein release *per se*. Nothing is known about the mechanism of scaffolding protein exit. The fact that the coat protein apparently must undergo a conformational change during expansion, and that the surface area of the coat protein shell increases by 23% during this process, make quite attractive the idea that transitory holes form in the shell and then each scaffolding molecule leaves by its own exit hole.

4. Prohead Assembly and Regulation of Scaffolding Protein Synthesis

Unlike that of most other viruses, the P22 assembly process has been found to affect the synthesis of one of the assembly components, namely, scaffolding protein. Expression of the scaffolding protein gene *in vivo* is turned down about six-fold by unassembled scaffolding protein, while scaffolding protein assembled into proheads has no such effect.[91-93] This regulation seems appropriate since scaffolding protein in effect catalyzes coat protein shell assembly and DNA packaging, but is still required in fairly large amounts. The result of this regulation is that just enough scaffolding protein is made to assemble with the available coat protein since any excess unassembled scaffolding protein subunits depress further synthesis. This modulation of expression is post-transcriptional and the stability of the scaffolding protein mRNA is lowered when the gene is being turned off by unassembled scaffolding protein.[47]

D. INITIATION OF DNA PACKAGING
1. The Role of gp2 and gp3

The products of genes *2* and *3* are "accessory" DNA packaging proteins. *In vivo* experiments have clearly shown that they are required for successful DNA packaging; in their absence, fully functional proheads and concatemeric DNA molecules accumulate.[11,94] Immediately after DNA packaging, a few molecules (<20) of each accessory protein are bound to the particle;[21,95] however, they are not present in mature phage particles (Figure 3).[11,43] It is not known if gp2 and/or gp3 actually act catalytically. Poteete and Botstein[58] purified gp2 by its ability to promote *in vitro* DNA packaging in a gene *2*-defective, infected-cell extract and found that gp3 copurified with gp2. It is thus thought that they form a complex; however, it has been argued that either gp2 or gp3 has a function outside that which it performs in this complex.[54] The biochemical properties of these proteins have not yet been studied in detail.

Genetic studies have implicated gp3 in the recognition of DNA that is to be packaged. Schmieger[96,97] isolated mutants of P22 with altered generalized transduction properties, which also affect initiation of packaging (see below). One of these, called HT12/4, has been shown to be a single nucleotide change in gene *3*.[56,57] This mutant fails to initiate packaging series at the normal location and initiates at two new locations on the P22 chromosome.[56,59] A *cis/trans* test showed that the mutant gene *3* protein has altered properties and that this phenotype is not due to an altered target site on the DNA.[59] A mutation that alters DNA target specificity is likely to affect amino acids which contact the DNA in the recognition process.[98-101] Thus, this gene *3* mutation, which changes a glutamic acid (amino acid 83) to a lysine, strongly suggests that gp3 is directly involved in the DNA recognition process and, furthermore, is likely to identify the portion of gp*3* which contacts the DNA target site. The glutamic acid 83 region of gp*3* contains neither a canonical "helix-turn-helix" nor a "Zn^{2+} finger" DNA-binding sequence motif.[27,56] In addition to the roles of gp*2* and gp*3* in initiation, the *in vitro* packaging experiments of Schmieger and co-workers[60,101,102] show that even when mature DNA molecules are being packaged, and apparently no specific target *(pac)* site is required, gp*2* and gp*3* are still required for packaging. It seems possible, therefore, that these two proteins may be required in some packaging step after initiation as well as being involved in initiation.

2. The DNA Target (*Pac*) Site

Since phage P22 packages P22 DNA much more efficiently than host DNA,[103] and since packaging is unidirectional along the DNA,[53,54,104] it seems very likely that a specific nucleotide sequence on the P22 chromosome, called the *pac* site, marks it for preferential packaging. Experiments — in which portions of the P22 chromosome were integrated into the host chromosome[104,105] or into an F'-factor[106] and packaging starts from the integrated P22 DNA were monitored after infection by P22 — have shown that the *pac* site must lie within or very near gene *3*. This agrees well with the observation that the left ends of mature chromosomes generated in the first events of packaging series lie near the middle of gene *3* (Figure 4).[27] The actual sequence of the *pac* site is not yet known with certainty, but there does exist a "consensus" sequence at the two locations where the phage mentioned above with the mutant gp*3* initiates packaging series (Figure 8).[56] It is unlikely that a single amino acid change could alter recognition of more than a few nucleotides and, indeed, most of this consensus sequence is also found in the region of wild-type series initiation. There are two parts to the consensus sequence, called the "+10" and the "+110" portions (Figure 8). It seems likely that this sequence is the target recognized by the packaging apparatus (most likely gp*3*). This is supported by an unpublished observation reported by Backhaus; a 114-bp fragment of DNA which includes the "+10 consensus" sequence increases the frequency of packaging of a plasmid under certain conditions.[27] Single nucleotide changes within the +10 and +110 consensus regions give the phage a high transduction frequency phenotype (packaging events initiate on the host chromosome about 100 times more frequently than with wild-type phage), but both these changes also alter the amino acid sequence of gp*3*.[107] It is not yet known if the phenotype of the nucleotide changes is caused by the nucleotide change in the putative target site change or in the amino acid sequence of the recognition protein.

3. The Left End of Headful Number One

Although it is thought that packaging initiates from a specific *pac* site, the locations of the left ends of the first chromosomes of packaging series are, in fact, not uniquely positioned on the DNA sequence. The left ends of these chromosomes fall into six regions called *end* regions (Figures 5 and 8). These end regions are separated by approximately 20 or 40 bp. In addition, there is microheterogeneity of left ends in each end region over about 10 bp (Figure 8).[27,61] This microheterogeneity has some specificity in that only certain nucleotides are chosen as ends, but no clear relationship of the exact endpoints to the nucleotide sequence in the region has been found.[27] Furthermore, there is no apparent repeat in the sequence that correlates with end region locations. One is left with the idea (without direct proof) that perhaps the recognition proteins initially bind the *pac* consensus site and that the positions of the physical ends are somehow determined by their spatial relationship to that site. It is interesting to note that the ends have a similar, but not identical, relationship at the two new series initiation locations used by the gene *3* high-transduction mutant mentioned above. In one case, there are six and in the other seven *end* regions, all of which are separated by about 20-bp intervals, and the relative frequency of use of the individual end regions is not

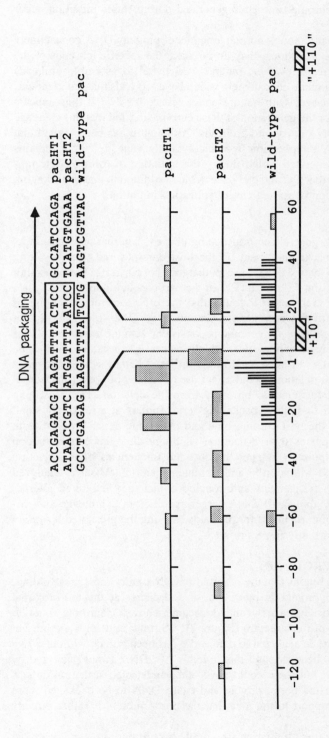

FIGURE 8.   A consensus P22 packaging recognition (*pac*) site. Nucleotide sequences are shown above for the regions thought to contain the wild-type *pac* site, *pac*HT1 and *pac*HT2 (the latter two are recognized by the HT12/4 mutant gp3), with the homologies boxed. Below, a larger region is shown schematically aligned by the "+10" portion of the consensus *pac* sequence (bounded by the stippled vertical lines). The vertical lines above the wild-type map indicate the approximate frequency of use of endpoints at precisely determined nucleotides of *end* regions 2 (centered at position −20), 3 (−1), 4 (17), and 5 (35).[27] The stippled rectangles indicate the approximate location and frequency of use of the *end* regions within which the exact endpoints have not yet been positioned on the nucleotide sequence.[56,61] The cross-hatched horizontal bars below indicate the positions of the "+10" and "+110" consensus sequences (see text).[56]

the same in the two cases, suggesting that local sequence must have some influence on this parameter. In wild-type and the mutant, the putative *pac* site lies in the middle of the 120 to 160 bp regions in which the chromosome ends generated during series initiation occur (Figure 8).[56]

Laski and Jackson[108] found that although normal amounts of progeny DNA concatemers are found in gene *2* or *3* defective infections, they do not have the specific left ends of the first molecule packaged in a series. However, mutants which fail to assemble proheads (defective in gene *1, 8,* or *5*) accumulate concatemers which have left ends that are identical, including frequency of end site usage, with mature phage left ends.[108,109] It thus appears that gp2 and gp3 are responsible for the generation of these ends *in vivo* and that this cleavage is not necessarily coupled with DNA entry into proheads. Although it has been argued that these ends are generated after DNA replication by nucleolytic cleavage,[56,104,108] it remains formally possible that they are generated earlier during the initiation of rolling circle replication. The latter model requires that gp2 and gp3 have a role in addition to that in packaging (see above) in dictating the position of rolling circle replication initiation.

4. Generalized Transduction

As mentioned above, P22 is a generalized transducing phage[1,110] in that on occasion its packaging apparatus initiates a packaging event on the host chromosome. The resulting particles are apparently identical to actual phage, except that they contain a P22 chromosome-sized fragment of the host chromosome.[103] About 2% of the particles in a normal P22 lysate are such transducing particles.[103] Schmieger[111] found that the efficiency of packaging of sequences from various regions of the host genome varies up to 1000-fold. His plot of transducing efficiency across the *Salmonella* genome is consistent with the existence of less than ten high-efficiency packaging initiation sites, which in turn is consistent with a *pac* site of approximately ten base pairs.[56] Chelala and Margolin[62] and Krajewska-Grybkiewicz and Klopotowski[63] found that host deletions could affect the frequency of cotransduction of two markers up to several hundred thousand bp away from the deletion. The packaging series model easily explains such findings if packaging series initiate at a relatively small number of *pac*-like locations on the host chromosome and packaging series proceed from those points. Thus, if a series passes over a deletion, it might alter the cotransduction frequencies (packaged DNA endpoints), but not the absolute frequencies of transduction beyond the deletion. Chelala and Margolin[112] and Schmieger and Backhaus[113] analyzed mutants with raised transduction frequencies and concluded that they exhibited altered, presumably relaxed (rather than abolished) DNA target specificity. Schmieger and co-workers[114,115] devised a method of enrichment for plasmids carrying the presumed host *pac*-like sites, and their analysis is currently in progress.

E. DNA ENTRY INTO THE PROHEAD

Proheads contain only three proteins that are required for DNA entry (coat, scaffolding, and portal proteins). The portal protein is present as a dodecamer at the tail-proximal vertex,[18,45,83] and computer enhanced images of this dodecamer show a clear hole about 30 Å in diameter through the center of the structure (Figure 7).[83] Similar structures are present in all dsDNA bacteriophage proheads analyzed to date and it has been argued without *direct* evidence that DNA enters the prohead through this hole.[42,116-120] DNA is not packaged by proheads lacking portal protein,[11] but these could have other undetected structural defects as well. P22 mutations that alter the portal protein and cause DNA to be packaged more tightly within the particle lend support to the idea that the portal protein is rather directly involved in DNA entry.[120]

According to this model, once an "initiation structure" is built with gp2 and/or gp3 at the *pac* site, the DNA is somehow inserted through the hole in the portal structure. Then the DNA is moved into the prohead through that hole. The rate of DNA entry is >3000 bp/s.[90] The condensation of DNA within the prohead must require energy to overcome charge repulsion and partial dehydration within the virion.[121,122] P22 DNA packaging in crude extracts, although quite inefficient, requires ATP cleavage as an energy source. How this energy might be used to move the DNA is not yet clear. The history of the search for the elusive physicochemical mechanism of DNA entry into dsDNA phage proheads has been reviewed[2,3] and will not be reiterated in detail here. Suffice it to say that a number of potential mechanisms have been ruled out and current interest is focused on three possible mechanisms: (1) a "DNA translocase" or "motor" moves the DNA into the prohead,[2,41,123,124] (2) prohead expansion draws in DNA,[125,126] and (3) DNA enters the prohead because of a "favorable" internal environment. Although they have not been ruled out, the latter two models may be less likely for the following reasons: (model 2) the increase in volume during shell expansion is less than the volume of the DNA being packaged[2] and in bacteriophages λ and T3, expansion occurs before 50% of the DNA is packaged;[127,128] (model 3) unless stabilizing proteins are added after packaging, DNA is spontaneously released *in vivo*.[84] However, the putative P22 DNA translocase remains unidentified and uncharacterized. The portal protein, gp*1*, and the "accessory" proteins, gp*2* and gp*3*, are reasonable candidates for participants in such a device. Work with *in vitro* packaging systems in two other dsDNA bacteriophages, φ29 and T3, make it seem unlikely that host proteins play any major role in DNA entry;[129,130] however, basic *E. coli* DNA-binding proteins may be involved in the initiation of lambdoid phage DNA packaging.[131,132] A report that the φ29 portal protein may bind DNA[133] supports the idea that portal proteins may be part of the hypothetical DNA translocase.

F. TERMINATION OF DNA PACKAGING
1. The Headful Nuclease

After sufficient DNA is packaged within the coat protein shell, a nuclease cleaves the packaged chromosome from the concatemer. This style of packaging has been called "headful" packaging since cleavage of the DNA occurs *only* when the head is full of DNA.[41] A mechanism for sensing when the head is full of DNA is not unique to viruses with circularly permuted chromosomes. Herpesviruses and bacteriophages such as lambda do not have circularly permuted chromosomes and require a specific sequence at the point of cleavage of the mature chromosome from the concatemeric product of replication. Yet they do not perform this cleavage unless a sufficient amount of DNA has been packaged within the capsid.[134-138] Thus, a "headful sensing device" appears to be a common feature of the large dsDNA viruses.

When the P22 head is sensed to be full of DNA, the headful nuclease cleaves the DNA. The genetic identity of this nuclease is unknown. The proteins gp*1*, gp*2*, and gp*3* are the candidates for the nuclease since gp*1* is thought to be at the point of entry, and although gp*3* and gp*2* are involved in initiation (above), they are also present in particles which have finished packaging DNA[21,95] and are probably present when the headful cleavage takes place. In a recent study, we analyzed the locations where the headful nuclease performed cleavages to generate mature chromosomes *in vivo*.[26] It was found that the headful sensing device is imprecise in that headful cleavages occur when a chromosome length of 103.8 ± 1.7% of the genome is packaged. Mature chromosomes are thus 43,400 ± 750 bp in length, where the uncertainty is the actual variation in length (thus, the length of the terminal redundancy varies from 850 to 2450 bp in different virions). These studies also showed that within the regions of headful cleavage, only one in about every 20 to 25 phosphodiester bonds is a substrate for efficient cleavage, suggesting that the headful nuclease has some DNA sequence/ structure specificity that is superimposed on the headful requirement.

2. Stabilization of Packaged DNA

After P22 DNA is packaged and the headful DNA cleavage is made, the particles are unstable and DNA is spontaneously lost from them both *in vitro*[11,16,139] and *in vivo*.[84] Three additional P22 gene products, gp*4*, gp*10*, and gp*26*, must add in this order to the particles (2 to 20 molecules of each) to stabilize the DNA within the particles (Figure 3).[139] These proteins are present in the resulting particle at the portal vertex[18] and, hence, likely block DNA passage out through the portal. Finally gp*9*, the tail protein, joins to complete the phage particle, but it does not have a strong additional effect on the particle stability.[11,16] Emptied phage heads isolated from infections lacking gp*4*, gp*10*, or gp*26* contain small amounts (5 to 15 molecules per particle) of the DNA packaging accessory proteins, gp*3* and gp*2*.[21,95] It thus appears that (1) these proteins can bind directly to the head proteins since these heads contain no DNA and (2) gp*26* appears to be required for their release. Thus, completion of the packaging process involves stabilization of the newly packaged DNA by the addition of several specific proteins to the portal vertex and release of the DNA packaging accessory proteins.

3. Termination of Packaging Series

Although normal packaging series have a 40 to 50% probability of terminating after each packaging event,[26] it is not known with certainty what limits series length. Two extreme possibilities exist: series length is limited by (1) the physical length of the DNA concatemers or by (2) the rate of initiation at *pac* sites relative to the number of DNA concatemers present. Existing evidence tends to support the latter possibility since lowering the *in vivo* concentration of gp*3*, a protein involved in series initiation, can cause the average series length to increase dramatically without altering the burst size.[55]

IV. RELATIONSHIP TO OTHER SYSTEMS

In all complex dsDNA viruses with preformed proheads, packaging events must have initiation, entry, and termination stages. Among the dsDNA phages studied, there are striking similarities in the overall strategy of DNA packaging. All have two DNA packaging "accessory proteins" (required in small numbers which may or, like P22, may not be prohead or virion components), all have preassembled proheads constructed with similar design principles, all require ATP as an energy source, all coat proteins undergo a conformational change (shell expansion) during packaging, and all except φ29 cleave the DNA during the packaging reaction (reviewed in References 2, 3, and 42). These similarities lend support to the idea that all these phage DNA packaging proteins derive from a common ancestor. However, there is substantial variation in the details of packaging among the different phages. For example, in φ29 one of the "accessory" DNA packaging proteins is covalently attached to the DNA ends, triangulation numbers of the coat protein shells vary from 1 to 13, and T4 has at least three different scaffolding proteins, all of which are removed by proteolysis, while P22 has one scaffolding protein that exits the structure intact (reviewed by Casjens and Hendrix[50]). Comparisons of the DNA-sequence-derived amino acid sequences among analogous proteins have not yet revealed recognizable similarities. For example, when the amino acid sequences of the portal proteins of phages P22, T7, φ29, and lambda are compared, even though these four portal structures are strikingly similar morphologically,[83,116-119] no amino acid sequence homology has yet been found.[35] At present, we cannot determine whether the similarities in strategy of DNA packaging by these different phages are the result of spectacular convergent evolution or whether they have descended so far from a common ancestor that the proteins involved have lost recognizable amino acid sequence homology. As mentioned above, herpesvirus and iridovirus DNA packaging have similarities in strategy with dsDNA bacteriophages, but a more detailed knowledge of these animal viruses is required before we can conclude that there are also detailed mechanistic similarities.

REFERENCES

1. **Zinder, N. and Lederberg, J.,** Genetic exchange in *Salmonella, J. Bacteriol.,* 64, 679, 1952.
2. **Earnshaw, W. and Casjens, S.,** DNA packaging by the double-stranded DNA bacteriophages, *Cell,* 21, 319, 1980.
3. **Casjens, S.,** Nucleic acid packaging by viruses, in *Virus Structure and Assembly,* Casjens, S., Ed., Jones and Bartlett, Boston, 1985, 75.
4. **Matthews, R.,** *A Critical Appraisal of Viral Taxonomy,* CRC Press, Boca Raton, FL, 1983.
5. **Susskind, M. and Botstein, D.,** Molecular genetics of bacteriophage P22, *Microbiol. Rev.,* 42, 385, 1978.
6. **Susskind, M. and Youderian, P.,** Bacteriophage P22 antirepressor and its control, in *Lambda II,* Hendrix, R., Roberts, J., Stahl, F., and Weisberg, R., Eds., Cold Spring Harbor Laboratory, Cold Spring Harbor, NY, 1983, 347.
7. **Poteete, A.,** Bacteriophage P22, in *The Bacteriophages,* Vol. 1, Calendar, R., Ed., Plenum Press, New York, 1988, 647.
8. **Anderson, T.,** On the fine structures of the temperate bacteriophages P1, P2 and P22, in *Proceedings of the European Regional Conference on Electron Microscopy,* Vol. 2, Delft, The Netherlands, 1960, 1008.
9. **Yamamoto, N. and Anderson, T.,** Genomic masking and recombination between serologically unrelated phages P22 and P221, *Virology,* 14, 430, 1961.
10. **Earnshaw, W., Casjens, S., and Harrison, S.,** Assembly of the head of bacteriophage P22: X-ray diffraction from heads, proheads and related structures, *J. Mol. Biol.,* 104, 387, 1976.
11. **Botstein, D., Waddell, C., and King, J.,** Mechanism of head assembly and DNA encapsulation in *Salmonella* phage P22. I. Genes, proteins, structures, and DNA maturation, *J. Mol. Biol.,* 80, 669, 1973.
12. **Israel, V., Anderson, T., and Levine, M.,** *In vitro* morphogenesis of phage P22 from heads and base-plate parts, *Proc. Natl. Acad. Sci. U.S.A.,* 57, 284, 1967.
13. **Casjens, S. and King, J.,** Virus Assembly, *Annu. Rev. Biochem.,* 44, 55, 1975.
14. **Casjens, S.,** Molecular organization of the bacteriophage P22 coat protein shell, *J. Mol. Biol.,* 131, 1, 1979.
15. **Earnshaw, W.,** Modelling of the small-angle x-ray diffraction arising from the surface lattices of phages lambda and P22, *J. Mol. Biol.,* 131, 14, 1979.
16. **King, J., Lenk, E., and Botstein, D.,** Mechanism of head assembly and DNA encapsidation in *Salmonella* phage P22. II. Morphogenetic pathway, *J. Mol. Biol.,* 80, 697, 1973.
17. **Israel, V., Rosen, H., and Levine, M.,** Binding of bacteriophage P22 tail parts to cells, *J. Virol.,* 10, 1152, 1972.
18. **Hartweig, E., Bazinet, C., and King, J.,** DNA injection apparatus of phage P22, *Biophys. J.,* 49, 24, 1986.
19. **Israel, V.,** E proteins of bacteriophage P22. I. Identification and ejection from wild-type and defective particles, *J. Virol.,* 23, 91, 1977.
20. **Hoffman, B. and Levine, M.,** Bacteriophage P22 virion protein which performs an essential early function. II. Characterization of the gene 16 function, *J. Virol.,* 16, 1547, 1975.
21. **Poteete, A. and King, J.,** Functions of two new genes in *Salmonella* phage P22 assembly, *Virology,* 76, 725, 1977.
22. **Rhoades, M., MacHattie, L., and Thomas, C.,** The P22 bacteriophage DNA molecule. I. The mature form, *J. Mol. Biol.,* 37, 21, 1968.
23. **Tye, B., Chan, R., and Botstein, D.,** Packaging of an oversize transducing genome by *Salmonella* phage P22, *J. Mol. Biol.,* 85, 485, 1974.
24. **Suzuki, K., Mise, K., and Nakaya, R.,** Electron microscopic observation of new transposable elements inserted into P22 phage genome from R plasmids, *Microbiol. Immunol.,* 24, 309, 1980.
25. **Jackson, E., Miller, H., and Adams, M.,** *Eco* RI restriction endonuclease cleavage site map of bacteriophage P22 DNA, *J. Mol. Biol.,* 118, 347, 1978.
26. **Casjens, S. and Hayden, M.,** Analysis *in vivo* of the bacteriophage P22 headful nuclease, *J. Mol. Biol.,* 199, 467, 1988.
27. **Backhaus, H.,** DNA packaging initiation of *Salmonella* bacteriophage P22: determination of cut sites within the DNA sequence coding for gene 3, *J. Virol.,* 55, 458, 1985.
28. **Earnshaw, W. and Harrison, S.,** DNA arrangement in isometric phage heads, *Nature (London),* 268, 598, 1977.
29. **Kosturko, L., Hogan, M., and Dattagupta, N.,** Structure of DNA within three isometric bacteriophages, *Cell,* 16, 515, 1979.
30. **Richards, K., Williams, R., and Calendar, R.,** Mode of DNA packing within bacteriophage heads, *J. Mol. Biol.,* 78, 255, 1973.
31. **Earnshaw, W., King, J., Harrison, S., and Eiserling, F.,** The structural organization of DNA packaged within heads of T4 wild-type, isometric and giant bacteriophages, *Cell,* 14, 559, 1978.

32. **Black, L., Newcomb, W., Boring, J., and Brown, J.,** Ion etching of bacteriophage T4: support for a spiral-fold model of packaged DNA, *Proc. Natl. Acad. Sci. U.S.A.,* 82, 7960, 1985.

33. **Serwer, P.,** Arrangement of double-stranded DNA packaged in bacteriophage capsids. An alternative model, *J. Mol. Biol.,* 190, 509, 1986.

34. **Lepault, J., Dubochet, J., Baschong, W., and Kellenberger, E.,** Organization of double-stranded DNA in bacteriophages: a study by cryo-electron microscopy of vitrified particles, *EMBO J.,* 6, 1507, 1987.

35. **Casjens, S.,** unpublished data, 1988.

36. **Padmanabhan, R., Wu, R., and Bode, V.,** Arrangement of DNA in lambda bacteriophage heads. III. Location and number of nucleotides cleaved from λ DNA by micrococcal nuclease attack on heads, *J. Mol. Biol.,* 69, 201, 1972.

37. **Thomas, J.,** Chemical linkage of the tail to the right-hand end of bacteriophage lambda DNA, *J. Mol. Biol.,* 87, 1, 1974.

38. **Chattoraj, D. and Inman, R.,** Location of DNA ends in P2, 186, P4 and lambda bacteriophage heads, *J. Mol. Biol.,* 87, 11, 1974.

39. **Saigo, K. and Uchida, H.,** Connection of the right-hand terminus of DNA to the proximal end of the tail in bacteriophage lambda, *Virology,* 61, 524, 1974.

40. **Inman, R., Schnos, M., and Howe, M.,** Location of the "variable end" of Mu DNA within the bacteriophage particle, *Virology,* 72, 393, 1976.

41. **Streisinger, G., Emrich, J., and Stahl, M.,** Chromosome structure in phage T4. III. Terminal redundancy and length determination, *Proc. Natl. Acad. Sci. U.S.A.,* 57, 292, 1967.

42. **Bazinet, C. and King, J.,** The DNA translocating vertex of dsDNA bacteriophage, *Annu. Rev. Microbiol.,* 39, 109, 1985.

43. **Casjens, S. and King, J.,** P22 morphogenesis: catalytic scaffolding protein in capsid assembly, *J. Supramol. Struct.,* 2, 202, 1974.

44. **Bazinet, C. and King, J.,** A late gene product of phage P22 affecting virus infectivity, *Virology,* 143, 368, 1985.

45. **Casjens, S., Wyckoff, E., Parr, R., Goates, J., Sampson, L., and Eppler, K.,** unpublished data, 1988.

46. **Weinstock, G., Riggs, P., and Botstein, D.,** Genetics of bacteriophage P22. III. The late operon, *Virology,* 106, 82, 1980.

47. **Casjens, S. and Adams, M.,** Posttranscriptional modulation of bacteriophage P22 scaffolding protein gene expression, *J. Virol.,* 53, 185, 1985.

48. **King, J.,** Regulation of structural protein interactions as revealed in phage morphogenesis, in *Biological Regulation and Development,* Vol. 2, Goldberger, R., Ed., Plenum Press, New York, 1980, 101.

49. **Berget, P.,** Pathways in viral morphogenesis, in *Virus Structure and Assembly,* Casjens, S., Ed., Jones and Bartlett, Boston, 1985, 149.

50. **Casjens, S. and Hendrix, R.,** Control mechanisms in dsDNA bacteriophage assembly, in *The Bacteriophages,* Vol. 1, Calendar, R., Ed., Plenum Press, New York, 1988, 15.

51. **Botstein, D. and Matz, M.,** A recombination function essential to the growth of bacteriophage P22, *J. Mol. Biol.,* 54, 417, 1970.

52. **Botstein, D.,** Synthesis and maturation of phage P22 DNA. I. Identification of intermediates, *J. Mol. Biol.,* 43, 621, 1968.

53. **Tye, B., Huberman, J., and Botstein, D.,** Non-random circular permutation of phage P22 DNA, *J. Mol. Biol.,* 85, 501, 1974.

54. **Jackson, E., Jackson, D., and Deans, R.,** *Eco* RI analysis of bacteriophage P22 DNA packaging, *J. Mol. Biol.,* 118, 365, 1978.

55. **Adams, M., Hayden, M., and Casjens, S.,** On the sequential packaging of bacteriophage P22 DNA, *J. Virol.,* 46, 673, 1983.

56. **Casjens, S., Huang, W., Hayden, M., and Parr, R.,** Initiation of bacteriophage P22 DNA packaging series. Analysis of a mutant that alters the DNA target specificity of the packaging apparatus, *J. Mol. Biol.,* 194, 411, 1987.

57. **Raj, A. and Schmieger, H.,** Phage genes involved in the formation of generalized transducing particles in *Salmonella*-phage P22, *Mol. Gen. Genet.,* 135, 175, 1974.

58. **Poteete, A. and Botstein, D.,** Purification and properties of proteins essential to DNA encapsulation by phage P22, *Virology,* 95, 565, 1979.

59. **Jackson, E., Laski, F., and Andres, C.,** P22 mutants which alter the specificity of DNA packaging, *J. Mol. Biol.,* 154, 551, 1982.

60. **Strobel, E., Behnisch, W., and Schmieger, H.,** *In vitro* packaging of mature phage DNA by *Salmonella* phage P22, *Virology,* 133, 158, 1984.

61. **Casjens, S. and Huang, W.,** Initiation of sequential packaging of bacteriophage P22 DNA, *J. Mol. Biol.,* 157, 287, 1982.

62. **Chelala, C. and Margolin, P.,** Effects of deletions on cotransduction linkage in *Salmonella typhimurium:* evidence that bacterial chromosome deletions affect the formation of transducing DNA fragments, *Mol. Gen. Genet.,* 131, 97, 1974.

63. **Krajewska-Grynkeiwicz, K. and Klopotowski, T.,** Altered linkage values in phage P22-mediated transduction caused by distant deletions or insertions in donor chromosomes, *Mol. Gen. Genet.,* 176, 87, 1979.

64. **Hayden, M., Adams, M., and Casjens, S.,** Bacteriophage L: chromosome physical map and structural proteins, *Virology,* 147, 431, 1985.

65. **MacHattie, L. and Gill, G.,** DNA maturation by the "headful" mode in bacteriophage T1, *J. Mol. Biol.,* 110, 441, 1977.

66. **Bachi, B. and Arber, W.,** Physical mapping of *BglII, BamHI, EcoRI, HindIII* and *PstI* restriction fragments of bacteriophage P1, *Mol. Gen. Genet.,* 153, 311, 1977.

67. **Sternberg, N. and Coulby, J.,** Recognition and cleavage of the bacteriophage P1 packaging site *(pac).* I. Differential processing of the cleaved ends, *in vivo, J. Mol. Biol.,* 194, 453, 1986.

68. **Morelli, G., Fisseau, C., Beheren, B., Trautner, T., Luh, J., Ratcliff, S., Allison, D., and Ganesan, A.,** The genome of *B. subtilis* phage SPP1, *Mol. Gen. Genet.,* 168, 153, 1979.

69. **Pretorius, G. and Coetzee, W.,** *Proteus mirabilis* phage 5006M: a physical characterization, *J. Gen. Virol.,* 45, 389, 1979.

70. **Jones, P. and Pretorius, G.,** *Achromobacter* sp. 2 phage α3: a physical characterization, *J. Gen. Virol.,* 53, 275, 1981.

71. **Schnaubel, H., Schramm, E., Schnabel, R., and Zillig, W.,** Structural variability in the genome of bacteriophage φH of *Halobacterium halobium, Mol. Gen. Genet.,* 188, 370, 1982.

72. **Johnson, L. and Schievert, P.,** A physical map of the group A streptococcal pyrogenic exotoxin bacteriophage T12 genome, *Mol. Gen. Genet.,* 189, 251, 1983.

73. **Guidolin, A., Morelli, G., Kamke, M., and Manning, P.,** *Vibrio cholerae* bacteriophage CP-T1: characterization of bacteriophage DNA and restriction analysis, *J. Virol.,* 51, 163, 1984.

74. **Sengupta, A., Ray, P., and Das, J.,** Characterization and physical map of choleraphage φ149 DNA, *Virology,* 140, 217, 1985.

75. **Stellwag, E., Fink, J., and Zissler, J.,** Physical characterization of the genome of the *Myxococcus xanthus* bacteriophage MX-8, *Mol. Gen. Genet.,* 199, 123, 1985.

76. **Chung, S. and Thompson, S.,** Physical and genetic characterization of actinophage SF1, a P1-like phage isolated from *Streptomyces fradiae,* in *Microbiology — 1985,* Levine, L., Ed., American Society for Microbiology, Washington, D.C., 1985, 431.

77. **Chowdhuey, R. and Das, J.,** Infection by choleraphage φ138: bacteriophage DNA and replicative intermediates, *J. Virol.,* 57, 960, 1986.

78. **Verma, M., Rao, A., and Chakravorty, M.,** Isolation of temperature-sensitive mutants of bacteriophage MB78 and correlation between the physical and genetic maps, *Virology,* 151, 274, 1986.

79. **Grund, A. and Hutchinson, C.,** Bacteriophages of *Saccharapolyspora erythrae, J. Bacteriol.,* 169, 3013, 1987.

80. **Murti, K., Goorha, R., and Granoff, A.,** Structure of frog virus 3 genome: size and arrangement of nucleotide sequences as determined by electron microscopy, *Virology,* 116, 275, 1982.

81. **Guo, P., Erickson, S., and Anderson, D.,** A small viral RNA is required for *in vitro* packaging of bacteriophage φ29 DNA, *Science,* 236, 690, 1987.

82. **King, J. and Casjens, S.,** Catalytic head assembling protein in virus morphogenesis, *Nature (London),* 251, 112, 1974.

83. **Bazinet, C., Benbasat, J., King, J., Carazo, J., and Carrascosa, J.,** Purification and organization of the gene 1 portal protein required for phage P22 DNA packaging, *Biochemistry,* 27, 1849, 1988.

84. **Lenk, E., Casjens, S., Weeks, J., and King, J.,** Intracellular visualization of precursor capsids in phage P22 mutant infected cells, *Virology,* 68, 182, 1975.

85. **Earnshaw, W. and King, J.,** Structure of phage P22 protein aggregates formed in the absence of the scaffolding protein, *J. Mol. Biol.,* 126, 721, 1978.

86. **Fuller, M. and King, J.,** Regulation of coat protein polymerization by the scaffolding protein of bacteriophage P22, *Biophys. J.,* 23, 381, 1980.

87. **Fuller, M. and King, J.,** Purification of the coat and scaffolding proteins from procapsids of bacteriophage P22, *Virology,* 112, 529, 1981.

88. **Fuller, M. and King, J.,** Assembly *in vitro* of bacteriophage P22 procapsids from purified coat and scaffolding subunits, *J. Mol. Biol.,* 156, 633, 1982.

89. **Tilly, K. and Georgopoulos, C.,** Viral-host interactions in virus morphogenesis, in *Virus Structure and Assembly,* Casjens, S., Ed., Jones and Bartlett, Boston, 1985, 269.

90. **Gope, R. and Serwer, P.,** Bacteriophage P22 *in vitro* packaging monitored by agarose gel electrophoresis: rate of DNA entry into capsids, *J. Virol.,* 47, 96, 1983.

91. **King, J., Hall, C., and Casjens, S.,** Control of the synthesis of phage P22 scaffolding protein is coupled to capsid assembly, *Cell,* 15, 551, 1978.

92. **Casjens, S., Adams, M., Hall, C., and King, J.,** Assembly-controlled autogenous modulation of bacteriophage P22 scaffolding protein gene expression, *J. Virol.,* 53, 174, 1985.

93. **Wyckoff, E. and Casjens, S.,** Autoregulation of the bacteriophage P22 scaffolding protein gene, *J. Virol.,* 53, 192, 1985.

94. **Poteete, A., Jarvick, V., and Botstein, D.,** Encapsidation of P22 DNA *in vitro, Virology,* 95, 550, 1979.

95. **Casjens, S. and Adams, M.,** unpublished data, 1988.

96. **Schmieger, H.,** A method for detection of phage mutants with altered transducing ability, *Mol. Gen. Genet.,* 110, 378, 1971.

97. **Schmieger, H.,** Phage P22 mutants with increased or decreased transduction abilities, *Mol. Gen. Genet.,* 119, 75, 1972.

98. **Youderian, P., Vershon, A., Bouvier, S., Sauer, R., and Susskind, M.,** Changing the DNA-binding specificity of a repressor, *Cell,* 35, 777, 1983.

99. **Ebright, R.,** Evidence for contact between glutamine-18 of *lac* repressor and base pair 7 of *lac* promoter, *Proc. Natl. Acad. Sci. U.S.A.,* 83, 303, 1986.

100. **Wharton, R. and Ptashne, M.,** Changing the binding specificity of a repressor by redesigning an α-helix, *Nature (London),* 316, 601, 1985.

101. **Strobel, E. and Schmieger, H.,** *In vitro* packaging of exogenous DNA by *Salmonella* phage P22, *J. Gen. Virol.,* 45, 291, 1979.

102. **Benisch, W. and Schmieger, H.,** *In vitro* packaging of plasmid DNA oligomers by *Salmonella* phage P22: independence of the *pac* site, and evidence for the termination cut *in vitro, Virology,* 144, 310, 1985.

103. **Ebel-Tsipis, J., Botstein, D., and Fox, M.,** Generalized transduction by phage P22 in *Salmonella typhimurium.* I. The molecular origin of transducing DNA, *J. Mol. Biol.,* 71, 433, 1972.

104. **Weaver, S. and Levine, M.,** Replication *in situ* and DNA encapsulation following induction of an excision-defective lysogen of *Salmonella* bacteriophage P22, *J. Mol. Biol.,* 118, 389, 1978.

105. **Kufer, B., Backhaus, H., and Schmieger, H.,** The packaging initiation site of phage P22. Analysis of packaging events by transduction, *Mol. Gen. Genet.,* 187, 510, 1982.

106. **Schmieger, H.,** *Pac* sites are indispensable for *in vivo* packaging of DNA by phage P22, *Mol. Gen. Genet.,* 195, 252, 1984.

107. **Casjens, S., Randall, S., Eppler, K., and Schmieger, H.,** unpublished data, 1988.

108. **Laski, F. and Jackson, E.,** Maturation cleavage of bacteriophage P22 DNA in the absence of packaging, *J. Mol. Biol.,* 154, 565, 1982.

109. **Casjens, S. and Hayden, M.,** unpublished data, 1988.

110. **Margolin, P.,** Generalized transduction, in *Escherichia coli* and *Salmonella typhimurium,* Neidhardt, F., Ed., American Society for Microbiology, Washington, D.C., 1986, 1154.

111. **Schmieger, H.,** Packaging signals for phage P22 on the chromosome of *Salmonella typhimurium, Mol. Gen. Genet.,* 187, 516, 1982.

112. **Chelala, C. and Margolin, P.,** Evidence that HT mutant strains of bacteriophage P22 retains an altered form of substrate specificity in the formation of transducing particles in *Salmonella typhimurium, Genet. Res.,* 27, 315, 1976.

113. **Schmieger, H. and Backhaus, H.,** Altered cotransduction frequencies exhibited by HT-mutants of *Salmonella*-phage P22, *Mol. Gen. Genet.,* 143, 307, 1976.

114. **Schmidt, C. and Schmieger, H.,** Selective transduction of recombinant plasmids with cloned *pac* sites by *Salmonella* phage P22, *Mol. Gen. Genet.,* 196, 123, 1984.

115. **Vogel, W. and Schmieger, H.,** Selection of bacterial *pac* sites recognized by *Salmonella* phage P22, *Mol. Gen. Genet.,* 205, 563, 1986.

116. **Driedonks, R., Engel, A., ten Heggler, B., and van Driel, R.,** Gene 20 product of bacteriophage T4. Its purification and structure, *J. Mol. Biol.,* 152, 641, 1981.

117. **Carazo, J., Santisteban, A., and Carrascosa, J.,** Three-dimensional reconstruction of the bacteriophage φ29 neck particles at 2.2 nm resolution, *J. Mol. Biol.,* 183, 79, 1985.

118. **Kochan, J., Carrascosa, J., and Murialdo, H.,** Bacteriophage lambda preconnectors: purification and structure, *J. Mol. Biol.,* 174, 433, 1984.

119. **Carazo, J., Fujisawa, H., Nakasu, S., and Carrascosa, J.,** Bacteriophage T3 gene 8 product oligomer structure, *J. Ultrastruct. Mol. Struct. Res.,* 96, 105, 1987.

120. **Casjens, S., Hayden, M., Eppler, K., and Wyckoff, E.,** unpublished data, 1988.

121. **Riemer, S. and Bloomfield, V.,** Packaging of DNA in bacteriophage heads: some considerations on energetics, *Biopolymers,* 17, 784, 1978.

122. **Rau, D., Lee, B., and Parsegian, V.,** Measurement of the repulsive force between polyelectrolyte molecules in ionic solution: hydration forces between parallel DNA double helices, *Proc. Natl. Acad. Sci. U.S.A.,* 81, 2621, 1984.

123. **Hendrix, R.,** Symmetry mismatch and DNA packaging in large DNA bacteriophages, *Proc. Natl. Acad. Sci. U.S.A.,* 75, 4779, 1978.

124. **Black, L. and Silverman, D.,** Model for DNA packaging into bacteriophage T4 heads, *J. Virol.,* 28, 643, 1978.

125. **Hohn, B., Wurtz, M., Klein, B., Lustig, A., and Hohn, T.,** Phage lambda DNA packaging, *in vitro, J. Supramol. Struct.,* 2, 302, 1974.

126. **Serwer, P.,** A metrizamide-impermeable capsid of the DNA packaging pathway of bacteriophage T7, *J. Mol. Biol.,* 138, 65, 1980.

127. **Hohn, B.,** DNA sequences necessary for packaging of bacteriophage λ DNA, *Proc. Natl. Acad. Sci. U.S.A.,* 80, 7456, 1983.

128. **Shibata, H., Fujisawa, H., and Minagawa, T.,** Characterization of the bacteriophage T3 DNA packaging reaction *in vitro* in a defined system, *J. Mol. Biol.,* 196, 845, 1987.

129. **Guo, P., Grimes, S., and Anderson, D.,** A defined system for *in vitro* packaging of DNA-gp3 of the *Bacillus subtilis* bacteriophage φ29, *Proc. Natl. Acad. Sci. U.S.A.,* 83, 3505, 1986.

130. **Hamada, K., Fujisawa, H., and Minagawa, T.,** A defined *in vitro* system for packaging of bacteriophage T3 DNA, *Virology,* 151, 119, 1986.

131. **Bear, S., Court, D., and Friedman, D.,** An accessory role for *Escherichia coli* integration host factor: characterization of a lambda mutant dependent upon integration host factor for DNA packaging, *J. Virol.,* 52, 966, 1984.

132. **Gold, M. and Parris, W.,** A bacterial protein requirement for the bacteriophage λ terminase reaction, *Nucleic Acids Res.,* 14, 9797, 1986.

133. **Herranz, L., Salas, M., and Carrascosa, J.,** Interaction of bacteriophage φ29 connector protein with the viral DNA, *Virology,* 155, 289, 1986.

134. **Emmons, S.,** Bacteriophage lambda derivatives carrying two copies of the cohesive end site, *J. Mol. Biol.,* 83, 511, 1974.

135. **Feiss, M., Fisher, D., Siegele, D., Nichols, B., and Egner, C.,** Packaging of the bacteriophage λ chromosome: effect of chromosome length, *Virology,* 77, 281, 1977.

136. **Feiss, M. and Becker, A.,** DNA packaging and cutting, in *Lambda II,* Hendrix, R., Roberts, J., Stahl, F., and Weisberg, R., Eds., Cold Spring Harbor Laboratory, Cold Spring Harbor, NY, 1983, 305.

137. **Vlazny, D., Kwong, A., and Frenkel, N.,** Site specific cleavage/packaging of herpes simplex virus DNA and the selective maturation of nucleocapsids containing full-length viral DNA, *Proc. Natl. Acad. Sci. U.S.A.,* 79, 1423, 1982.

138. **Deiss, L. and Frenkel, N.,** Herpes simplex virus amplicon: cleavage of concatemeric DNA is linked to packaging and involves amplification of the terminally reiterated α sequence, *J. Virol.,* 57, 933, 1986.

139. **Strauss, H. and King, J.,** Steps in stabilization of newly packaged DNA during phage P22 morphogenesis, *J. Mol. Biol.,* 173, 523, 1984.

Index

INDEX

A

Accessory DNA packaging proteins, phage P22, 251, 256

Adenine methylation, P1 DNA packaging, 233—234

Adenosine diphosphate, dnaA protein binding, 46

Adenosine triphosphate

dnaA protein binding, 46

in vitro initiation of replication at *oriC,* 48

phage packaging, 209—210, 256

procapsid binding of DNA, 215

T antigen binding, 124

A enhancer, polyoma, 124

Aggregation sensitivity, bacterial nucleoids, 12—14

Agnoprotein, 117—118

Annuli, 81

AP2, 135—136

AP3, 136

AP4, 136

ArgF, 70

Asparagine synthetase *(asn)* operon, 69

Aspartic acid protease, 152

Asymmetric chromosomes, termination of replication, 96—97

Asynchrony of initiation, 44

AT-rich regions

oriC, 29

phage lambda DNA, 165—166, 173

polyoma and SV40, 118, 120

AUG codons

cauliflower mosaic virus, 151—152

polyoma and SV40, 114, 117—118

B

Bacillus subtilis

bacteriophage φ29, 255—256

chromosomal separation model, 81

dnaA proteins, 46

dnaA region, 52

oriC sequence comparison, 30—31, 52

origin, 51—53

termination of replication

asymmetric chromosomes, 96—97

blocked replication forks, 97—99

replication pattern, 96

termination sequence, 105—107

Bacterial chromatin

replication, origin of, see *OriC*

replication, termination of

Bacillus subtilis, 96—99

Escherichia coli, 99—105

structure-function relationships, 4—21

aggregation sensitivity of DNA plasm of nucleoids, 12—14

fine structure of DNA-containing plasm of nucleoid, 14—17

immunocytochemical localization histone-like protein HU, 17—19

nucleoid in thin sections, 7—13

special features of prokaryotic chromatin, 4—7

Bacterial loop model, 16

Bacteriophage lambda

DNA packaging, see Bacteriophages, double-stranded DNA packaging

reinitiation of DNA replication in, 164—175

abnormal reinitiation events, 167—169

effect of SOS on regulation of initiation, 173—175

first round of DNA replication, 164

initiation of replication, 166

regulation of initiation of replication, 169—171

requirement for negative superhelical tension at origin, 171—173

unusual reinitiation of replication, 166—169

Bacteriophage P1 DNA packaging, 226—238

comparative biology: P22 vs. P1, 238

genes involved in head morphogenesis and packaging, 227—228

genes involved in vegetative replication, 228—229

host DNA packaging into generalized transducing particles, 236—238

pac processing genes, 230—232

pac site, 232—235

physical structure, 226—227

processive headful packaging, 229—230

Bacteriophage P2 DNA packaging, see Bacteriophages, double-stranded DNA packaging

Bacteriophage P22 DNA packaging, 242—256, see also Bacteriophages, double-stranded DNA packaging

mechanisms, 243—256

DNA entry into prohead, 254—255

DNA metabolism, 245—248

general pathway, 243—245

initiation of packaging, 251—254

proheads, 248—251

termination of packaging, 255—256

vs. phage p1, 238

virion, 242—243

Bacteriophage φ29, 255—256, see also Bacteriophages, double-stranded DNA packaging

Bacteriophages, double-stranded DNA packaging, 204—218

conformation, 205—208

analogies with condensation of chromatin, 208

model comparisons, 206—208

orientation of packaged DNA segments, 206

sites of perturbed duplex, 205

types of nucleotide sequence, 205

energetics, 208—211

additional hypotheses, 211

distribution of segments of packaged DNA during entry, 211